Lithium Ion Rechargeable Batteries

Edited by

Kazunori Ozawa

Related Titles

Mitsos, A. / Barton, P. I. (eds.)

Microfabricated Power Generation Devices

Design and Technology

2009

ISBN: 978-3-527-32081-3

Liu, Hansan / Zhang, Jiujun (eds.)

Electrocatalysis of Direct Methanol Fuel Cells

From Fundamentals to Applications

2010

ISBN: 978-0-471-68958-4

Sundmacher, K., Kienle, A., Pesch, H. J., Berndt, J. F., Huppmann, G. (eds.)

Molten Carbonate Fuel Cells

Modeling, Analysis, Simulation, and Control

2007

ISBN: 978-3-527-31474-4

Zuttel, A. / Borgschulte, A. / Schlapbach, L. (eds.)

Hydrogen as a Future Energy Carrier

2008

ISBN: 978-3-527-30817-0

Lithium Ion Rechargeable Batteries

Edited by
Kazunori Ozawa

WILEY-VCH Verlag GmbH & Co. KGaA

The Editor

Dr. Kazunori Ozawa
Enax, Inc.
President & CEO
2-11-19 Otowa, Bunkyo-ku
Tokyo 112-0013
Japan

Library of Congress Card No.:
applied for

British Library Cataloguing-in-Publication Data
A catalogue record for this book is available from the British Library.

Bibliographic information published by the Deutsche Nationalbibliothek
The Deutsche Nationalbibliothek lists this publication in the Deutsche Nationalbibliografie; detailed bibliographic data are available on the Internet at http://dnb.d-nb.de.

Printed in the Federal Republic of Germany
Printed on acid-free paper

Typesetting Laserwords Private Limited, Chennai, India
Printing Strauss GmbH, Mörlenbach
Binding Litges & Dopf GmbH, Heppenheim
Cover Design Adam-Design, Weinheim

ISBN: 978-3-527-31983-1

Contents

Preface *XI*

List of Contributors *XIII*

1 General Concepts *1*
Kenzo Matsuki and Kazunori Ozawa
1.1 Brief Outline of Batteries *1*
1.1.1 Galvanic Cell System – Aqueous Electrolyte System *2*
1.1.2 Lithium-Cell System – Nonaqueous Electrolyte System *4*
1.2 Early Development of Lithium-Ion Batteries *5*
1.2.1 Ceramics Production Capability *5*
1.2.2 Coating Technology *6*
1.2.3 $LiPF_6$ as a Salt for Electrolytes *6*
1.2.4 Graphite Conductor in the Cathode *6*
1.2.5 Using Hard Carbon for the Anode *6*
1.2.6 Nonwoven Shut-down Separator *6*
1.2.7 Ni-Plated Fe Can *7*
1.3 Toward a Realistic Goal *7*
References *9*

2 Lithium Insertion Materials Having Spinel-Framework Structure for Advanced Batteries *11*
Kingo Ariyoshi, Yoshinari Makimura, and Tsutomu Ohzuku
2.1 Introduction *11*
2.2 Structural Description of Spinel *12*
2.3 Derivatives of Spinel-Framework Structure *15*
2.3.1 Superlattice Structures Derived from "Spinel" *15*
2.3.2 Examples of Superstructure Derived from "Spinel" *20*
2.4 Electrochemistry of Lithium Insertion Materials Having Spinel-Framework Structure *24*
2.4.1 Lithium Manganese Oxides (LMO) *24*
2.4.2 Lithium Titanium Oxide (LTO) *27*

Lithium Ion Rechargeable Batteries. Edited by Kazunori Ozawa

ISBN: 978-3-527-31983-1

2.4.3 Lithium Nickel Manganese Oxide (LiNiMO) *28*
2.5 An Application of Lithium Insertion Materials Having Spinel-Framework Structure to 12 V "Lead-Free" Accumulators *29*
2.5.1 Twelve-Volt Batteries Consisting of Lithium Titanium Oxide (LTO) and Lithium Manganese Oxide (LMO) *32*
2.5.2 Twelve-Volt Batteries Consisting of Lithium Titanium Oxide (LTO) and Lithium Nickel Manganese Oxide (LiNiMO) *34*
2.6 Concluding Remarks *36*
References *37*

3 Overlithiated $Li_{1+x}(Ni_z Co_{1-2z}Mn_z)_{1-x}O_2$ as Positive Electrode Materials for Lithium-Ion Batteries *39*
Naoaki Kumagai and Jung-Min Kim
3.1 Introduction *39*
3.2 Co-Free $Li_{1+x}(Ni_{1/2}Mn_{1/2})_{1-x}O_2$ *40*
3.3 $Li_{1+x}(Ni_{1/3}Co_{1/3}Mn_{1/3})_{1-x}O_2$ *44*
3.4 Other $Li_{1+x}(Ni_zCo_{1-2z}Mn_z)_{1-x}O_2$Materials *48*
3.5 Conclusion *50*
References *51*

4 Iron-Based Rare-Metal-Free Cathodes *53*
Shigeto Okada and Jun-ichi Yamaki
4.1 Introduction *53*
4.2 2D Layered Rocksalt-Type Oxide Cathode *54*
4.3 3D NASICON- Type Sulfate Cathode *55*
4.4 3D Olivine-Type Phosphate Cathode *58*
4.5 3D Calcite-Type Borate Cathode *62*
4.6 3D Perovskite-Type Fluoride Cathode *64*
4.7 Summary *65*
References *65*

5 Thermodynamics of Electrode Materials for Lithium-Ion Batteries *67*
Rachid Yazami
5.1 Introduction *67*
5.2 Experimental *71*
5.2.1 The ETMS *71*
5.2.2 Electrochemical Cells: Construction and Formation Cycles *73*
5.2.3 Thermodynamics Data Acquisition *73*
5.3 Results *74*
5.3.1 Carbonaceous Anode Materials *74*
5.3.1.1 Pre-coke (HTT $<$ 500 C) *77*
5.3.1.2 Cokes HTT 900–1700 °C *79*
5.3.1.3 Cokes HTT 2200 and 2600 °C *80*
5.3.1.4 Natural Graphite *82*
5.3.1.5 Entropy and Degree of Graphitization *84*

5.3.2 Cathode Materials 86
5.3.2.1 $LiCoO_2$ 86
5.3.2.2 $LiMn_2O_4$ 90
5.3.2.3 Effect of Cycling on Thermodynamics: 93
5.4 Conclusion 94
References 96

6 Raman Investigation of Cathode Materials for Lithium Batteries 103
Rita Baddour-Hadjean and Jean-Pierre Pereira-Ramos
6.1 Introduction 103
6.2 Raman Microspectrometry: Principle and Instrumentation 104
6.2.1 Principle 104
6.2.2 Instrumentation 105
6.3 Transition Metal-Oxide-Based Compounds 106
6.3.1 $LiCoO_2$ 107
6.3.2 $LiNiO_2$ and Its Derivative Compounds $LiNi_{1-y}Co_yO_2$ $(0 < y < 1)$ 113
6.3.3 Manganese Oxide-Based Compounds 114
6.3.3.1 MnO_2-Type Compounds 114
6.3.3.2 Ternary Lithiated Li_xMnO_y Compounds 117
6.3.4 V_2O_5 127
6.3.4.1 V_2O_5 Structure 127
6.3.4.2 Structural Features of the $Li_xV_2O_5$ Phases 131
6.3.5 Titanium Dioxide 143
6.4 Phospho-Olivine $LiMPO_4$ Compounds 149
6.5 General Conclusion 156
References 157

7 Development of Lithium-Ion Batteries: From the Viewpoint of Importance of the Electrolytes 163
Masaki Yoshio, Hiroyoshi Nakamura, and Nikolay Dimov
7.1 Introduction 163
7.2 General Design to Find Additives for Improving the Performance of LIB 166
7.3 A Series of Developing Processes to Find Novel Additives 169
7.4 Cathodic and the Other Additives for LIBs 172
7.5 Conditioning 174
References 177

8 Inorganic Additives and Electrode Interface 179
Shinichi Komaba
8.1 Introduction 179
8.2 Transition Metal Ions and Cathode Dissolution 180
8.2.1 Mn(II) Ion 181
8.2.2 Co(II) Ion 184
8.2.3 Ni(II) Ion 186

8.3 How to Suppress the Mn(II) Degradation *187*
8.3.1 LiI, LiBr, and NH_4I *188*
8.3.2 2-Vinylpyridine *190*
8.4 Alkali Metal Ions *197*
8.4.1 Na^+ Ion *197*
8.4.2 K^+ Ion *204*
8.5 Alkali Salt Coating *207*
8.6 Summary *209*
References *210*

9 Characterization of Solid Polymer Electrolytes and Fabrication of all Solid-State Lithium Polymer Secondary Batteries *213*
Masataka Wakihara, Masanobu Nakayama, and Yuki Kato
9.1 Molecular Design and Characterization of Polymer Electrolytes with Li Salts *213*
9.1.1 Introduction *213*
9.1.2 Solid Polymer Electrolytes with Plasticizers *217*
9.1.3 Preparation of SPE Films with B-PEG and Al-PEG Plasticizers *217*
9.1.4 Evaluation of SPE Films with B-PEG Plasticizers *219*
9.1.5 Ionic Conductivity of SPE Films with B-PEG Plasticizers *223*
9.1.6 Transport Number of Lithium Ions *227*
9.1.7 Electrochemical Stability *229*
9.1.8 Summary *230*
9.2 Fabrication of All-Solid-State Lithium Polymer Battery *231*
9.2.1 Introduction *231*
9.2.2 Required Ionic Conductivity of SPE *231*
9.2.3 Difference between Conventional Battery with Liquid Electrolyte and All-Solid-State LPB *232*
9.2.4 Fabrication and Electrochemical Performance of LPBs Using SPE with B-PEG and/or Al-PEG Plasticizers *235*
9.2.5 Fabrication of a Nonflammable Lithium Polymer Battery and its Electrochemical Evaluation *243*
9.2.6 Summary *250*
References *251*

10 Thin-Film Metal-Oxide Electrodes for Lithium Microbatteries *257*
Jean-Pierre Pereira-Ramos and Rita Baddour-Hadjean
10.1 Introduction *257*
10.2 Lithium Cobalt Oxide Thin Films *259*
10.2.1 Sputtered $LiCoO_2$ Films *259*
10.2.1.1 Liquid Electrolyte *259*
10.2.1.2 Solid-State Electrolyte *262*
10.2.2 PLD $LiCoO_2$ Films *265*
10.2.3 CVD $LiCoO_2$ Films *269*
10.2.4 $LiCoO_2$ Films Prepared by Chemical Routes *269*

10.2.5 Conclusion 271
10.3 $LiNiO_2$ and Its Derivatives Compounds $LiNi_{1-x}MO_2$ 272
10.3.1 Solid-State Electrolyte 273
10.3.2 Liquid Electrolyte 274
10.3.3 Li – Ni – Mn Films 274
10.3.4 Conclusion 275
10.4 $LiMn_2O_4$ Films 275
10.4.1 Sputtered $LiMn_2O_4$ Films 276
10.4.2 PLD $LiMn_2O_4$ Films 277
10.4.3 ESD $LiMn_2O_4$ Films 281
10.4.4 $LiMn_2O_4$ Films Prepared Through Chemical Routes 282
10.4.5 Substituted $LiMn_{2-x}M_xO_4$ Spinel Films 283
10.4.6 Conclusion 283
10.5 V_2O_5 Thin Films 285
10.5.1 Sputtered V_2O_5 Thin Films 286
10.5.1.1 Liquid Electrolyte 286
10.5.1.2 Solid-State Electrolyte 294
10.5.2 PLD V_2O_5 Thin Films 296
10.5.3 CVD V_2O_5 Films 297
10.5.4 V_2O_5 Films Prepared by Evaporation Techniques 297
10.5.5 V_2O_5 Films Prepared by Electrostatic Spray Deposition 298
10.5.6 V_2O_5 Films Prepared via Solution Techniques 299
10.5.7 Conclusion 300
10.6 MoO_3 Thin Films 301
10.6.1 Liquid Electrolyte 301
10.6.2 Solid State Electrolyte 302
10.6.3 Conclusion 303
10.7 General Conclusions 303
References 305

11 Research and Development Work on Advanced Lithium-Ion Batteries for High-Performance Environmental Vehicles 313
Hideaki Horie
11.1 Introduction 313
11.2 Energy Needed to Power an EV 313
11.3 Quest for a High-Power Characteristic in Lithium-Ion Batteries 315
11.4 Cell Thermal Behavior and Cell System Stability 322
Further Reading 326

Index 329

Preface

Lithium ion battery has become the basis of the huge market for cellular phones and lap top computers, and these mobile communication market continues to grow at a rapid rate, supported by the demand all over the world. Even so, intensive efforts are still under way to further improve the technology. The main target of the effort is not only the automobile industry by achieving higher energy and higher power, but also the energy storage market supplementing environmentally friendly power source such as solar energy and wind turbine.

Though the lithium ion technology is so wide this book can include only a few topics, I believe the readers can find an indicator to do the research.

Chapter one covers the basic concepts of electrochemical devices and lithium ion battery.

From Chapter two to Chapter four cathode materials are described, and Chapter two especially proposes new application such as an accumulator.

The basic thoughts of the materials are mentioned in Chapter five and Chapter six. These two Chapters are so new , then they may give a big impact to the readers.

Chapter seven and Chapter eight focus on the solid electrolyte interface, so called SEI which is important to develop high performance lithium ion batteries.

Solid state batteries are discussed in Chapter nine and Chapter ten. These batteries may show the big business chance in the future.

Las but not least, Chapter eleven explain advanced lithium ion batteries for high performance environmental vehicles.

The substantial contribution of each of the authors to this book is gratefully acknowledged, as well as their cooperation in preparing their manuscripts in the style and format selected. I also wish to express my appreciation to the companies, associations who supported the contributing authors and willingly provided their technical information and data permitted its use in this book.

January, 2009 *Kazunori Ozawa*

Lithium Ion Rechargeable Batteries. Edited by Kazunori Ozawa

ISBN: 978-3-527-31983-1

List of Contributors

Kingo Ariyoshi
Osaka City University (OCU)
Graduate School of Engineering
Department of Applied Chemistry
Sugimoto 3-3-138
Sumiyoshi, Osaka 558-8585
Japan

Rita Baddour-Hadjean
Institut de Chimie et des Matériaux
Paris-Est, ICMPE/GESMAT
UMR 7182 CNRS et Université Paris
XII, CNRS, 2 rue Henri Dunant
94320 Thiais
France

Nikolay Dimov
Saga University
Advanced Research Center
IM & T Inc.
1341 Yoga-machi, Saga 840-0047
Japan

Hideaki Horie
Nissan Motor Co., Ltd.
Nissan Research Center
1, Natsushima-cho, Yokosuka-shi
Kanagawa, 237-85223
Japan

Yuki Kato
University of Tokyo
Institute of Industrial Science
4-6-1 Komaba, Meguro-ku
Tokyo 153-8505
Japan

Jung-Min Kim
Iwate University
Graduate School of Engineering
Department of Frontier Materials
and Functional Engineering
4-3-5 Ueda, Morioka
Iwate 020-8551
Japan

Shinichi Komaba
Tokyo University of Science
Department of Applied Chemistry
1-3 Kagurazaka, Shinjuku
Tokyo 162-8601
Japan

Naoaki Kumagai
Iwate University
Graduate School of Engineering
Department of Frontier Materials
and Functional Engineering
4-3-5 Ueda, Morioka
Iwate 020-8551
Japan

Lithium Ion Rechargeable Batteries. Edited by Kazunori Ozawa

ISBN: 978-3-527-31983-1

Yoshinari Makimura
Toyota Central Research and
Development Laboratories, Inc.
Nagakute, Aichi 480-1192
Japan

Kenzo Matsuki
Yamagata University
4-3-16 Jonan, Yonezawa
Yamagata 992-8510
Japan

Hiroyoshi Nakamura
Saga University
Department of Applied Chemistry
1 Honjyo, Saga, 840-8502
Japan

Masanobu Nakayama
Nagoya Institute of Technology
Graduate School of Engineering
Department of Materials and
Science and Engineering
Gokiso-cho, Syowa-ku
Nagoya 466-8555
Japan

Tsutomu Ohzuku
Osaka City University (OCU)
Graduate School of Engineering
Department of Applied Chemistry
Sugimoto 3-3-138
Sumiyoshi, Osaka 558-8585
Japan

Shigeto Okada
Kyushu University
Institute for Materials Chemistry and
Engineering, 6-1, Kasuga Koen
Kasuga, 816-8580
Japan

Kazunori Ozawa
Enax, Inc.
8F Otowa KS Bldg.
2-11-19 Otowa
Bunkyo-ku, Tokyo 112-0013
Japan

Jean-Pierre Pereira-Ramos
Institut de Chimie et des
Matériaux Paris-Est
ICMPE/GESMAT, UMR 7182
CNRS et Université Paris XII, CNRS
2 rue Henri Dunant
94320 Thiais
France

Masataka Wakihara
Tokyo Institute of Technology
Office Wakihara:
Dear City Akasaka W-403
2-12-21 Akasaka Minato-ku
Tokyo 107-0052
Japan

Jun-ichi Yamaki
Kyushu University
Institute for Materials
Chemistry and Engineering
6-1, Kasuga Koen
Kasuga 816-8580
Japan

Rachid Yazami
California Institute of Technology
(CALTECH)
International Associated Laboratory
on Materials for Electrochemical
Energetics (LIA-ME^2), MC 138-78
Pasadena, CA 91125
USA

Masaki Yoshio
Saga University
Advanced Research Center
1341 Yoga-machi, Saga, 840-0047
Japan

1
General Concepts

Kenzo Matsuki and Kazunori Ozawa

1.1
Brief Outline of Batteries

The first practical battery is the generally known *Volta cell* (also called the *Galvanic cell*). Its invention, over two centuries ago, spawned the invention of a variety of batteries based principally on the Volta cell. However, interestingly, during the last century, only three batteries, namely, the MnO_2 primary battery and the secondary batteries of lead/acid or nickel have been in use. Knowing why such batteries continue to be used would give us some important pointers toward the development of new technology in this line. These old batteries are close to reaching their technical limit. Recently, however, new concepts have been used in the development of lithium-ion secondary batteries with higher ability.

A battery generally provides two functions – the ability to supply power over a duration of time and the ability to store power. These are defined by two operations, charge/discharge (progress of the reaction) and storage/stop (termination of the reaction), that is, a battery is a device that provides two functions, namely, energy storage and energy conversion (from chemical to electrical, and vice versa). As shown in Figure 1.1, the field of energy conversion is a multiphase system that is composed of positive/negative terminals and positive/negative active materials and electrolyte; the ions and electrons transfer through their interfaces. The interfaces reflect the nature of each phase. In addition, the state of these interfaces changes over time with the operation of the battery. The cell voltage is supported by an electric double layer with a remarkably high electric field between the electrodes and the electrolytic solution in which the electrode reactions take place. It should be emphasized that battery technology is essentially the same as the technology that controls these interfaces.

Lithium Ion Rechargeable Batteries. Edited by Kazunori Ozawa

ISBN: 978-3-527-31983-1

Phase (I)		Phase (II)		Phase (III)		Phase (IV)		Phase (V)
e−	⇄	e− ion	⇄	ion	⇄	e− ion	⇄	e−
Electronic conductor		Metal electrode		Ionic conductor		Solid matrix		Electronic conductor
Negative terminal		Negative(anode) active material		Electrolyte		Positive(cathode) active material		Positive terminal

Fig. 1.1 Multilayer system composed of five phases and four interfaces.

1.1.1
Galvanic Cell System – Aqueous Electrolyte System

To understand the cell structure and its reaction, the well-known Daniel cell is schematically shown in Figure 1.2.

The two half cells of $Zn|Zn^{2+}$ and $Cu|Cu^{2+}$ are combined and a separator is placed between them so that they are not miscible with each other. The formula that shows the principle and the structure of the Daniel cell is as follows:

$$(-)\ Zn|Zn^{2+}, SO_4{}^{2-}||Cu^{2+}, SO_4{}^{2-}|Cu\ (+)$$

where the symbols | and || show the interface of different phases and the liquid–liquid junction (separator), respectively.

The cathode (positive electrode) active material of the Daniel cell is the Cu^{2+} ion in the electrolyte, while the Zn anode (negative electrode) dissolves to form the Zn^{2+} ion. The drop in voltage of the cell occurs because of self-discharge of

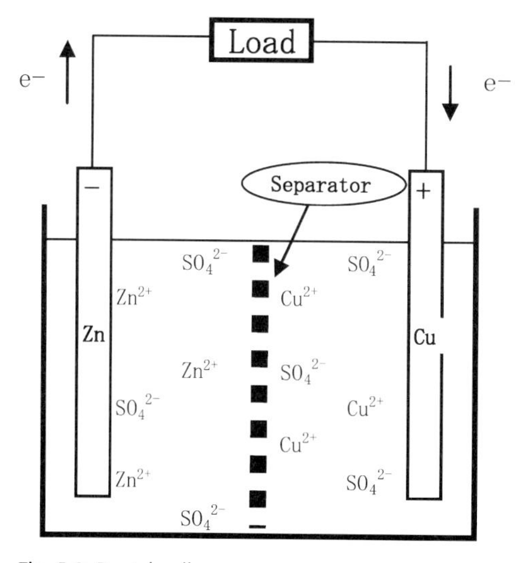

Fig. 1.2 Daniel cell.

the active materials. Generally, a self-discharge tends to occur when the dissolved chemical species such as Cu^{2+} ion are used as the cathode-active material. This is one of the reasons that the Daniel cell was not used for practical purposes.

Cathode- and anode-active materials in the Leclanche cell are MnO_2 (solid) and Zn metal, respectively. These electrode reactions are as follows:

$$\text{Cathode:} \quad 2MnO_2 + 2H^+ + 2e \longrightarrow 2MnOOH$$

$$\text{Anode:} \quad Zn + 2NH_4{}^+ \longrightarrow Zn(NH_3)_2{}^{2+} + 2H^+ + 2e$$

When the discharge reaction takes place, the Zn anode dissolves to form a complex ion. Since MnO_2 has a depolarizing ability that reduces the potential drop produced, the Leclanche type battery has been improved progressively to produce several kinds of batteries for commercial use, such as the manganese dry cell, the $ZnCl_2$ cell, and the alkaline MnO_2 cell.

In 1859, Plante invented the lead acid storage battery. This battery has been improved over the years and is now industrially mass-produced. The electrode reactions in the lead storage battery are described as follows:

$$\text{Cathode:} \quad PbO_2 + H_2SO_4 + 2H^+ + 2e = PbSO_4 + 2H_2O$$

$$\text{Anode:} \quad Pb + H_2SO_4 = PbSO_4 + 2H^+ + 2e$$

During discharge, a secondary solid phase of $PbSO_4$ is formed on both the anode and the cathode. Moreover, sulfuric acid in the aqueous solution – which is another active species – and water also participate in the charge/discharge reactions. These factors cause some polarizations that lower the cell performance.

Electrode reactions in Ni–Cd cell are as follows:

$$\text{Cathode:} \quad 2NiOOH + 2H_2O + 2e = 2Ni(OH)_2 + 2OH^-$$

$$\text{Anode:} \quad Cd + 2OH^- = Cd(OH)_2 + 2e$$

The cathode reaction involves the insertion of an H^+ ion into the solid NiOOH, which is similar to the cathode reaction of MnO_2 in the manganese battery, while the anode reaction is the formation of a secondary solid phase $Cd(OH)_2$ on the Cd anode. This prevents a smooth reaction as the Cd anode is covered with $Cd(OH)_2$.

The cathode-active material of nickel–metal hydride (Ni–MH) battery is the H species, which is adsorbed by the hydrogen-adsorbing alloy (MH) instead of the Cd anode of the Ni–Cd battery; the cell reaction is very simple because only hydrogen participates in the charge/discharge reaction. The Ni–MH battery has almost same voltage and larger electric capacity when compared with that of the Ni–Cd battery; moreover, it is free from environmental contamination. Therefore, the industrial production of Ni–MH battery has increased rapidly in recent years.

1.1.2
Lithium-Cell System – Nonaqueous Electrolyte System

To realize a battery with high potential of 3 V, batteries using lithium metal as the anode-active material and a powerful oxidizing agent as the cathode-active material were considered to be ideal. One such promising cathode-active material was MnO_2; the development of lithium battery using this commenced in 1962. Fortunately, at that time, substantial amount of basic and application data with MnO_2 was available. About 10 years later, an Li–MnO_2 battery with a lithium metal anode was made available by SANYO Inc.; this became the first representative primary lithium battery.

Since then, considerable research and development has taken place in the design and manufacture of rechargeable lithium batteries. Many cathode-active materials such as TiSe, NbSe, MoS_2, and MnO_2 were studied. For example, rechargeable batteries based on a lithium metal anode and a molybdenum sulfide cathode (Li insertion electrode) were developed by MOLI Energy, Inc. in 1985. This battery system was abandoned owing to safety problems. Lithium batteries based on Li metal anodes and commonly used electrolyte systems revealed the thermal runaway of these systems, which can lead to their explosion; this was almost inevitable in abuse cases such as short circuit, overheating, and overcharging. Although the highest energy density available for Li batteries is achieved by a battery system that can use Li metal anode, a solution to safety issues needs to be found.

Active materials with good reversibility for the Li intercalation/deintercalation and low charge/discharge voltage were used as anode materials instead of Li metal. A carbon material was found to meet these requirements, and a rechargeable Li battery based on a carbon anode and $LiCoO_2$ (layered lithium cobalt oxide) cathode was developed, mass-produced, and commercialized by Sony Inc. in 1991; this lithium-ion battery was capable of high performance as well as a high voltage of 4 V. As shown schematically in Figure 1.3, lithium-ion rechargeable batteries are charged and discharged through the transport of Li^+ ions between anode and cathode, with electron exchange as a result of insertion (doping) and extraction (undoping). Both anode and cathode materials are layered compounds, and, as a result, the battery reaction is very simple because only Li^+ ions participate in the charge/discharge reactions.

The features of the Li-ion batteries, compared with the other rechargeable batteries, can be summarized as follows: (i) Charge and discharge reactions

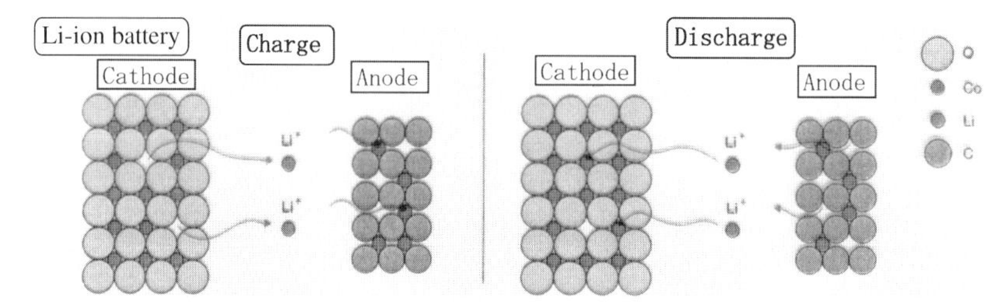

Fig. 1.3 Schematic illustration of the reaction in a lithium-ion battery.

transfer Li^+ ions between cathode and anode with minimal side reactions; (ii) The electrolytes work only as a path for the Li^+ ions; and (iii) The volume of the electrolyte between cathode and anode will not be required.

1.2 Early Development of Lithium-Ion Batteries

The UK Atomic Energy Authority showed in their patent [1] that the intercalation and deintercalation of A_x-ion of the compound $A_xM_yO_2$ reversibly occurs, where A_x is an alkaline metal and M_y is a transition metal. In 1990, Sony used this patent to first produce Li-ion batteries for a cellular phone HP-211. The cell sizes were 14500 and 20500, where 20 refers to the diameter and 50 the length in mm. The chemistry was $LiCoO_2$/soft carbon system, and the capacities of the 14500 and 20500 models were only 350 and 900 mAh, respectively. The production was on a pilot scale. However, the naming of the *lithium-ion rechargeable battery* [2] was a marketing success. It was a controversial issue whether Li existed as ion or metal in the carbon anode. By the measurement using NMR, it was revealed that some part of the Li could exist as ion [3].

The actual mass production of the lithium-ion cell was carried out for a camcorder TR-1 in 1991. The cell size was 18650, which has the same volume as the 20500 cell. The chemistry was $LiCoO_2$/hard-carbon system [4]. Figure 1.4 displays the inside structure of a 18650 cell.

The reasons for Sony's success as the first producer of lithium-ion batteries is explained in the following section.

1.2.1 Ceramics Production Capability

Sony were already one of the biggest Mn–Zn ferrites producers in Japan. They also had considerable experience in the production of $LiCoO_2$.

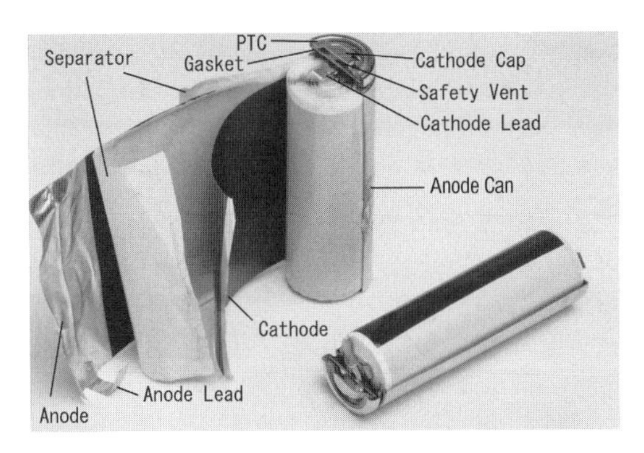

Fig. 1.4 The inside structure of a 18650 lithium-ion cell.

1.2.2
Coating Technology

Sony had been producing magnetic tapes for audio- and videotape recorders. The coating technology for magnetic tapes was very useful and very important in making cathode and anode electrodes, especially in making good slurry and performing intermittent coating.

1.2.3
$LiPF_6$ as a Salt for Electrolytes

$LiPF_6$ is unstable and easily decomposes with heat and moisture. Since the acid HF is produced in the presence of water, numerous arguments opposing the use of $LiPF_6$ were put forth in conferences and seminars. However, it was revealed that a small amount of HF increased the life cycle, because of the formation of a very strong passive layer such as AlF_3 on the surface of an aluminum cathode collector [5]. $LiPF_6$ is also easily soluble in the solvent, making it a good solid electrolyte interface (SEI) on the surface of anode materials.

1.2.4
Graphite Conductor in the Cathode

Synthesized graphite KS-15 was added as a conductor in the cathode. Since the cathode material is an oxide, which is nonconductive, a conductor has to be added. Metallic materials were also considered as conductors, but only carbon was effective for cycle performance because it acts as a reduction material giving a conductive path on the surface of the material.

1.2.5
Using Hard Carbon for the Anode

It is possible to use hard carbon and soft carbon in the form of propylene carbonate (PC) as solvent for the electrolyte. At first, soft carbon was used for the 14500 and 20500 cells, but it was changed to hard carbon in the 18650 cell after considering the float-charge stability – soft carbon is graphitizable carbon, whereas hard carbon is nongraphitizable carbon when it is heated to high temperatures of 2800–3000°C.

1.2.6
Nonwoven Shut-down Separator

Nonwoven polyethylene (PE) separators made by Tonen could be used for the 18650 cell. This separator melted at ~140°C and gave good safety results by shutting down the rush current in the case of abuse.

1.2.7 Ni-Plated Fe Can

Since the acid HF is produced inside the cell, a can should be made with stainless steel. However, the conductivity of stainless steel is so low that it is not suitable for the can, which requests low electroresistance. Ni-plated Fe can was then used for producing cylindrical cells.

After Sony's commercialization, various modifications have been made to develop advanced Li^+ ion batteries with higher energy density, retaining their good safety characteristics. These efforts achieved energy density of $200\,W\,h\,kg^{-1}$ and $500\,W\,h\,l^{-1}$, values of which would be close to those postulated earlier.

The shapes of cells have been widely expanded from cylindrical to prismatic and laminated. The applications of cells have also widely expanded from cellular phones and camcorders to laptop computers, power tools, and light electric vehicles.

Not only $LiCoO_2$ but also $LiMn_2O_4$, $LiNi_{1/3}Co_{1/3}Mn_{1/3}O_2$, $LiNiAlO_2$, etc. have been developed as cathode materials.

It has become possible to use graphite for the anode by controlling the SEI layer. However, $LiPF_6$ is interestingly still preferred as a salt in the electrolyte, and components of the solvent are usually still cyclic carbonates such as PC, ethylene carbonate (EC), and linear carbonates such as dimethyl carbonate (DMC), diethyl carbonate (DEC), ethyl–methyl carbonates (EMC), and/or their combinations. Lithium-ion batteries are considered to be good in electric vehicles (EVs), hybrid electric vehicles (HEVs), and plug-in hybrid vehicles. In tune with the increased and diverse applications, a strong demand for higher power density (kilowatts per kilograms) and higher energy density (kilowatt hours per kilogram) will become the target of the further research and development activities.

1.3 Toward a Realistic Goal

One of our final goals is a high-energy and high-power battery that can replace the lead/acid battery of automobiles satisfying EU RoHS (EU Directive: Restrictions on Hazardous Substances) instructions. To realize lead-free accumulators (engine starters), it is important to study the materials from their basic properties. Lithium insertion materials without the destruction core structures, called *topotactic reactions*, are classified into three categories:

1. Layer structure: $LiCoO_2$, $LiNiO_2$, $LiNi_{1/2}Mn_{1/2}O_2$, $LiCo_{1/3}Ni_{1/3}Mn_{1/3}O_2$, $LiAl_{0.05}Co_{0.15}Ni_{0.8}O_2$, etc. [6];
2. Spinel-frame work structure: $LiMn_2O_4$, $Li[Ni_{1/2}Mn_{3/2}]O_4$, LiV_2O_4, $Li[Li_{1/3}Ti_{5/3}]O_4$, etc. [7, 8];
3. Olivine structure: $LiFePO_4$, $LiMnPO_4$, $LiCoPO_4$, etc. [9].

Table 1.1 Redox Potential.

Electrode reaction	E^0 (V)	Electrode reaction	E^0 (V)
$Li^+ + e \rightleftharpoons Li$	−3.01	$Co^{2+} + 2e \rightleftharpoons Co$	−0.27
$K^+ + e \rightleftharpoons K$	−2.92	$Ni^{2+} + 2e \rightleftharpoons Ni$	−0.23
$Ba^{2+} + 2e \rightleftharpoons Ba$	−2.92	$Sn^{2+} + 2e \rightleftharpoons Sn$	−0.14
$Sr^{2+} + 2e \rightleftharpoons Sr$	−2.89	$Pb^{2+} + 2e \rightleftharpoons Pb$	−0.13
$Na^+ + e \rightleftharpoons Na$	−2.71	$H^+ + e \rightleftharpoons 1/2H_2$	0.00
$Mg^{2+} + 2e \rightleftharpoons Mg$	−2.38	$Cu^{2+} + 2e \rightleftharpoons Cu$	0.34
$Ti^{2+} + 2e \rightleftharpoons Ti$	−1.75	$Cu^+ + e \rightleftharpoons Cu$	0.52
$Al^{3+} + 3e \rightleftharpoons Al$	−1.66	$Ag^+ + e \rightleftharpoons Ag$	0.80
$Mn^{2+} + 2e \rightleftharpoons Mn$	1.05	$O_2 + 4H^+ + 4e \rightleftharpoons 2H_2O$	1.23
$Zn^{2+} + 2e \rightleftharpoons Zn$	−0.76	$Cl_2 + 2e \rightleftharpoons 2Cl^-$	1.36
$Fe^{2+} + 2e \rightleftharpoons Fe$	−0.44	$F_2 + 2e \rightleftharpoons 2F^-$	2.87

$LiMn_2O_4$ is a very attractive material as a cathode, but is reported to have a poor cycle performance. Measurement of the entropy of lithiation may indicate a solution [10], which is described in Chapter 5. Olivine structure, also called *chrysoberyl structure*, has a space group system of Pnma and shows excellent stability, but very low electric conductivity.

As for an accumulator of a consumer car, the battery has to have very high energy that resists overdischarging during parking. For the higher energy density, the species that have the biggest redox potential should be considered. The term *redox* is obtained from a contraction of the words *reduction* and *oxidation*. Table 1.1 shows the standard potentials of electrode reaction at room temperature. Lithium has the maximum potential on the negative side, and fluorine has the maximum on the positive side. These suggest our future target.

It is also necessary to consider the fact that the key elements of lithium-ion batteries are facing the crisis of exhaustion. Co metal used for the cathode has been in great shortage for quite some time. Element resources in the earth were estimated as the Clarke number, which was presented about 80 years ago. These values are not fully accepted, but they suggest our standpoint.

Focusing on the transition metals, they are ranked as follows in weight:

$$Fe > Ti > Mn > Cr > V > Ni > Cu > \cdots$$

Lithium-ion batteries are still one of the most promising storage devices. To realize the ideal battery, further development efforts are continuing all over the world.

References

1 Goodenough, J.B. and Mizushima, K. (1979) c/o United Kingdom Atomic Energy Authority, United Kingdom Patent GB 11953/79, April 1979.

2 Nagaura, T. and Tozawa, K. (1990) *Prog. Batteries Sol. Cells*, **9**, 209.

3 Tanaka, K., Itabashi, M., Aoki, M., Hiraka, S., Kataoka, M., Sataori, K., Fujita, S., Sekai, K. and Ozawa, K. (1993) 184th ECS Fall Meeting, New Orleans, Louisiana.

4 Ozawa, K. (1994) *Solid State Ionics*, **69**, 212.

5 Tachibana, K., Nishina, T., Endo, T., and Matsuki, K. (1999) The 1999 Joint International meeting, 196th Meeting of the Electrochemical Society, 1999 Fall Meeting of The Electrochemical Society of Japan with Technical cosponsorship of the Japan Society of Applied Physics, Honolulu, Abstract No. 381 (1999).

6 Mizushima, K., Jopnes, P.C., Wiseman, P.J., and Goodenough, J.B. (1984) *Mater. Res. Bull.*, **19**, 170.

7 Hunter, J.C. (1981) *J. Solid State Chem.*, **39**, 142.

8 Thackeray, M.M., Johnson, P.J., de Piciotto, L.A., Bruce, P.G., and Goodenough, J.B. (1984) *Mater. Res. Bull.*, **19**, 179.

9 Phadhi, A.K., Nanjundaswang, K.S., and Goodenough, J.B. (1997) *J. Electrochem. Soc.*, **144**, 1188.

10 Yazami, R., Reynier, Y., and Fultz, B. *IMLB 2006*, Biarritz, France.

2
Lithium Insertion Materials Having Spinel-Framework Structure for Advanced Batteries

Kingo Ariyoshi, Yoshinari Makimura, and Tsutomu Ohzuku

2.1
Introduction

Lithium-ion batteries are popular worldwide as power sources for wireless telephones, laptop computers, and other electronic devices. Current lithium-ion batteries mostly consist of $LiCoO_2$ and graphite, which are the layered structures. Possible alternatives to $LiCoO_2$ and graphite have been intensively investigated by many research groups over the past 15 years, and many lithium insertion materials have been reported [1–4]. However, $LiCoO_2$ and graphite continue to be used in lithium-ion batteries because of their high-energy density. Current lithium-ion batteries have about 550 Wh dm^{-3} of volumetric energy density. Because graphite shows the lowest operating voltage against lithium with the high rechargeable capacity of more than 350 mAh g^{-1} and $LiCoO_2$ shows a flat operating voltage of about 4 V versus Li with rechargeable capacity of about 140 mAh g^{-1}, it is very difficult to replace $LiCoO_2$ or graphite with other lithium insertion materials with respect to energy density. Recently, possible applications of batteries have diversified from high energy density to high power density and long life owing to environmental issues. This new direction gives us a chance to promote research on brand-new batteries and materials. Lithium insertion materials having spinel-framework structure described in this chapter fit applications that require high power and long life.

We review the crystal structure of spinel in Section 2.2, in which the structural feature of spinel in relation to a lithium insertion scheme is described. Derivatives of spinel-framework structure are explained by lowering the crystal symmetry of cubic spinel and a series of crystal structures in terms of superstructure due to cation ordering are discussed in Section 2.3.

The structural chemistry and electrochemistry of lithium insertion materials having spinel-framework structure are described in Section 2.4; these include the 4-V lithium insertion material of $Li[Li_xMn_{2-x}]O_4$, 5-V material of $Li[Ni_{1/2}Mn_{3/2}]O_4$ (LiNiMO), and the zero-strain insertion material of $Li[Li_{1/3}Ti_{5/3}]O_4$. We focus on each characteristic value in this section.

Lithium Ion Rechargeable Batteries. Edited by Kazunori Ozawa

ISBN: 978-3-527-31983-1

In Section 2.5, we discuss how lithium insertion materials having spinel-framework structure can be used in batteries in a fruitful and practical manner. We introduce the new concept of 12 V "lead-free" accumulators. Although the combination of lithium titanium oxide (LTO) and lithium manganese oxide (LMO) (or LiNiMO) does not give energy densities higher than the current lithium-ion batteries of $LiCoO_2$ and graphite, it has been found that they show three to five times higher energy density than current 12-V lead-acid batteries. Power sources for high-power and long-life applications, such as hybrid electric vehicles (HEVs) or pure electric vehicles (PEVs), are also highlighted in Section 2.5. The final section, Section 2.6 contains a summary and the prospects of the future in lithium insertion materials based on "spinel."

2.2
Structural Description of Spinel

The aim of this chapter is to review the crystal structure of spinel for lithium insertion materials. The mineral spinel is known as a *gemstone*, which is composed of $MgAl_2O_4$. The crystal structure of spinel, $MgAl_2O_4$, is shown in Figure 2.1. Such an illustration with emphasis on the tetrahedral and octahedral coordination around Mg^{2+} and Al^{3+} ions, respectively, is usually found in textbooks. Spinel has a cubic close-packed (ccp) oxygen array in which Mg^{2+} ions occupy one-eighth of the tetrahedral sites and Al^{3+} ions occupy one-half of the octahedral sites. More specifically, oxygen ions reside at 32(e) sites to form ccp structure where Mg^{2+} ions are located at tetrahedral 8(a) sites and Al^{3+} ions are at octahedral 16(d) sites in the space group symmetry of $Fd\bar{3}m$. The meaning of 8(a) and 16(d) is the number of equivalent sites in the unit cell combined with the Wyckoff letter. In describing a lithium insertion scheme, we use this symbol with the Hermann–Mauguin space group symbols. There are many materials isostructural

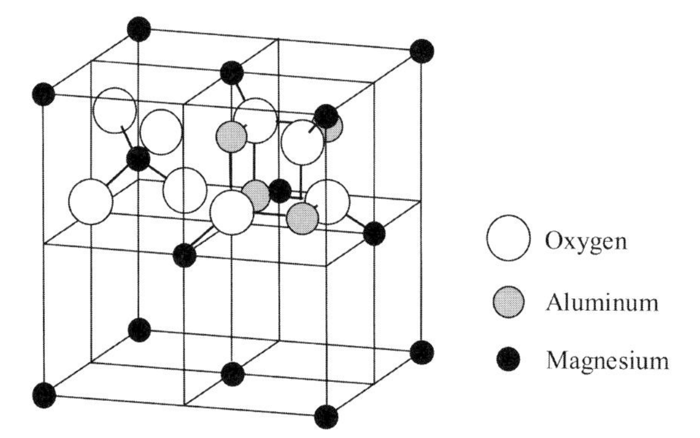

Fig. 2.1 Schematic illustration of the crystal structure of spinel, $MgAl_2O_4$. The illustration is depicted with emphasis on the MgO_4 tetrahedral and the AlO_6 octrahedral coordination.

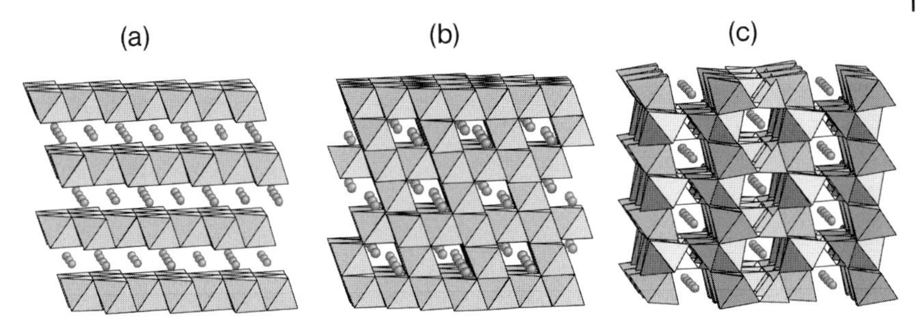

Fig. 2.2 The crystal structures of (a) α-$NaFeO_2$-type layered structure, (b) spinel structure, and (c) olivine structure represented by MO_6 polyhedra.

with the mineral spinel. In describing the spinel, the composition is generally represented as AB_2O_4 where A ion is surrounded by four oxygen ions, forming a unit tetrahedron, usually in its divalent (or monovalent) state, and B ion is surrounded by six oxygen ions, forming a unit octahedron, usually in trivalent (and/or tetravalent) states. Although the illustration in Figure 2.1 is useful to understand the coordination around the A and B ions in the spinel structure, the way to link BO_6 octahedra and AO_4 tetrahedra is not well illustrated in this figure.

Figure 2.2 shows the polyhedral representation of layered (a), spinel (b), and olivine structures (c). In the materials under consideration, lithium ions are assumed to be mobile while transition metal ions are immobile or fixed at octahedral sites, so that lithium ions are illustrated as spheres in the framework of such octahedral linkage. The linkage of FeO_6-octahedra and PO_4-tetrahedra is used in drawing the olivine structure ($LiFePO_4$). The olivine structure has hexagonally close-packed (HCP) oxygen array, while α-$NaFeO_2$-type layered and spinel structures have ccp oxygen arrays. Although the oxygen packing is different between the olivine and spinel structures, similarity in crystal structure is seen between the two structures. In spinel-related $LiMn_2O_4$, lithium and manganese ions are located at tetrahedral 8(a) and octahedral 16(d) sites, respectively, in the space group of $Fd\bar{3}m$. If $LiMn_2O_4$ is described as $Li[Mn_2]O_4$, $LiFePO_4$ can be represented as $P[LiFe]O_4$ since phosphorus ions are located at tetrahedral sites and lithium and iron ions are at octahedral sites. Actually, olivine $(Mg, Fe)_2SiO_4$ transforms to the spinel structure at high temperature under extremely high pressure. This method to illustrate the structural features of lithium insertion materials shown in Figure 2.2 is useful for material researches. As clearly seen in Figure 2.2b, the spinel-framework structure has a three-dimensional (3D) channel consisting of one-dimensional (1D) tunnels with respect to lithium-ion transportation in contrast to two-dimensional (2D) networks in the layered structure or 1D tunnels in the olivine structure.

The spinel structure can also be described as being composed of single chains consisting of edge-shared BO_6-octahedra. The layer consisting of BO_6-octahedral chains parallel to each other pile up along the c axis with 90° rotation to form the spinel structure. Because the single chains in the neighboring layers are perpendicular to each other, the chains in the spinel structure run in two perpendicular directions. For example, stoichiometric $LiMn_2O_4$ consists of single chains of edge-shared MnO_6-octahedra, i.e., $Mn^{3+}-Mn^{4+}-Mn^{3+}-Mn^{4+}-Mn^{3+}-Mn^{4+}$ in its ideal charge order. At room temperature, stoichiometric $LiMn_2O_4$ is cubic. When it is cooled to low temperature, phase transition from cubic to tetragonal (or orthorhombic) is observed, which reminds us of a Peierls transition coupled with localized/delocalized electrons associated with Jahn–Teller effect although this is not well supported by the experimental facts. Cubic to tetragonal transformation during the reduction of $LiMn_2O_4$ to $Li_2Mn_2O_4$, which is described in Section 2.4, is also explained using this structural description.

The general composition for the spinel structure described above is represented as $A[B_2]O_4$ in which the number of B ions is twice that of the A ions. "A" and "B" sites are usually used in describing structural chemistry of spinel. The A sites are tetrahedral 8(a) sites and B sites are octahedral 16(d) sites in the space group symmetry of $Fd\bar{3}m$. $A[B_2]O_4$ is called a *normal spinel*. However, the spinel structure is flexible with regard to the distribution of A and B ions at tetrahedral and octahedral sites. When A ions occupy one-half of the octahedral 16(d) sites and B ions occupy the other half of the octahedral 16(d) sites as well as all of the tetrahedral 8(a) sites, the structure is called an *inverse spinel*. The inverse spinel is represented by $B[AB]O_4$ in which A and B ions in the square brackets occupy the octahedral 16(d) sites. Some examples of "normal" and "inverse" spinel in cubic symmetry are listed in Table 2.1. The spinel structure can hold cation vacancies in a regular part of the crystal, called a *defect spinel*. γ-Fe_2O_3 is a defect spinel represented by Fe^{3+} $[\square_{1/3}Fe^{3+}{}_{5/3}]O_4$ in which one-sixth of Fe^{3+} at the 16(d) sites is vacant to compensate charge, where open square indicates vacant octahedral sites. γ-Al_2O_3 is also a defect spinel, while the distribution of cation vacancies together with Al^{3+} has not yet been determined. Cation vacancies are probably distributed randomly at both tetrahedral

Table 2.1 "Normal" and "inverse" spinel in cubic symmetry.

Composition	Structure	Lattice parameter, cubic (Å)
$MgAl_2O_4$	Normal : $Mg^{2+}[Al^{3+}{}_2]O_4$	8.08
$MnCr_2O_4$	Normal : $Mn^{2+}[Cr^{3+}{}_2]O_4$	8.44
$LiMn_2O_4$	Normal : $Li^+[Mn^{3+}Mn^{4+}]O_4$	8.25
$LiTi_2O_4$	Normal : $Li^+[Ti^{3+}Ti^{4+}]O_4$	8.42
$Li[CrTi]O_4$	Normal : $Li^+[Cr^{3+}Ti^{4+}]O_4$	8.32
$CoFe_2O_4$	Inverse : $Fe^{3+}[Co^{2+}Fe^{3+}]O_4$	8.41
$NiFe_2O_4$	Inverse : $Fe^{3+}[Ni^{2+}Fe^{3+}]O_4$	8.36
γ-Al_2O_3	Defect : $\square_{1/3}Al^{3+}{}_{8/3}O_4$	7.86
γ-Fe_2O_3	Defect : $Fe^{3+}[\square_{1/3}Fe^{3+}{}_{5/3}]O_4$	8.33
$Li[Li_{1/3}Ti_{5/3}]O_4$	Defect : $Li^+[Li^+{}_{1/3}Ti^{4+}{}_{5/3}]O_4$	8.36

and octahedral sites. Among these materials in Table 2.1, LTOs, LMOs, and their derivatives are important for batteries. Because lithium insertion materials extract and insert lithium ions during charge and discharge, their crystal structures must have lithium-ion conduction paths. Lithium transition metal oxides having an inverse spinel structure are not appropriate at present, because the transition metal ions at tetrahedral 8(a) sites block lithium-ion transfer in a solid matrix. The spinel-framework structure based on a normal spinel has been accepted for lithium-ion technology because of its 3D channels consisting of 1D tunnels of vacant octahedral sites, which make lithium-ion transportation possible throughout the solid matrix.

2.3 Derivatives of Spinel-Framework Structure

The previous section reviewed the structural description of spinel. To extend the knowledge obtained from the "spinel" structure and, hopefully, to give some insight into material design for advanced batteries, derivatives of spinel-framework structure are described in this section. Although considerable amount of experimental and theoretical work has been done for predicting electrochemical behavior of lithium insertion materials, it is difficult to predict the electrochemical performances in nonaqueous lithium cells. Therefore, in this section, "material design" means the construction of the appropriate crystal structure by using geometrical methods; the optimum chemical composition is then determined from the structure under the restricted conditions discussed previously [1] for lithium insertion materials.

One of the methods to design the crystal structure based on "spinel" is to use the concept of "superlattice." A superlattice is derived from cation ordering, which results in the materials having the space group symmetry lower than that of $Fd\bar{3}m$. Materials design based on the superlattice allows us to expect something new for lithium insertion materials, and therefore this should stimulate us to find such superlattice materials. In this section, superlattice structures (or superstructures) derived from "spinel" are surveyed, and the possibility of superstructural lithium insertion materials using the oxidation states and ionic radii of transition metal ions to form cation ordering in spinel-framework structure is discussed.

2.3.1 Superlattice Structures Derived from "Spinel"

Possible superlattice structures derived from "spinel", which have space group symmetry lower than $Fd\bar{3}m$, can be found by using group–subgroup relations [5–7]. For example, space group of $Fd\bar{3}m$ is reduced to $F\bar{4}3m$ by eliminating some symmetry operations; tetrahedral A sites, 8(a) sites in $Fd\bar{3}m$, can be divided into two tetrahedral 4(a) and 4(c) sites in $F\bar{4}3m$. Consequently, superlattice structures with

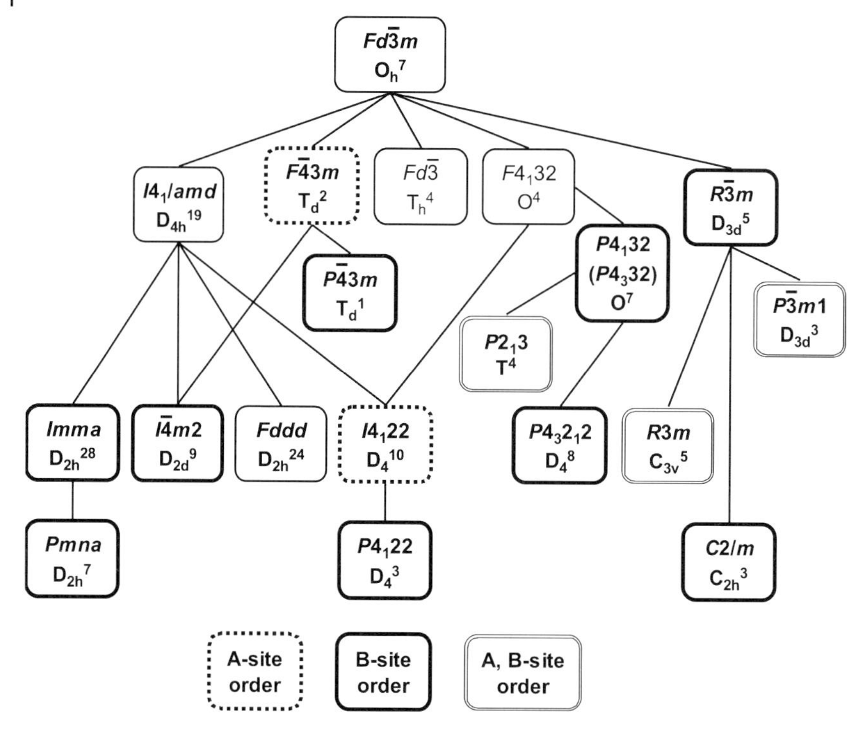

Fig. 2.3 Group–subgroup relations obtained by the reduction of the space group symmetry of $Fd\bar{3}m$. Superstructures derived from "spinel" are distinguished in terms of cation-ordering site, i.e., A-site, B-site, or both A- and B-site order (see in text).

1 : 1 ratio of cation ordering at A sites can be expected, so that A sites can be occupied by two kinds of cations or by cations and vacant sites, which restrict the chemical composition. By applying group–subgroup relations for finding superlattice, many possible structures can be found. Figure 2.3 shows the group–subgroup relations obtained by the reduction of the space group symmetry of $Fd\bar{3}m$. Lines connecting two space groups indicate the super- or subgroup relation between two space groups. Geometrically, many types of superlattice structures can be obtained from "spinel." Table 2.2 summarizes the superlattice structures derived from "spinel" together with some examples of superlattice materials prepared and examined so far. As seen in Table 2.2, superlattice structures are classified as (i) cation ordering at tetrahedral A site, (ii) cation ordering at octahedral B site, and (iii) cation ordering at both tetrahedral A and octahedral B sites.

When 1 : 3 order on B sites occurs, there are three types of superstructures due to the different cation ordering. The crystal structures of these superstructures can be illustrated using space group symmetries of $P4_332$ (or $P4_132$), $R\bar{3}m$, and *Pmna*. These three structures derived from spinel are illustrated in Figure 2.4. The difference among these structures is the distribution of two cations in the octahedral

Table 2.2 Superstructures due to cation ordering in "spinel".

Order	Space Group	Formulation	Example
	$Fd\bar{3}m$-$O_h{}^7$	$A[B_2]O_4$	$Mg[Al_2]O_4$
1 : 1 order at A sites	$F\bar{4}3m$-$T_d{}^2$	$A_{1/2}A'_{1/2}[B_2]O_4$	$Li_{1/2}Fe_{1/2}[Cr_2]O_4$
1 : 2 order at A sites	$I4_122$-$D_4{}^{10}$	$A_{1/3}A'_{2/3}[B_2]O_4$	$In_{2/3}\square_{1/3}[In_2]S_4$
1 : 1 order at B sites	$P4_322$-$D_4{}^7$	$A[BB']O_4$	$Zn[LiNb]O_4$
	$Imma$-$D_{2h}{}^{28}$	$A[BB']O_4$	$V[LiCu]O_4$
1 : 3 order at B sites	$P4_332$-O^6	$A[B_{1/2}B'_{3/2}]O_4$	$Fe[Li_{1/2}Fe_{3/2}]O_4$
			$Al[Li_{1/2}Al_{3/2}]O_4$
	$R\bar{3}m$-$D_{3d}{}^5$	$A[B_{1/2}B'_{3/2}]O_4$	Unknown
	$Pmna$-$D_{2h}{}^7$	$A[B_{1/2}B'_{3/2}]O_4$	Unknown
1 : 5 order at B sites	$P4_12_12$-$D_4{}^4$	$A[B_{1/3}B'_{5/3}]O_4$	$Fe[\square_{1/3}Fe_{5/3}]O_4$
1 : 1 order at A sites	$P2_13$-T^4	$A_{1/2}A'_{1/2}[B_{1/2}B'_{3/2}]O_4$	$Li_{1/2}Zn_{1/2}[Li_{1/2}Mn_{3/2}]O_4$
1 : 3 order at B sites			

16(d) sites in $Fd\bar{3}m$. To illustrate cation ordering at octahedral sites, we usually use the triangular lattice of sites. Figure 2.5 shows how spinel-framework structure is illustrated in the triangular lattice of sites. The spinel-framework structure consists of the $\square_{3/4}M_{1/4}O_2$ and $\square_{1/4}M_{3/4}O_2$ layers and they piled up alternately. One is the transition metal-poor sheet $\square_{3/4}M_{1/4}O_2$, in which cations are [2 × 2] ordering, and the other is the transition metal-rich sheet $\square_{1/4}M_{3/4}O_2$, in which vacant octahedral sites are also [2 × 2]-ordering. Therefore, the spinel-framework structure is the [2 × 2]-superstructure with respect to a layered formulation. As clearly seen in Figure 2.6, two kinds of cations are distributed in both layers for

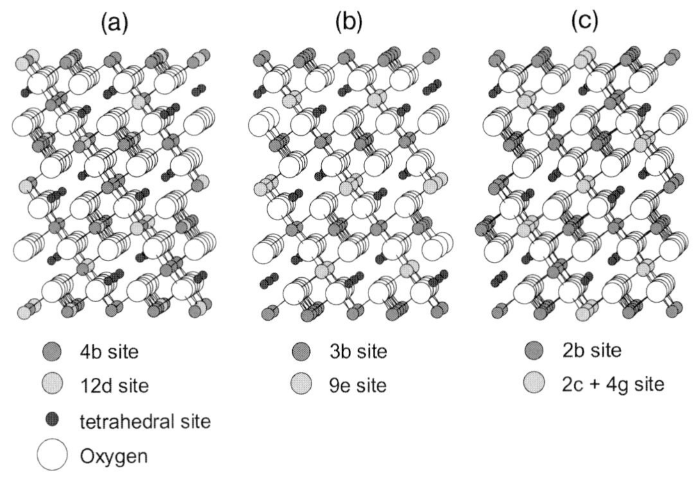

Fig. 2.4 The crystal structures on three types of superstructures of $Li[M'_{1/2}M_{3/2}]O_4$ having space group symmetry of (a) $P4_332$, (b) $R\bar{3}m$, and (c) $Pmna$, where M′ and M denote metal ions.

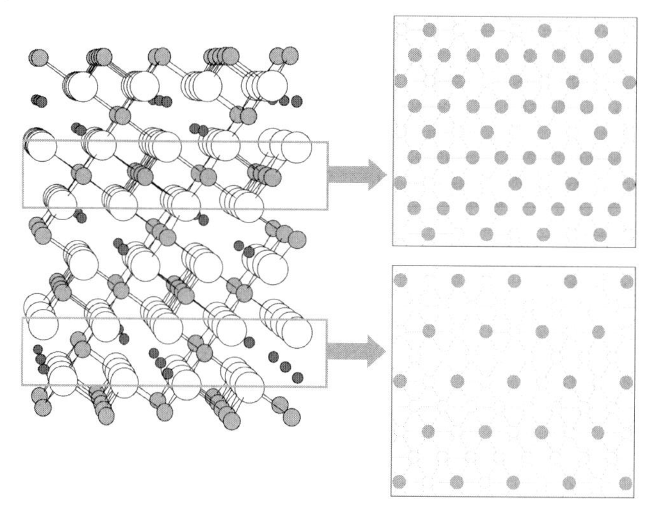

Fig. 2.5 Schematic illustration of the triangular lattice of sites for the spinel-framework structure; Transition metal-rich sheet ($\square_{1/4}M_{3/4}O_2$) with [2 × 2] ordering and transition metal-poor sheet ($\square_{3/4}M_{1/4}O_2$) also with [2 × 2] ordering can be seen in spinel-framework structure.

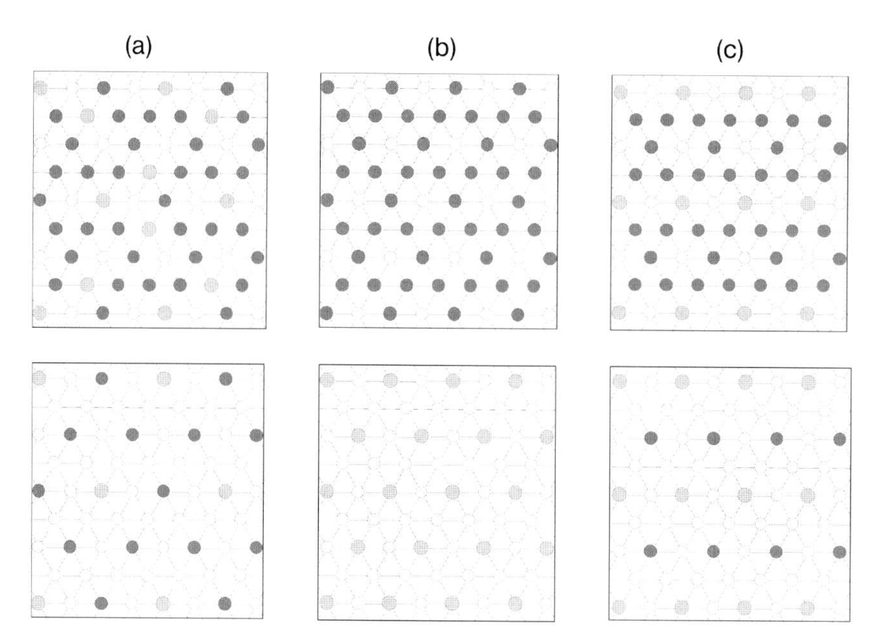

Fig. 2.6 The cation distribution of three types of superstructures of $Li[M'_{1/2}M_{3/2}]O_4$ having space group symmetry of (a) $P4_332$, (b) $R\bar{3}m$, and (c) *Pmna*. Transition metal-rich sheet is illustrated at the top and transition metal-poor sheet at the bottom.

Table 2.3 Possible chemical composition and species to form superstructures due to cation ordering.

Charges for cations	Formulation	order on B-site	
1–4	$Li[M^{+}{}_{1/3}\ M'^{4+}{}_{5/3}]O_4$	1 : 5 order	$M^{+} = Li^{+}, Na^{+}, K^{+}, Cu^{+}, \ldots$
1–5	$Li[M^{+}{}_{3/4}\ M'^{5+}{}_{5/4}]O_4$	3 : 5 order ?	
1–6	$Li[M^{+}\ M'^{6+}]O_4$	1 : 1 order	
2–4	$Li[M^{2+}{}_{1/2}\ M'^{4+}{}_{3/2}]O_4$	1 : 3 order	$M^{2+} = Mg^{2+}, Ni^{2+}, Cu^{2+}, Zn^{2+}, \ldots$
2–5	$Li[M^{2+}\ M'^{5+}]O_4$	1 : 1 order	
2–6	$Li[M^{2+}{}_{5/4}\ M'^{6+}{}_{3/4}]O_4$	5 : 3 order ?	
3–4	$Li[M^{3+}\ M'^{4+}]O_4$	1 : 1 order	$M^{3+} = Al^{3+}, Cr^{3+}, Fe^{3+}, Co^{3+}, \ldots$
3–5	$Li[M^{3+}{}_{3/2}\ M'^{5+}{}_{1/2}]O_4$	3 : 1 order	
3–6	$Li[M^{3+}{}_{5/3}\ M'^{6+}{}_{1/3}]O_4$	5 : 1 order	

$P4_332$, separated into two sheets for $R\bar{3}m$, and distributed periodically, drawing stripes for $Pmna$.

To find possible space group lower than $Fd\bar{3}m$ is easy. However, chemical species to construct superlattice structure are limited in considering lithium insertion materials. Table 2.3 summarizes possible chemical compositions for $Li[M_xM'_{2-x}]O_4$. For manganese-based spinel, three types of possible superstructures can be found, i.e., 1 : 5 order on B sites for Li^{+} and Mn^{4+}, 1 : 3 order for divalent cation and Mn^{4+}, and 1 : 1 order between trivalent cation and Mn^{4+}. Chemical compositions of the structures are linked with the oxidation state of chemical species due to the conservation of charge. From the electrostatic point of view, difference in oxidation states between different cations is preferable to form the superlattice. Oxidation states are closely related to ionic radii [8]. When two different cations having quite different ionic radii are mixed, phase separation or the formation of superlattice is expected in order to reduce internal stress derived from the difference in size. Figure 2.7 shows ionic radii of the 3d-transition metal ions together with Li^{+}, Mg^{2+}, and Al^{3+} [8]. The values are selected to be at octahedral sites, i.e., coordination number (CN) = 6. Upper and lower dashed lines in Figure 2.7 indicate ionic radii of Mn^{3+} (HS; 0.645 Å) and Mn^{4+} (0.53 Å). HS and LS stand for high-spin and low-spin states, respectively. Cations having almost the same ionic radii as that of manganese ions tend to form substitutional solid solution, not superlattice structure. In other words, if one intends to make superlattice structure of $LiM_xMn_{2-x}O_4$, cations having larger or smaller ionic radius than manganese ions should be selected. Actually, Mg^{2+}-substituted $LiMn_2O_4$ forms the superlattice in which Mg^{2+} and Mn^{4+} are ordered in the ratio of 1 : 3 at the octahedral B sites, because of big difference in ionic radii, i.e., 0.72 Å for Mg^{2+} and 0.53 Å for Mn^{4+}. Divalent ions, Ni^{2+}, Cu^{2+}, and Zn^{2+} in Figure 2.7 are possible for superlattice formation of the same kind [9]. The crystal structure and electrochemical properties for these materials are described in the next sections.

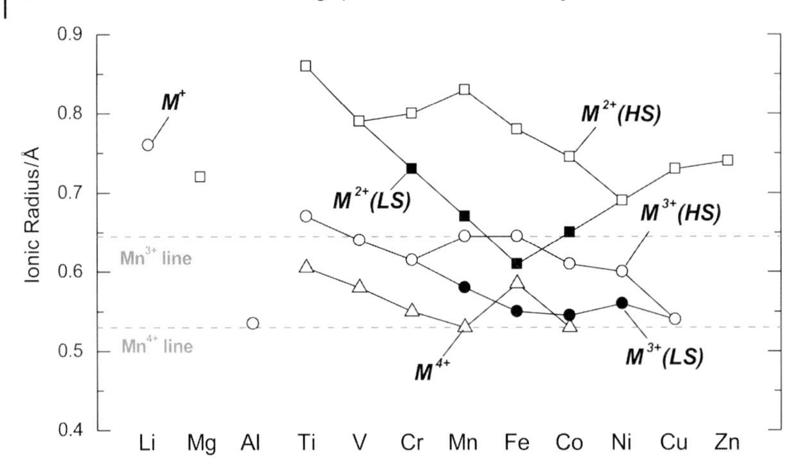

Fig. 2.7 Ionic radii of the 3d-transition metal ions together with Li^+, Mg^{2+}, and Al^{3+}. Ionic radii given in this figure are taken from [8] with coordination number (CN) = 6. HS and LS stand for high-spin and low-spin states, respectively.

2.3.2
Examples of Superstructure Derived from "Spinel"

Because Zn^{2+} ions prefer to occupy the tetrahedral A sites, lithium zinc manganese oxide of $LiZn_{1/2}Mn_{3/2}O_4$ is an inverse spinel, i.e., $Li_{1/2}Zn_{1/2}[Li_{1/2}Mn_{3/2}]O_4$. Figure 2.8 shows the X-ray diffraction (XRD) patterns of $LiZn_{1/2}Mn_{3/2}O_4$ prepared by different conditions: (a) two-step solid-state reaction of oxidation at 650°C for 12 hours following the heating process at 1000°C for 12 hours, (b) heating at 900°C and then cooling, and (c) heating at 1000°C and then cooling. All diffraction lines can be indexed by assuming a cubic lattice. The strong (220) diffraction lines in Figure 2.8 are indicative of inverse spinel structure. The locations and intensities of main diffraction lines, i.e., (111), (220), (311), (400), and (440), are almost the same. However, extra diffraction lines are observed in Figure 2.8(a) and (b). This suggests that there are three types of $Li_{1/2}Zn_{1/2}[Li_{1/2}Mn_{3/2}]O_4$ with different cation distribution.

The first possible structure is $Li_{1/2}Zn_{1/2}[Li_{1/2}Mn_{3/2}]O_4$ with random distribution at both the tetrahedral A and the octahedral B sites, but this is not a superstructure. The second one is the superlattice $Li_{1/2}Zn_{1/2}[Li_{1/2}Mn_{3/2}]O_4$ with cation ordering at the tetrahedral A sites with 1 : 1 ratio of Li and Zn while Li and Mn randomly distribute at the octahedral B sites. The third is superlattice $Li_{1/2}Zn_{1/2}[Li_{1/2}Mn_{3/2}]O_4$ with cation ordering at the octahedral B sites with 1 : 3 ratio of Li and Mn while Li and Zn randomly distribute at the tetrahedral A sites. The fourth one is the superstructural $Li_{1/2}Zn_{1/2}[Li_{1/2}Mn_{3/2}]O_4$ with 1 : 1 ordering of Li and Zn at the tetrahedral A sites and the 1 : 3 ordering of Li and Mn at the octahedral B sites. Figure 2.9 shows the calculated XRD patterns of possible superstructural $LiZn_{1/2}Mn_{3/2}O_4$ with different cation ordering: (i) space group of $Fd\bar{3}m$ with random distribution

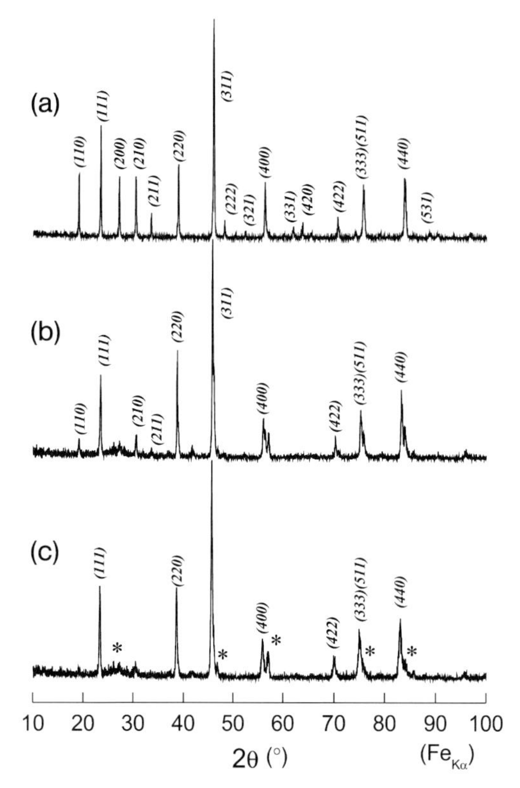

Fig. 2.8 X-ray diffraction patterns of $LiZn_{1/2}Mn_{3/2}O_4$ prepared by heating a reaction mixture of LiOH, ZnO, and MnOOH (manganite) at (a) 1000°C for 12 hours in air followed by heating at 650°C for 12 hours in air ($a = 8.20$ Å), (b) 900°C for 12 hours in air ($a = 8.25$ Å), and (c) 1000°C for 12 hours in air ($a = 8.27$ Å). Asterisks (*) indicate the diffraction line for Li_2MnO_3.

of Li and Zn at both tetrahedral and octahedral sites; (ii) $F\bar{4}3m$ with 1 : 1 ordering of Li and Zn at the tetrahedral A sites; (iii) $P4_332$ with 1 : 3 ordering of Li and Mn at the octahedral B sites; and (iv) $P2_13$ with 1 : 1 ordering of Li and Zn at the tetrahedral A sites and 1 : 3 ordering of Li and Mn at the octahedral B sites. When cations are only ordered at the tetrahedral sites, (200) line is visible, as seen in Figure 2.9(b) for $F\bar{4}3m$. Superlattice lines of (110), (210), and (211) due to the cation ordering at the octahedral B sites are seen for $P4_332$. Consequently, $P2_13$ superstructure shows both superlattice lines. A comparison of the XRD patterns in Figure 2.8 with those in Figure 2.9 shows that $LiZn_{1/2}Mn_{3/2}O_4$ is crystallized as $Li_{1/2}Zn_{1/2}[Li_{1/2}Mn_{3/2}]O_4$ having space group symmetry of $P2_13$ [10].

Although $LiZn_{1/2}Mn_{3/2}O_4$ provides useful information on cation ordering among lithium, divalent, and tetravalent ions at tetrahedral and/or octahedral sites in the spinel-framework structure, $LiZn_{1/2}Mn_{3/2}O_4$ is not good lithium insertion material [11, 12] partly because Zn^{2+} at the tetrahedral sites interferes with lithium-ion migration and partly because Zn^{2+} at the octahedral sites is not a good

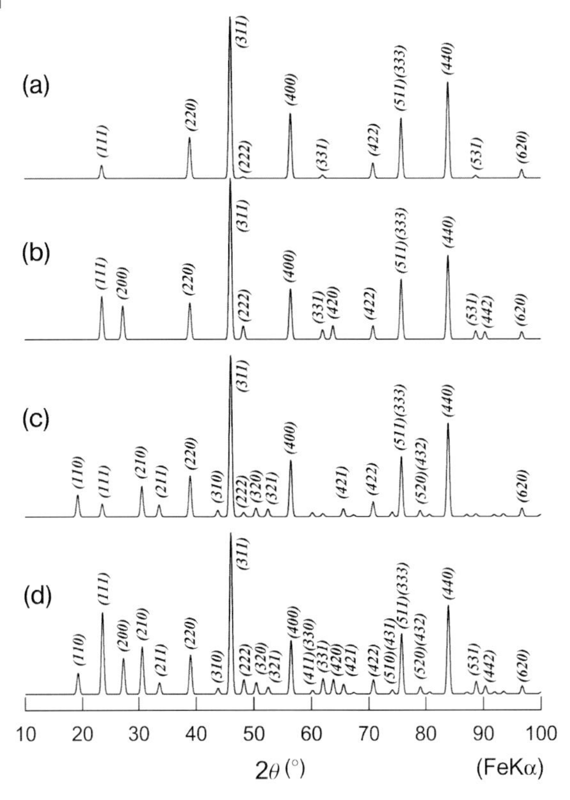

Fig. 2.9 Calculated XRD patterns of four structural models on $Li_{1/2}Zn_{1/2}[Li_{1/2}Mn_{3/2}]O_4$ ($a = 8.19$ Å) having the space group symmetry of (a) $Fd\bar{3}m$, (b) $F\bar{4}3m$, (c) $P4_332$, and (d) $P2_13$. Superlattice lines can be seen for diffraction angles below 40° in 2θ.

redox center among 3d-transition metals. The chemistry is not well suited for the lithium insertion materials. When Zn^{2+} is replaced with Ni^{2+}, cation ordering of Ni^{2+} and Mn^{4+} at octahedral sites can be expected, whereas conservation of charge in $LiNi_{1/2}Mn_{3/2}O_4$ requires the combination of either Ni^{3+} and Mn^{3+} or Ni^{2+} and Mn^{4+}. Actually, $LiNi_{1/2}Mn_{3/2}O_4$ crystallizes as superlattice structure of $Li[Ni^{2+}_{1/2}Mn^{4+}{}_{3/2}]O_4$ having the space group symmetry of $P4_332$ (or $P4_132$) [13–15], and it is widely accepted as one of the 5-V lithium insertion materials as is described in Section 2.4.

Although the symmetry of crystal structures and the local environment of cations must be understood to deal with phase changes during electrochemical reaction, intuitive understanding of crystal structures using pictures and illustrations is important to consider their lithium insertion schemes. Figure 2.10 illustrates the beautiful cation ordering of Ni^{2+} and Mn^{4+} in superstructural $Li[Ni_{1/2}Mn_{3/2}]O_4$. When the electrochemical reaction takes place, bond distances extend and shrink between transition metal ions and oxygen ions due to the change in their ionic

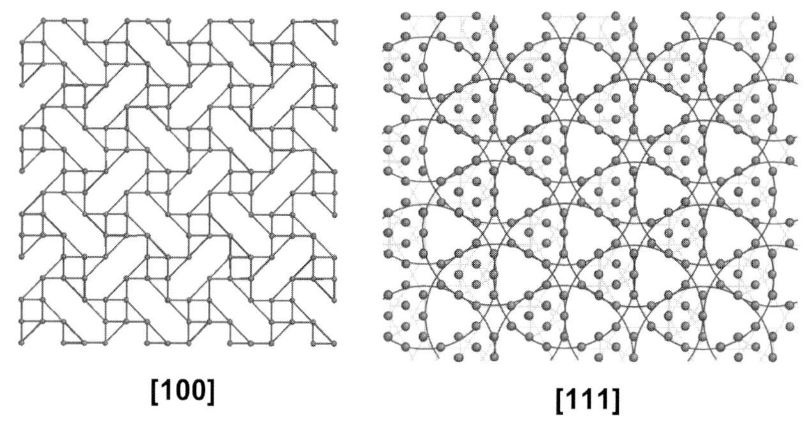

Fig. 2.10 Schematic illustration of Mn-sublattice for $Li[Ni_{1/2}Mn_{3/2}]O_4$ ($P4_332$) viewed along [100] and [111] directions. Cation ordering between Ni^{2+} and Mn^{4+} at the octahedral sites leads to a beautiful pattern that looks like an arabesque.

radii leading to the change in lattice parameters. The beautiful arabesques shown in Figure 2.10 suggest that $Li[Ni_{1/2}Mn_{3/2}]O_4$ minimizes internal stress during an electrochemical reaction because of symmetrical distribution of nickel and manganese ions in the solid matrix, which may be suitable for long cycling life from a microscopic viewpoint. Figure 2.11 shows scanning electron microscope (SEM) image of $Li[Ni_{1/2}Mn_{3/2}]O_4$ prepared at high temperature. From the viewpoint of particle morphologies, lithium insertion reaction and additional side reactions such as electrolyte decomposition take place at the interface between active material and electrolyte, and lithium ions migrate inside the particles during an electrochemical reaction. $Li[Ni_{1/2}Mn_{3/2}]O_4$ has an octahedral shape of particles with smooth (111) facets, which is ideal particle morphology for insertion materials. Such a smooth surface can minimize electrolyte decomposition and dissolution of materials, and a symmetrical octahedral shape can reduce the internal stress resulting from isotropic expansion or shrinkage of particles upon electrochemical reaction. In the octahedra, lithium-ion conduction paths run parallel to the edge of the octahedra

Fig. 2.11 Particle morphology observed by SEM for $Li[Ni_{1/2}Mn_{3/2}]O_4$ ($P4_332$) prepared at 1000°C for 12 hours and then oxidized at 700°C for 24 hours in air. Octahedral shape with smooth (111) facets can be seen.

corresponding to [110]-direction, which may be appropriate to fast lithium-ion transportation. Actually, $Li[Ni_{1/2}Mn_{3/2}]O_4$ exhibits excellent rate capability and cyclability for 2000-cycle test as described in a later section. Such an octahedral shape is characteristic of a single crystal for spinel, because, according to Wulff's theorem, the (111) facets may be a singular surface for spinel.

2.4
Electrochemistry of Lithium Insertion Materials Having Spinel-Framework Structure

As already described briefly in Section 2.2, lithium insertion materials having spinel-framework structure can be crystallized as $LiTi_2O_4$, LiV_2O_4, and $LiMn_2O_4$ by using 3d-transition elements. The reason that 3d-transition metal oxides are important for lithium insertion materials has already been discussed in a previous paper [1]. Among them, $LiTi_2O_4$ and $LiMn_2O_4$ are versatile nonstoichiometry between $LiTi_2O_4$ ($LiMn_2O_4$) and $Li[Li_{1/3}Ti_{5/3}]O_4$ ($Li[Li_{1/3}Mn_{5/3}]O_4$), and also titanium or manganese ions in spinel-framework structure can easily be replaced by other transition metal ions. LiV_2O_4 ($Fd\bar{3}m$; $a = 8.24$ Å) is an interesting material in its electronic structure. The reaction mechanism of LiV_2O_4 is known to be a topotactic two-phase reaction over the entire range of x in $Li_{1+x}V_2O_4$, i.e., LiV_2O_4 ($a = 8.24$ Å) + $Li^+ + e^- \rightarrow Li_2V_2O_4$ ($a = 8.29$ Å) [16, 17]. However, the reversible voltage of LiV_2O_4 is 2.43 V versus Li, meaning that LiV_2O_4 cannot be applied to a negative electrode for lithium-ion batteries and also to a positive electrode for rechargeable lithium batteries. LiV_2O_4 is neither one thing nor the other in terms of lithium insertion materials for batteries. In this section, we focus upon titanium and manganese systems.

2.4.1
Lithium Manganese Oxides (LMO)

Electrochemical reaction of stoichiometric $LiMn_2O_4$ ($a = 8.24$ Å) is known to proceed in a topotactic manner, which is devided into three parts. The reaction consists of cubic ($a = 8.03$ Å) – cubic ($a = 8.14$ Å), two phases at 4.11 V versus Li, cubic one phase ($a = 8.14$–8.24 Å) at 3.94 V of midpoint voltage, and cubic ($a = 8.24$ Å) – tetragonal two phases at 2.96 V [18]. Manganese dioxide having spinel-framework structure is called λ-MnO_2 [19]. Although stoichiometric $LiMn_2O_4$ shows high reversible voltage of about 4.0 V with 148 mAh g^{-1} of theoretical capacity, it shows poor cyclability in nonaqueous lithium cells. To extend cycle life with minimum sacrifice of rechargeable capacity, several trials have been carried out in the last 15 years [20, 21].

LMOs having spinel-framework structure show an interesting character – manganese ions at octahedral sites in $LiMn_2O_4$ can be replaced by lithium ions, forming $Li[Li_xMn_{2-x}]O_4$ without noticeable line broadening compared to stoichiometric $LiMn_2O_4$. Figure 2.12 shows the lattice constant a as a function of x in $Li[Li_xMn_{2-x}]O_4$ [22, 23]. Lattice dimension a shrinks linearly as x

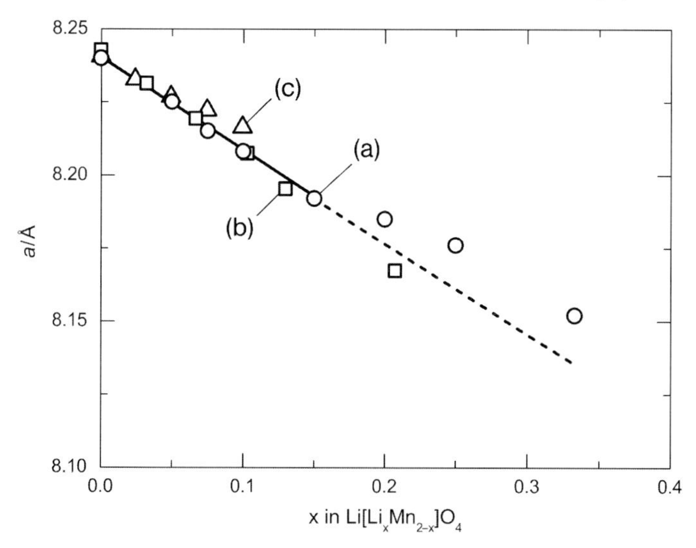

Fig. 2.12 The cubic lattice constant as a function of x in $Li[Li_xMn_{2-x}]O_4$. The values are taken from [22] (open squares (b)) and [23] (open triangles (c)) in addition to original data (a).

increases in $x = 0–0.2$, which is empirically represented as $a = 8.24 - 0.32x$ in $Li[Li_xMn_{2-x}]O_4$.

Charge and discharge curves of a series of LMOs are shown in Figure 2.13. Rechargeable capacity of $Li[Li_xMn_{2-x}]O_4$ decreases and operating voltage seemingly becomes sloping with increasing x in $Li[Li_xMn_{2-x}]O_4$. However, $Li[Li_xMn_{2-x}]O_4$ shows better cyclability and lower polarization than stoichiometric $LiMn_2O_4$. There is a trade-off relation between rechargeable capacity and cyclability. Charge and discharge capacities for $Li[Li_xMn_{2-x}]O_4$ at the first cycle as a function of x are shown in Figure 2.14. The solid line indicates theoretical capacity calculated using equation, $Li[Li_xMn^{3+}{}_{1-3x}Mn^{4+}{}_{1+2x}]O_4 \rightarrow \square_{1-3x}Li_{3x}[Li_xMn^{4+}{}_{2-x}]O_4 + (1-3x)Li^+ + (1-3x)e^-$, in which symbol for open square denotes the vacant site.

According to the equation, the rechargeable capacity at 4 V is limited by the availability of trivalent manganese ions, the so-called electron source-limited capacity [1]. As clearly seen in Figure 2.13, the rechargeable capacity of $Li[Li_xMn_{2-x}]O_4$ of about 100 mAh g^{-1} with good cyclability is attained, although the rechargeable capacity as close as 148 mAh g^{-1} of theoretical capacity for $LiMn_2O_4$ is difficult. For nearly stoichiometric samples of $x = 0$ and 0.05, small voltage steps at 4.5 V on charge and 4.3 and 3.3 V on discharge are seen in addition to 4 V, which are due to the reversible structural transformation from spinel to double hexagonal phases and vice versa [25, 26]. These voltage steps can be used as warning signals for capacity fading [27].

When $Li/Li[Li_xMn_{2-x}]O_4$ cells are operated at voltage above 5 V, a series of $Li[Li_xMn_{2-x}]O_4$ shows 5 V redox signals. Charge and discharge curves of $Li[Li_xMn_{2-x}]O_4$ in voltages of 1.0–5.2 V are shown in Figure 2.15. All $Li[Li_xMn_{2-x}]$

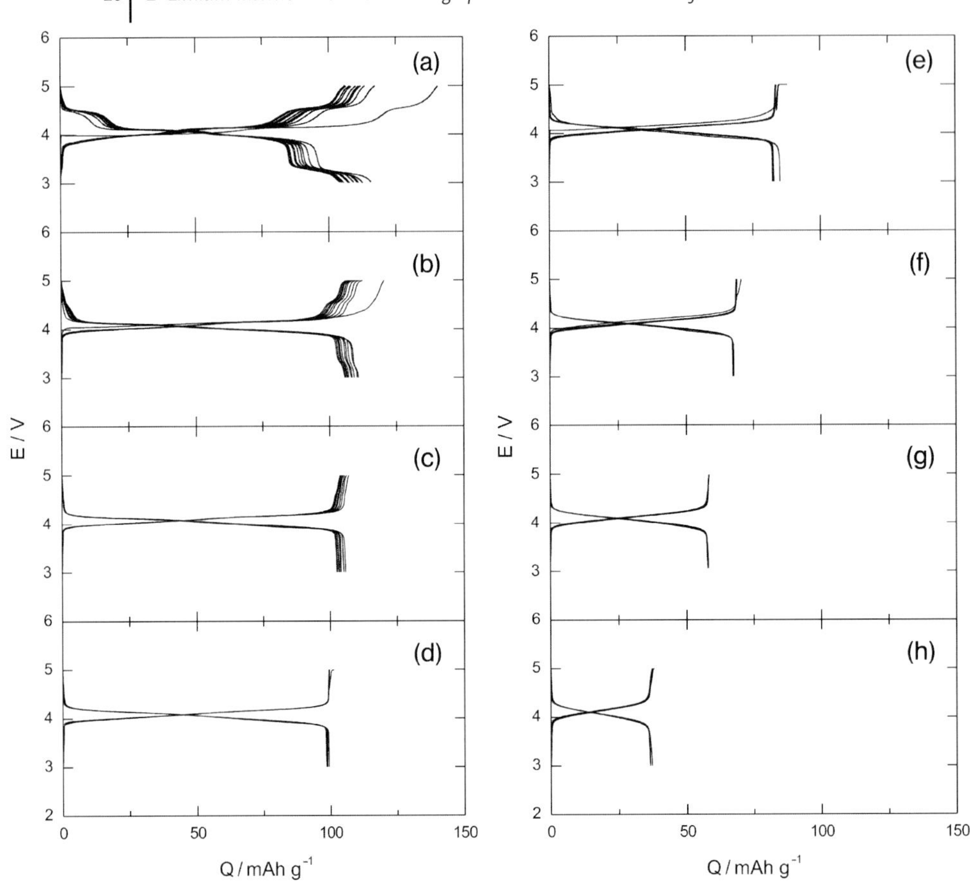

Fig. 2.13 Charge and discharge curves of $Li/Li[Li_xMn_{2-x}]O_4$ cells operated at a rate of 0.17 mA cm^{-2} in voltages of 3.0–5.0 V; (a) $x = 0$, (b) $x = 0.05$, (c) $x = 0.075$, (d) $x = 0.10$, (e) $x = 0.15$, (f) $x = 0.20$, (g) $x = 0.25$, and (h) $x = 0.33$.

O_4 ($0 \leq x < 1/3$) show almost the same rechargeable capacities in voltages of 3–5.2 V, i.e., 120–130 mAh g^{-1}, while the rechargeable capacities in the voltages of 3.5–4.5 V decreased as a function of x as seen in Figure 2.14. All samples give the same cubic lattice constant of $a \sim 8.0$ Å when the samples are potentiostatically charged at 5.0 or 5.2 V, suggesting that all samples are converted to λ-MnO_2, probably releasing oxygen ions [28].

Manganese dissolution from $Li[Li_xMn_{2-x}]O_4$ especially at 55 °C is another problem when the material is combined with the graphite-negative electrode. The addition of fluorine or other transition metal ions is reported to be effective to protect the manganese dissolution from the positive electrode [24]. As was reviewed by Amatucci and Tarascon [21], LMO has been examined as a possible alternative to $LiCoO_2$. However, LMO is hardly applied to high energy density lithium-ion

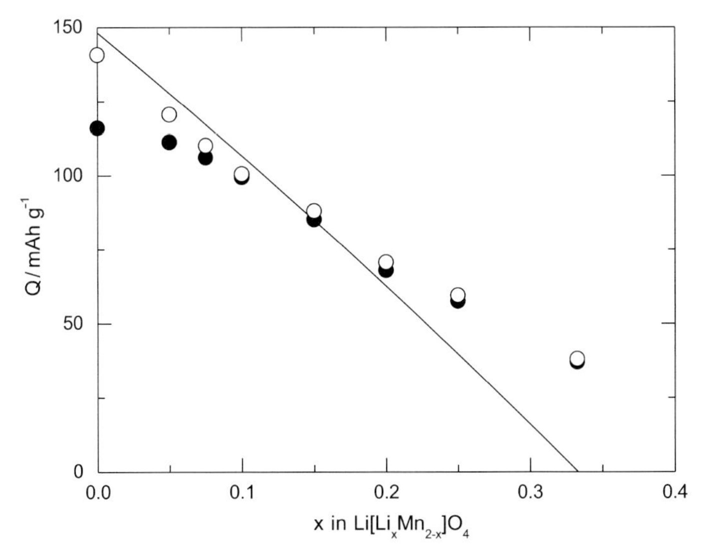

Fig. 2.14 Charge and discharge capacities as a function of x in $Li[Li_xMn_{2-x}]O_4$ observed during the first charge and discharge cycle. Solid line indicates theoretical capacity assuming electron source-limited capacity (see in text).

batteries with graphite-negative electrodes because $LiCoO_2$ is excellent in its performance. An application of LMO to high-power long-life batteries is described in Section 2.5.

2.4.2 Lithium Titanium Oxide (LTO)

The LTO $Li[Li_{1/3}Ti_{5/3}]O_4$ ($a = 8.365$ Å) has the unique characteristic that the lattice dimension does not change during the reduction to $Li_2[Li_{1/3}Ti_{5/3}]O_4$, so-called zero-strain lithium insertion material [29–32]. The reaction mechanism is shown to be topotactic and is represented as

$$Li_{8(a)}[Li_{1/3}Ti^{4+}{}_{5/3}]_{16(d)}O_4 + Li^+ + e^- \longrightarrow Li_{2\ 16(c)}[Li_{1/3}Ti^{3+}Ti^{4+}{}_{2/3}]_{16(d)}O_4 \qquad (2.1)$$

The reduction product still has tetravalent titanium ions to accept electrons, but no vacant octahedral sites to accommodate lithium ions, so-called the site-limited capacity [1]. The zero-strain insertion mechanism is also examined and explained in terms of "oxygen swing," with which the lattice dimension remains constant in spite of the change in Ti–O bond length during charge and discharge, i.e., change in the oxygen positional parameter [33].

The reversible voltage of $Li[Li_{1/3}Ti_{5/3}]O_4/Li_2[Li_{1/3}Ti_{5/3}]O_4$ is 1.55 V versus Li and the theoretical capacity is 175 mAh g^{-1} or 610 mAh cm^{-3}. Figure 2.16 shows the charge and discharge curves of LTO examined in a nonaqueous lithium cell. LTO shows no limitation of cycling life because of "zero-strain." Although rechargeable

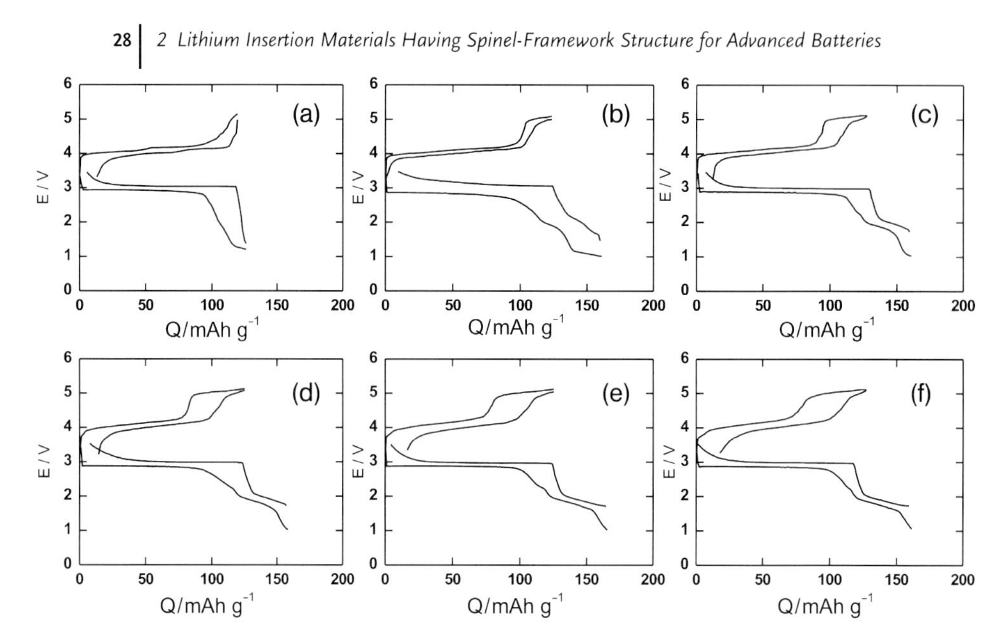

Fig. 2.15 Charge and discharge curves of $Li[Li_xMn_{2-x}]O_4$ in voltages of 1.0–5.2 V. Redox reaction at 5 V are commonly observed for all $Li[Li_xMn_{2-x}]O_4$ except $x = 0$ ($LiMn_2O_4$); (a) $x = 0$, (b) $x = 0.10$, (c) $x = 0.15$, (d) $x = 0.20$, (e) $x = 0.25$, and (f) $x = 0.33$.

gravimetric and volumetric capacities are relatively high together with excellent cyclability among lithium insertion materials, the operating voltage is about 1.5 V higher than that of current graphite-negative electrode currently used in lithium-ion batteries. This impedes the application of LTO to the negative electrode for high energy density lithium-ion batteries.

2.4.3
Lithium Nickel Manganese Oxide (LiNiMO)

As was described in Section 2.3, the lithium nickel manganese oxide (LiNiMO) of $Li[Ni_{1/2}Mn_{3/2}]O_4$ is crystallized as the superstructure having space group symmetry of $P4_332$. The reaction mechanism of LiNiMO consisting of Ni^{2+} and Mn^{4+} is already known to have three topotactic two-phase reactions [14], i.e., cubic ($a = 8.00$ Å)/cubic ($a = 8.09$ Å) two-phase reaction at 4.74 V versus Li in a range of 0–1/2 in $Li_x[Ni_{1/2}Mn_{3/2}]O_4$, cubic ($a = 8.09$ Å)/cubic ($a = 8.17$ Å) two-phase reaction at 4.72 V in $x = 1/2–1$, and cubic ($a = 8.17$ Å)/tetragonal ($a = 5.74$ Å, $c = 8.69$ Å) at 2.80 V in $x = 1–2$. The theoretical capacities of 147 mAh g^{-1} and 655 mAh cm^{-3} are almost the same as those of $LiMn_2O_4$ due to almost identical formula weight and lattice dimension. The difference is, however, seen in rechargeable capacity. LiNiMO shows a rechargeable capacity of more than 135 mAh g^{-1}, whereas LMO shows about 100 mAh g^{-1}. The charge and discharge curves of an Li/LiNiMO cell operated at a rate of 0.25 mA cm^{-2} are

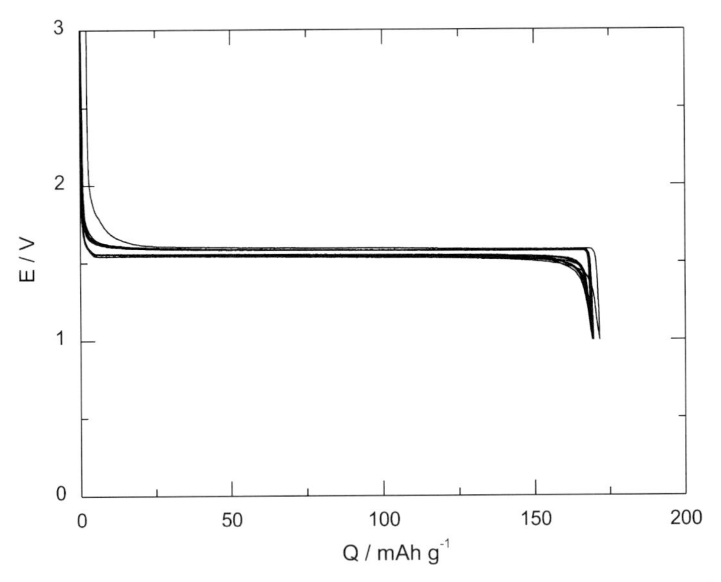

Fig. 2.16 Discharge and charge curves of a Li/LTO cell operated at 0.17 mA cm^{-2} in voltages of 1.0–3.0 V. The electrode consisted of 88 wt% of LTO, 6 wt% of acetylene black (AB), and 6 wt% of polyvinylidene fluoride (PVdF). The weight and thickness of electrode mix was 38.5 mg and 110 μm, respectively. The electrolyte used was 1 M $LiPF_6$ dissolved in EC/DMC solution (3/7 by volume). Two sheets of polypropylene membrane were used as a separator.

shown in Figure 2.17. The reversible voltage of 4.73 V versus Li with rechargeable capacity of about 135 mAh g^{-1} is the highest value among lithium insertion materials reported so far. The combination of LiNiMO and graphite gives the highest operating voltage of about 4.5 V with the highest energy density of 440 Wh kg^{-1} or 1500 Wh dm^{-3} calculated using the values of 135 mAh g^{-1} and 600 mAh cm^{-3} for LiNiMO and 350 mAh g^{-1} and 790 mAh cm^{-3} for $\Box C_6$. The volumetric energy density of more than 700 Wh dm^{-3} is estimated by analogy with the current lithium-ion batteries (18650) consisting of $LiCoO_2$ and graphite. Although LiNiMO potentially has superior properties to $LiCoO_2$ as a positive electrode in lithium-ion batteries, the practical reality of a lithium-ion battery consisting of LiNiMO and graphite is still an open question. We need electrolyte stability in voltages of −0.5 to +5.5 V versus Li to steadily operate the 4.5 V lithium-ion battery.

2.5
An Application of Lithium Insertion Materials Having Spinel-Framework Structure to 12 V "Lead-Free" Accumulators

As described in Section 2.4, LMO, LTO, and LiNiMO have their own characteristic values. However, they may not find application for high energy density

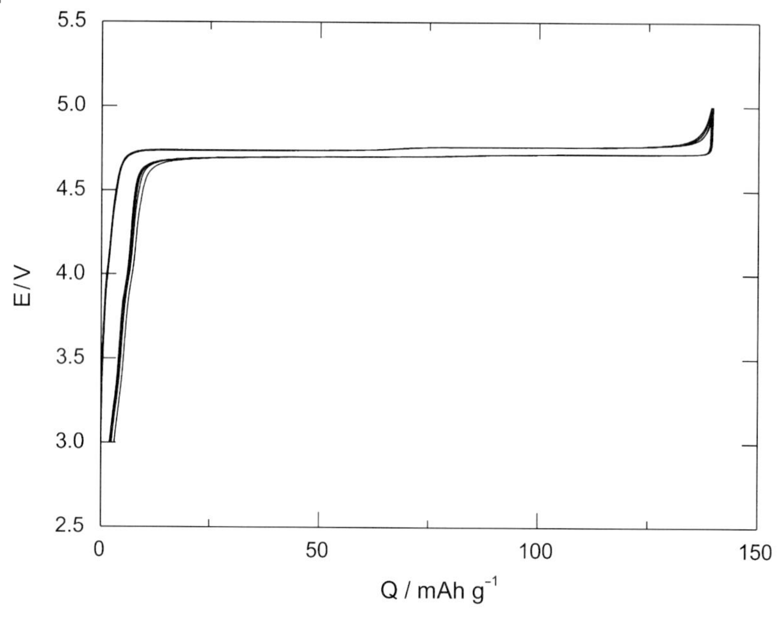

Fig. 2.17 Charge and discharge curves of a $Li/Li[Ni_{1/2}Mn_{3/2}]O_4$ cell operated at 0.25 mA cm^{-2} in voltages of 3.0–5.0 V. The electrode consisted of 88 wt% of $Li[Ni_{1/2}Mn_{3/2}]O_4$, 6 wt% of AB, and 6 wt% of PVdF. The weight and thickness of the electrode mix was 36.8 mg and 109 μm, respectively. The electrolyte used was 1 M $LiPF_6$ dissolved in EC/DMC solution (3/7 by volume). Two sheets of polypropylene membrane were used as a separator.

lithium-ion batteries because the combination of $LiCoO_2$ and graphite is excellent in its performance. Battery history has taught us that unless new applications of lithium insertion materials are proposed, designed, fabricated, and introduced for consumer use, the interest in basic and applied research on new materials will fade over the years [4]. Therefore, we need to find the application of LMO, LTO, and LiNiMO for the advancement of the science and technology of the lithium insertion materials having spinel-framework structure [34–36].

Twelve-volt lead-acid batteries are commonly used for automobile and stationary applications. Since 1859, lead-acid batteries have improved in their performance and have been established as among the most important secondary batteries. The energy density of 12-V lead-acid batteries ranges from 20 to 30 Wh kg^{-1} depending on the application. In 12-V lead-acid batteries, six cells are connected in series, since a single cell exhibits 2 V of operating voltage, which is the highest value among aqueous batteries. The cell reaction is usually described as the *double-sulfate* theory, i.e., $PbO_2 + Pb + 2H_2SO_4 \rightarrow 2PbSO_4 + 2H_2O$.

The discharge product is $PbSO_4$ at both positive and negative electrodes. Both electrodes expand upon discharge because of the destruction of the crystal structure

Table 2.4 Charge-end and discharge-end voltages together with operating voltage (nominal voltage for practical battery) of the unit cell required for the 12-V batteries.

Number of cells	Charge-end voltage (14.4 V)	Nominal voltage (12 V)	Discharge-end voltage (10.2 V)
3	4.8	4.0	3.40
4	3.6	3.0	2.55
5	2.9	2.4	2.04
6	2.4	2.0	1.70

of PbO_2 and Pb and the formation of $PbSO_4$. Such dimensional instability of both electrodes causes cycling failure, which is a key issue in lead-acid batteries. As regards the electrolyte in a lead-acid battery, H_2SO_4 is consumed on discharge and reproduced on charge. However, 100% H_2SO_4 cannot be used in lead-acid battery because of corrosion problems. Diluted H_2SO_4 having specific gravity of about 1.2, i.e., about 3.5 M H_2SO_4, which is called *flooded electrolyte,* is used in commercialized lead-acid batteries. Since sufficient amount of electrolyte is necessary to react with both electrodes in a cell, such diluted electrolyte largely reduces specific capacities both in ampere hours per kilogram and per cubic decimeter and, consequently, energy densities in watt hours per kilogram and per cubic decimeter.

As briefly described, lead-acid batteries may not be appropriate for advanced automobile and stationary applications, such as batteries for lightweight eco-cars, electric vehicles (EVs), uninterruptible power supply (UPS), and solar or wind power generation systems. There seems to be no room for improvement in the performance including energy density. The charge-end voltage of 12-V lead-acid batteries is typically 14.4 V (maximum 14.8 V). The discharge-end voltage is 10.2 V for 5-hour rate or 7.2 V when the burst of energy is supplied for short period of time. The 12-V lead-acid batteries are never fully discharged, which is the end of cell life. The charge and discharge at 5-hour rate are high for lead-acid batteries because of extremely slow convection and diffusion of sulfuric acid produced or consumed at the positive and negative electrodes. Table 2.4 summarizes the unit cell required for the construction of 12-V batteries. When the 12-V batteries are designed using five cells connected in series, an operating voltage ranging from 2.0 to 2.9 V is required with nominal voltage of 2.4 V, and an operating voltage of 2.55–3.6 V is required for the series connection of four cells. As clearly seen in Table 2.4, current lithium-ion battery consisting of $LiCoO_2$ and graphite cannot be applied to 12-V batteries, because the battery is operated in voltages of 3.0–4.2 V with a nominal voltage of 3.7 V. In this section, 12-V "lead-free" accumulators consisting of LTO and LMO (or LiNiMO) are described.

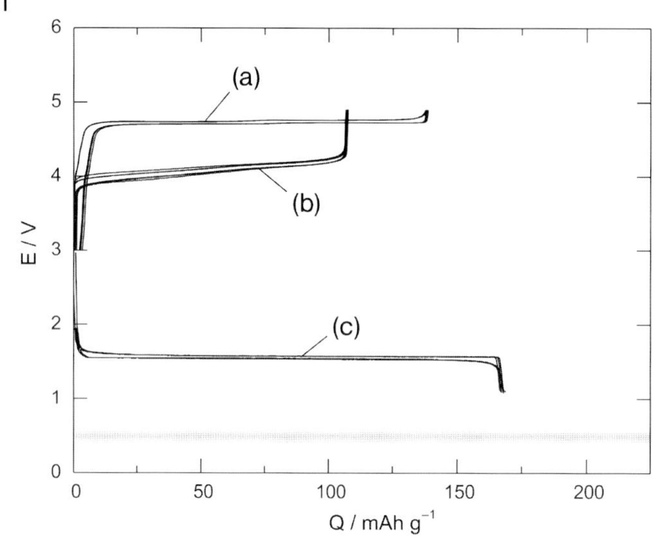

Fig. 2.18 Voltage profiles for the lithium insertion materials having spinel-framework structure: (a) LiNiMO, (b) LAMO, and (c) LTO. The diffused line at about 0.5 V is the targeted voltage of a negative electrode for 12-V batteries in which three cells are connected in series (see in text).

2.5.1
Twelve-Volt Batteries Consisting of Lithium Titanium Oxide (LTO) and Lithium Manganese Oxide (LMO)

Figure 2.18 shows the charge and discharge curves of LTO, LMO, and LiNiMO having spinel-framework structure. The combination of LTO and LMO gives a 2.5-V battery, which is well suited to the condition required for the construction of 12-V batteries. Figure 2.19 shows the charge and discharge curves of LTO/LAMO cell operated in voltage of 1.0–3.0 V. LAMO stands for lithium aluminum manganese oxide [37]. To stabilize LMO, aluminum is added in the spinel-framework structure. According to the results of prototype cells (31.95-mm diameter and 122.05-mm height), the cell delivers 6 Ah with nominal voltage of 2.5 V with excellent cycle life more than 3600 cycles. The calculated energy densities are 70 Wh kg^{-1} and 150 Wh dm^{-3}, which are far lower than those of current lithium-ion battery consisting of $LiCoO_2$ and graphite. However, the 12-V batteries consisting of LTO and LAMO have about three times larger gravimetric and volumetric energy densities than the 12-V lead-acid batteries. This means that one can apply the 12-V batteries of LTO and LAMO without the modification of current electronic circuits with reducing battery weight and volume by one-third while the ampere-hour capacity remains the same.

The operating voltage of LTO is about 1.5 V higher than that of lithium, so that there is no risk on lithium deposition even at high rate charge. In

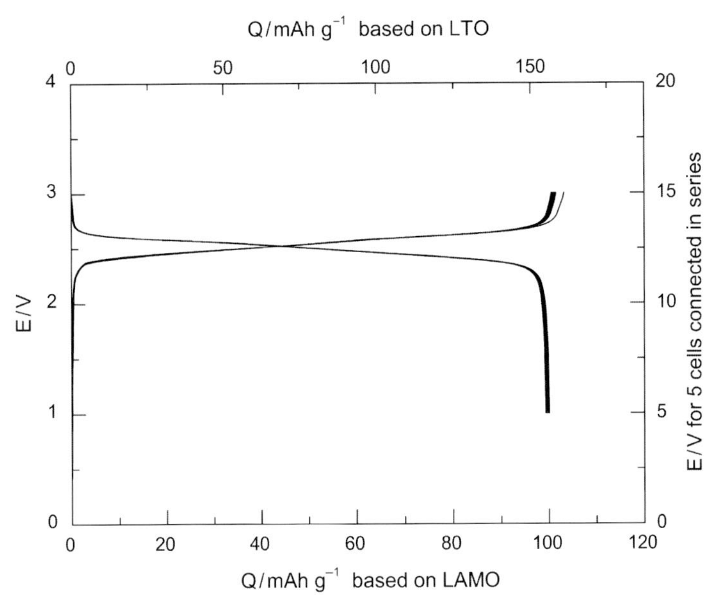

Fig. 2.19 Charge and discharge curves of an LTO/LAMO cell operated at 0.33 mA cm^{-2} in voltages of 1.0–3.0 V. The positive and negative electrodes consisted of 88 wt% of LAMO or LTO, 6 wt% of AB, and 6 wt% of PVdF. The weight and thickness of electrode mix was respectively 77.8 mg and 150 μm for the positive electrode and 49.9 mg and 139 μm for the negative electrode. Capacity ratio of Qp/Qn is 0.92, i.e., positive electrode-limited capacity. The electrolyte used was 1 M $LiPF_6$ dissolved in EC/DMC solution (3/7 by volume). Two sheets of polypropylene nonwoven clothes were used as a separator.

12-V batteries of LTO and LAMO, nonwoven cloth can be used as separator substituting for microporous membrane specially designed for lithium batteries. Aluminum substrate can be used for both positive and negative electrodes, which enable us to make lightweight batteries. While the 12-V lead-acid batteries cannot be discharged to 0 V, the 12-V batteries consisting of LTO and LAMO can be discharged to 0 V. Zero-volt storage is also possible for the 12-V batteries of LTO and LAMO if the material is well balanced for 0-V storage, which may help safe transport of 12-V batteries. The batteries can be operated in temperatures ranging from −30 to 85°C by selecting suitable electrolytes.

As described above, the 12-V batteries consisting of LTO and LAMO have several merits when compared to the 12-V lead-acid batteries. Figure 2.20 shows the pulse discharge curve of the LTO/LAMO cell. The cell delivers a burst of energy for a short period of time to start an engine at any state of charge. In addition to high power discharge, the 12-V batteries consisting of LTO and LAMO can accept high power input. Therefore, these 12-V batteries can also be used as power sources for HEVs or PEVs.

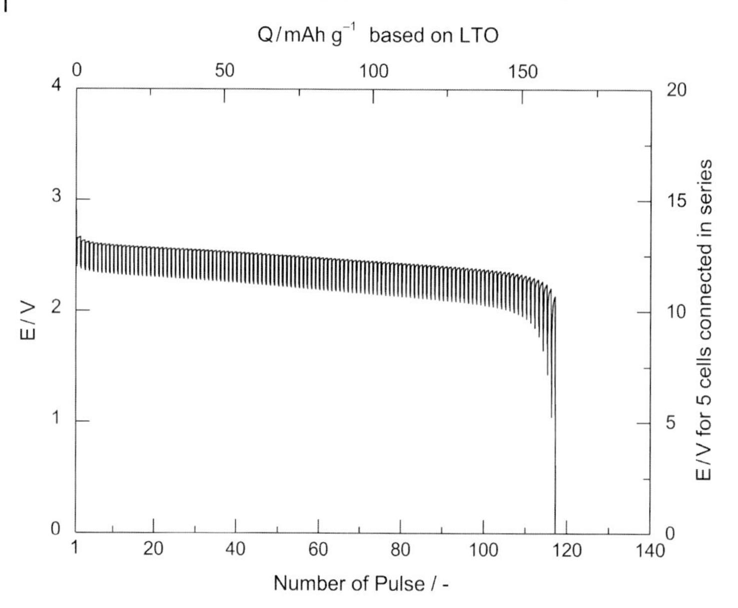

Fig. 2.20 The pulse discharge curve of an LTO/LAMO cell operated at 12.2 mA cm^{-2} (1 A g^{-1} based on LTO weight and 0.57 A g^{-1} based on LAMO) for 5 seconds current on and 25 seconds off. The electrodes consisted of 88 wt% of LAMO or LTO, 6 wt% of AB, and 6 wt% of PVdF. The weight of electrode mix was 73.2 mg for LAMO and 41.6 mg for LTO. The electrolyte used was 1 M $LiPF_6$ dissolved in ethylene carbonate (EC)/dimethyl carbonate (DMC) solution (3/7 by volume). Two sheets of polypropylene nonwoven cloth were used as a separator.

2.5.2
Twelve-Volt Batteries Consisting of Lithium Titanium Oxide (LTO) and Lithium Nickel Manganese Oxide (LiNiMO)

As seen in Figure 2.18, the combination of LTO and $Li[Ni_{1/2}Mn_{3/2}]O_4$ (LiNiMO) gives a 3.1-V battery [38–40] that is well suited to the requirement of the unit cell for 12-V batteries in Table 2.4. Figure 2.21 shows the charge and discharge curves of the LTO/LiNiMO cell. When four cells are connected in series, 12-V batteries can be made. A characteristic feature of the 12-V batteries consisting of LTO and LiNiMO is the flat operating voltage of about 12.5 V. Theoretical capacities of LTO and LiNiMO are respectively 175 mAh g^{-1} (610 mAh cm^{-3}) and 147 mAh g^{-1} (655 mAh cm^{-3}). Volumetric specific capacities of LTO and LiNiMO are almost the same because of the same lithium insertion scheme of spinel-framework structure. Electrodes with the same thickness between the positive and negative electrodes on an aluminum substrate may help improve cell performance. Electrode thickness of positive and negative electrodes is usually different because of the different volumetric specific capacities, so that electrode configuration is usually asymmetric in terms of thickness, leading to the limit

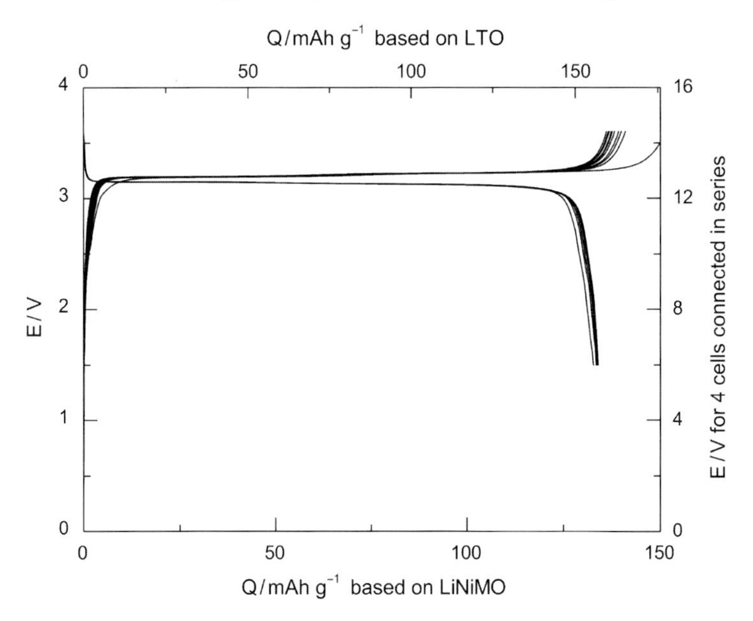

Fig. 2.21 Charge and discharge curves of an LTO/LiNiMO cell operated at 0.33 mA cm^{-2} in voltages of 1.0–3.6 V. The positive and negative electrodes consisted of 88 wt% of LiNiMO or LTO, 6 wt% of AB, and 6 wt% of PVdF. The weight and thickness of electrode mix were 73.1 mg and 138 μm for positive electrode, respectively, and 62.4 mg and 172 μm for negative electrode. Capacity ratio of Qp/Qn was 0.90, i.e., positive electrode-limited capacity. The electrolyte used was 1 M $LiPF_6$ dissolved in EC/DMC solution (3/7 by volume). Two sheets of polypropylene nonwoven clothes were used as a separator.

of lithium-ion migration and diffusion in pores inside the positive or negative electrode.

The calculated energy densities of 12-V batteries consisting of LTO and LiNiMO are 250 Wh kg^{-1} and 980 Wh dm^{-3}, which are higher than those for 12-V batteries consisting of LTO and LAMO. High energy density 12-V batteries can be expected. Because the operating voltage of LiNiMO is about 4.7 V versus Li, many people believe that there is no stable electrolyte to operate such a high-voltage positive electrode for thousands of cycles. The fact is that the LTO/LiNiMO cell with 1 M $LiPF_6$ ethylene carbonate (EC)/dimethyl carbonate (DMC) (3/7 by volume) can stably cycle for more than 2000 cycles at room temperature [40]. Figure 2.22 shows the charge and discharge curves of the LTO/LiNiMO cell operated at 3.33 mA cm^{-2}, corresponding to 259 mA g^{-1} based on LiNiMO and 216 mA g^{-1} based on LTO. The method of preparation of LiNiMO has been improved for long-life medium-power application [40]. Although the material and electrolyte have to be improved, the 12-V batteries consisting of LTO and LiNiMO can be used as high energy density 12-V batteries especially for medium-power application, such as wind or solar power generation system in off-grid remote areas.

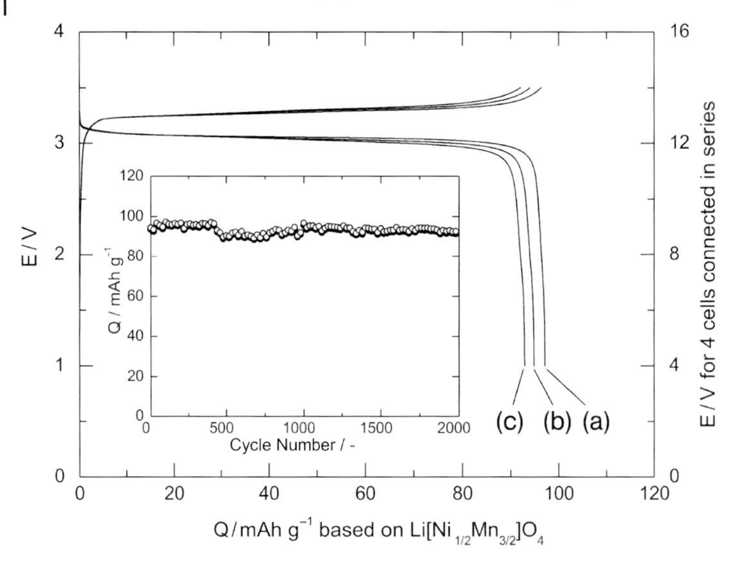

Fig. 2.22 Two thousand-cycle tests for an LTO/LiNiMO cell operated at 3.33 mA cm^{-2} corresponding to 259 mA g^{-1} based on LiNiMO and 216 mA g^{-1} based on LTO. Selected charge and discharge curves at (a) 100th, (b) 1000th, and (c) 2000th cycles are shown. Coulombs in and out for each cycle observed for the 2000-cycle tests of the lithium-ion battery of LiNiMO and LTO are also shown in inset of figure. The electrode mix consisted of 88 wt% of LiNiMO or LTO, 6 wt% of AB, and 6 wt% of PVdF. The electrolyte used was 1 M $LiPF_6$ dissolved in EC/DMC solution (3/7 by volume).

2.6 Concluding Remarks

In this chapter, we have dealt with lithium insertion materials having spinel-framework structure. Recent market demands for advanced batteries emphasize not only high energy density but also very high power density for both charge and discharge, especially in power tools, EV, and HEV applications. Although we described only the basic concept of 12-V batteries consisting of lithium insertion materials having spinel-framework structure, new concepts, and approaches from different angles have become more important than was previously recognized. There is a challenge for the materials chemist to develop innovative concepts to extend specific ampere-hour capacity with characteristic low or high operating voltage and to increase power density.

It should be noted here that 12-V batteries with different chemistries including 12-V lead-acid batteries should be appear on the market. The combination of LTO and LAMO reported in 2006 represents the first generation of 12-V lead-free accumulators, and the electrode couple of LTO and LiNiMO may be considered to be the second generation of the 12-V batteries. The negative electrode for both cells is based on the zero-strain lithium insertion material of the LTO $Li[Li_{1/3}Ti_{5/3}]O_4$. Positive-electrode material has been improved for 12-V batteries from LAMO to LiNiMO. The great success in handling a high-voltage (4.7 V versus Li) positive electrode of LiNiMO described here make the other 12-V batteries in which three cells connected in series possible, although we need a lithium insertion material whose operating voltage is about 0.5 V versus Li. Such a low operating voltage of 0.5 V versus Li, marked by pale black line in Figure 2.18, is usually neglected in research so far. We have to accept such a challenge to find the stable negative electrodes operated at around 0.5 V versus Li and to establish the science and technology of the third generation of 12-V batteries in which three cells are connected in series.

The consumers are not interested in chemistry inside the batteries. If the consumers accept 12-V batteries with a certain chemistry, the applications of such batteries will be extended. If not, such batteries will become obsolete. In our society, this depends on people's selection, which cannot be correctly predicted at present. A bright future in science and technology relating to advanced batteries can only be expected through continuing basic and applied research on lithium insertion materials, and, more generally, insertion materials.

Acknowledgments

In this chapter, we mostly describe the work done by us at Osaka City University (OCU). However, we owe our research to the results of other researchers via literature, e-mail, conference, or discussions at meetings, all of which have contributed to this chapter and deserve credit.

References

1 Ohzuku, T. and Ueda, A. (1994) *Solid State Ionics*, **69**, 201.
2 Whittingham, M.S. (2004) *Chem. Rev.*, **104**, 4271.
3 Ohzuku, T., Ariyoshi, K., Makimura, Y., Yabuuchi, N., and Sawai, K. (2005) *Electrochemistry (Tokyo, Japan)*, **73**, 2.
4 Ohzuku, T. and Brodd, R.J. (2007) *J. Power Sources*, **174**, 449.
5 Hass, C. (1965) *J. Phys. Chem. Solids*, **26**, 1225.
6 Wells, A.F. (1975) *Structural Inorganic Chemistry*, 4th edn., Oxford University Press, pp. 489–94.
7 Wondratschek, H. (1983) *International Tables for Crystallography*, Vol. **A**, D. Reidel Publishing Co., Dordrecht, pp. 726–28.
8 Shannon, R.D. (1976) *Acta Cryst.*, **A32**, 751.
9 Strobel, P., Ibarra-Palos, A., Anne, M., Poinsignon, C., and Crisci, A. (2003) *Solid State Sci.*, **5**, 1009.
10 Lee, Y.J., Park, S.-H., Eng, C., Parise, J.B., and Grey, C.P. (2002) *Chem. Mater.*, **14**, 194.
11 Ein-Eli, Y., Wen, W., and Mukerjee, S. (2005) *Electrochem. Solid-State Lett.*, **8**, A141.
12 Wen, W., Kumarasamy, B., Mulerjee, S., Auinat, M., and Ein-Eli, Y. (2005) *J. Electrochem. Soc.*, **152**, A1902.
13 Ooms, F.G.B., Kelder, E.M., Schoonman, J., Wagemaker, M., and Mulder, F.M. (2002) *Solid State Ionics*, **152**, 143.
14 Ariyoshi, K., Iwakoshi, Y., Nakayama, N., and Ohzuku, T. (2004) *J. Electrochem. Soc.*, **151**, A296.
15 Kim, J.-H., Myung, S.-T., Yoon, C.S., Kang, S.G., and Sun, Y.-K. (2004) *Chem. Mater.*, **16**, 906.
16 Guohua, L., Sakuma, K., Ikuta, H., Uchida, T., Wakihara, M., and Hetong, G. (1996) *Denki*

Kagaku (presently Electrochemistry, Tokyo, Japan), **64**, 202.

17 Ueda, A., Miyamoto, Y., and Ohzuku, T. (1995) "Extended Abstract No. 1A18". 36th Battery Symposium in Japan, Kyoto, September, 1995.

18 Ohzuku, T., Kitagawa, M., and Hirai, T. (1990) *J. Electrochem. Soc.*, **137**, 769.

19 Hunter, J.C. (1981) *J. Solid State Chem*, **39**, 142.

20 Gummow, R.J., De Kock, A., and Thackeray, M.M. (1994) *Solid State Ionics*, **69**, 59.

21 Amatucci, G. and Tarascon, J.M. (2002) *J. Electrochem. Soc.*, **149**, K31.

22 Ohzuku, T., Kitano, S., Iwanaga, M., Matsuno, H., and Ueda, A. (1997) *J. Power Sources*, **68**, 646.

23 Paulsen, J.M. and Dahn, J.R. (1999) *Chem. Mater.*, **11**, 3065.

24 Amatucci, G.G., Pereira, N., Zheng, T., and Tarascon, J.M. (2001) *J. Electrochem. Soc.*, **148**, A171.

25 Palacin, M.R., Chabre, Y., Dupont, L., Hervieu, M., Strobel, P., Rousse, G., Masquelier, C., Anne, M., Amatucci, G.G., and Tarascon, J.M. (2000) *J. Electrochem. Soc.*, **147**, 845.

26 Dupont, L., Hervieu, M., Rousse, G., Masquelier, C., Palacin, M.R., Chabre, Y., and Tarascon, J.M. (2000) *J. Solid State Chem.*, **155**, 394.

27 Gao, Y. and Dahn, J.R. (1996) *Solid State Ionics*, **84**, 33.

28 Iwata, E., Takeda, S., Iwanaga, M., and Ohzuku, T. (2003) *Electrochemistry (Tokyo, Japan)*, **71**, 1187.

29 Ohzuku, T., Ueda, A., and Yamamoto, N. (1995) *J. Electrochem. Soc.*, **142**, 1431.

30 Scharner, S., Weppner, W., and Schmid-Beurmann, P. (1999) *J. Electrochem. Soc.*, **146**, 857.

31 Albertini, V.R., Perfetti, P., Ronci, F., Reale, P., and Scrosati, B. (2001) *Appl. Phys. Lett.*, **79**, 27.

32 Ronci, F., Reale, P., Scrosati, B., Panero, S., Albertini, V.R., Perfetti, P., di Michiel, M., and Merino, J.M. (2002) *J. Phys. Chem. B*, **106**, 3082.

33 Ariyoshi, K., Yamato, R., and Ohzuku, T. (2005) *Electrochim. Acta*, **51**, 1125.

34 Ohzuku, T. and Ariyoshi, K. (2006) *Chem. Lett.*, **35**, 848.

35 Ariyoshi, K. and Ohzuku, T. (2007) *J. Power Sources*, **174**, 1258.

36 Thackeray, M.M. (1995) *J. Electrochem. Soc.*, **142**, 2558.

37 Ariyoshi, K., Iwata, E., Kuniyoshi, M., Wakabayashi, H., and Ohzuku, T. (2006) *Electrochem. Solid-State Lett.*, **9**, A557–A560.

38 Ohzuku, T., Ariyoshi, K., Yamamoto, S., and Makimura, Y. (2001) *Chem. Lett.*, **30**, 1270.

39 Ariyoshi, K., Yamamoto, S., and Ohzuku, T. (2003) *J. Power Sources*, **119–121**, 959.

40 Ariyoshi, K., Yamato, R., Makimura, Y., Amazutsumi, T., Maeda, Y., and Ohzuku, T. (2008) *Electrochemistry (Tokyo, Japan)*, **76**, 46.

3
Overlithiated $Li_{1+x}(Ni_zCo_{1-2z}Mn_z)_{1-x}O_2$ as Positive Electrode Materials for Lithium-Ion Batteries

Naoaki Kumagai and Jung-Min Kim

3.1
Introduction

Lithium-ion secondary batteries (LIBs) have become important power sources for portable electronics and electric vehicles because of their high energy densities. An LIB is mainly composed of a positive electrode, a nonaqueous electrolyte, and a negative electrode. Among them, the positive electrode material largely contributes to factors such as toxicity, cost, thermal safety, energy density, and power density, all of which affect the performance of the battery. Among various positive electrode materials, $LiCoO_2$ is most widely used owing to its ease of synthesis, an acceptable specific capacity, and high C rate performance. However, the relatively high cost of cobalt and concerns about toxicity and thermal safety of $LiCoO_2$ have lead to the development of other alternate materials.

Recently, $LiNi_zCo_{1-2z}Mn_zO_2$ materials have been investigated as promising positive electrode materials for LIBs in which cobalt ions are simultaneously substituted by nickel and manganese ions while preserving the α-$NaFeO_2$ structure [1–6]. The materials can be regarded as solid solutions of $LiNiO_2$, $LiCoO_2$, and $LiMnO_2$, in which the transition metals are trivalent in all three materials. The transition metals found in $LiNi_zCo_{1-2z}Mn_zO_2$ materials are Ni^{2+}, Co^{3+}, and Mn^{4+}. The electrochemical lithium deintercalation–intercalation processes of these materials are mainly accompanied by the $Ni^{2+/4+}$ and $Co^{3+/4+}$ redox reactions, whereas Mn^{4+} ions do not participate in the electrochemical redox reaction.

It is commonly believed that the double substitution of Ni and Mn for Co benefits material economy. However, the C rate performance of $LiNi_zCo_{1-2z}Mn_zO_2$ tends to degrade with decreasing Co content. Sun *et al.* [6] have reported that the rate performance of $LiNi_zCo_{1-2z}Mn_zO_2$ material rapidly falls with decreasing Co content, especially when it is <0.3 mol even though the electric conductivities of the materials are independent of the Co content. On the other hand, the Li^+ ion diffusion coefficient of $Li(Ni_zCo_{1-2z}Mn_z)O_2$ decreases with decreasing Co content [7]. One of the reasons for the degradation of the rate performance of $LiNi_zCo_{1-2z}Mn_zO_2$ with decreasing Co content (increasing Ni and/or Mn content)

Lithium Ion Rechargeable Batteries. Edited by Kazunori Ozawa

ISBN: 978-3-527-31983-1

is the structural disordering – the so-called cation mixing; Ni^{2+} can readily exchange its structural site for Li^+ owing to similar ionic radii of Ni^{2+} ($r = 0.69$ Å [8]) and Li^+ (0.74 Å [8]) in these materials. For example, about 11 and 6% cation mixing have been reported in $LiNi_{1/2}Mn_{1/2}O_2$ [9–11] and $LiNi_{1/3}Co_{1/3}Mn_{1/3}O_2$ [5], respectively. Whittingham has mentioned that the degree of the cation mixing in the $LiNi_yCo_{1-y-z}Mn_zO_2$ materials tends to increase as the ratio of Co/Ni decreases [12].

Recently, Li^+ substitution for transition metals in $Li_{1+x}(Ni_zCo_{1-2z}Mn_z)_{1-x}O_2$, so-called overlithiation, was introduced to improve their structural integrity and electrochemical performances [11, 13–18]. The overlithiation procedure was used to stabilize structures of positive electrode materials. For example, (i) Li^+ substitution for Mn in $Li_{1+x}Mn_{2-x}O_4$ increases the relative Mn^{4+} ion concentration in the spinel structure, which suppresses the manganese dissolution into electrolytes [19–21]; and (ii) the charge curves of $Li_{1+x}CoO_2$ (x and > 0) do not present the monoclinic distortion upon deintercalation, which is observed in lithium stoichiometric $LiCoO_2$ [22]. In this chapter, we introduce the overlithiation effects on the structure and electrochemical properties of the $Li_{1+x}(Ni_zCo_{1-2z}Mn_z)_{1-x}O_2$ materials (0 and $< z \leq 1/2$).

3.2 Co-Free $Li_{1+x}(Ni_{1/2}Mn_{1/2})_{1-x}O_2$

Recently, Co-free $Li(Ni_{1/2}Mn_{1/2})O_2$ has become attractive as positive electrode material for LIBs owing to its lower cost, greater stability at high voltage, and better thermal stability than $LiCoO_2$ [3, 9, 23–28]. $Li(Ni_{1/2}Mn_{1/2})O_2$ has a layered structure, similar to $LiCoO_2$, with complex cation orderings [9]. On the basis of theoretical calculations and spectroscopic studies, it is now widely accepted that the Ni and Mn ions have oxidation states of +2 and +4, respectively. Ohzuku [3] have reported that $Li(Ni_{1/2}Mn_{1/2})O_2$ electrodes deliver a reversible capacity of 150 mAh g^{-1} (2.5–4.3 V; 0.1 mA cm^{-2}) and show a milder thermal runaway than those of $LiCoO_2$ at fully charged state. Although the material shows a high reversible capacity at a very low rate, the high rate performance of the material does not meet commercial needs due to structural ordering (cation mixing). Kang *et al.* [9] have reported that the rate performance of $Li(Ni_{1/2}Mn_{1/2})O_2$ having a low cation mixing of 4.3% is superior to that of the material having a cation mixing of 10.9%. This result implies that the structural ordering of this material should be reduced to improve its electrochemical properties.

To improve structural integrity, overlithiation is introduced into the material. Figure 3.1 shows neutron diffraction data and Rietveld refinement results of the overlithiated $Li_{1+x}(Ni_{1/2}Mn_{1/2})_{1-x}O_2$ prepared by an emulsion drying method [10]. The refinement results show a good agreement between the observed and calculated patterns based on α-$NaFeO_2$ structure. It is found that the cation mixing is 7.3% for $Li(Ni_{1/2}Mn_{1/2})O_2$ and 2.6% for $Li_{1.06}(Ni_{1/2}Mn_{1/2})_{0.94}O_2$ and that the excess lithium is located in the transition metal layer. Recently, Kang *et al.* [16] have also reported that the cation mixing decreases from 10.4 to 2.4% when x in $Li_{1+x}(Ni_{1/2}Mn_{1/2})_{1-x}O_2$ increases from 0 to 0.15.

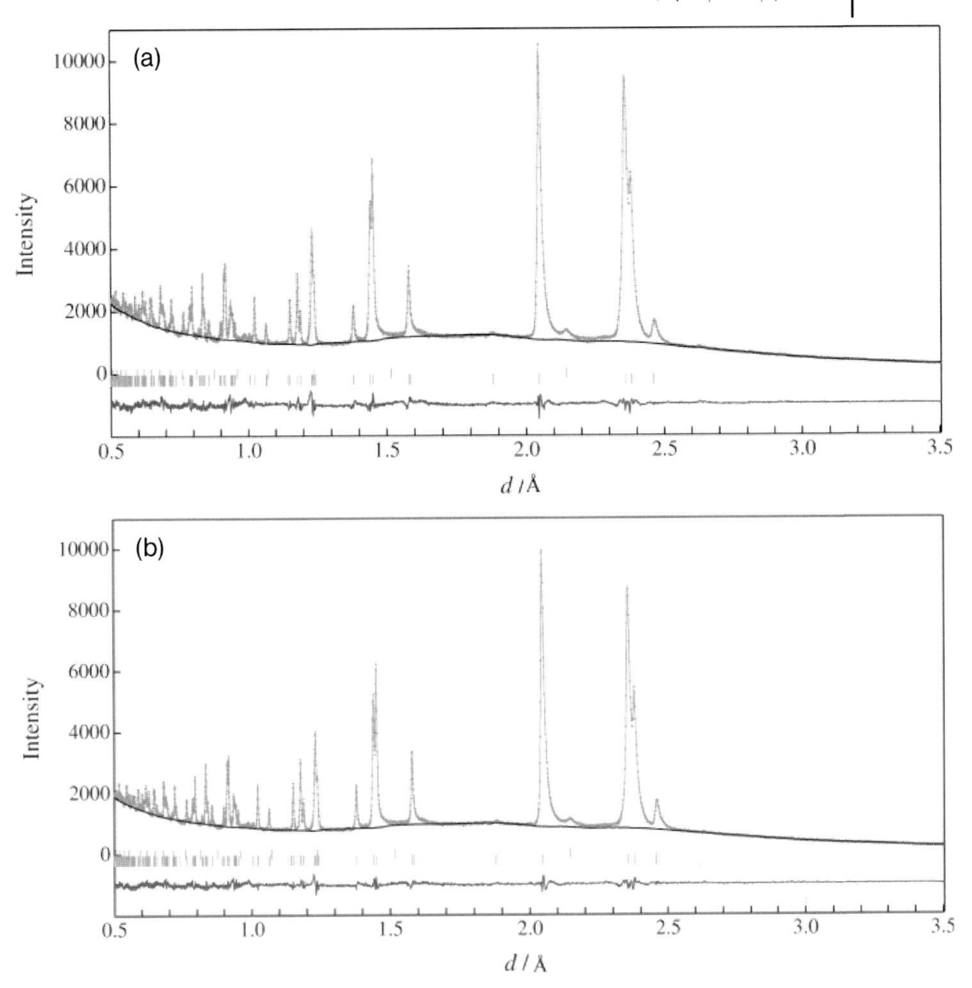

Fig. 3.1 Neutron diffraction patterns of (a) $Li(Ni_{1/2}Mn_{1/2})O_2$ and (b) $Li_{1.06}(Ni_{1/2}Mn_{1/2})_{0.94}O_2$.

If excess lithium substitutes for the transition metals in $Li(Ni_{1/2}Mn_{1/2})O_2$ and there is no oxygen deficiency, the oxidation of the transition metals would account for charge compensation. In this case, $Li_{1+x}(Ni_{1/2}Mn_{1/2})_{1-x}O_2$ could be rewritten as $Li[(Ni_{1/2}Mn_{1/2})_{1-x}Li_x]O_2$ in α-$NaFeO_2$ structure notation. Figure 3.2 shows the Ni and Mn K-edge X-ray absorption near edge spectroscopy (XANES) spectra of $Li(Ni_{1/2}Mn_{1/2})O_2$ and $Li[(Ni_{1/2}Mn_{1/2})_{0.94}Li_{0.06}]O_2$ to compare the oxidation states of the absorber atoms in both the materials. The Ni K-edge spectra of the materials are compared with those of the spinel $LiNi_{0.5}Mn_{1.5}O_4$ and layered $LiNiO_2$, where the average oxidation states of Ni are +2 and +3, respectively. The spectrum of Ni for $Li(Ni_{1/2}Mn_{1/2})O_2$ is close to that of $LiNi_{0.5}Mn_{1.5}O_4$, suggesting that the oxidation state of Ni in $Li(Ni_{1/2}Mn_{1/2})O_2$ is close to +2. However, the spectrum for $Li[(Ni_{1/2}Mn_{1/2})_{0.94}Li_{0.06}]O_2$ is shifted slightly toward a higher energy position

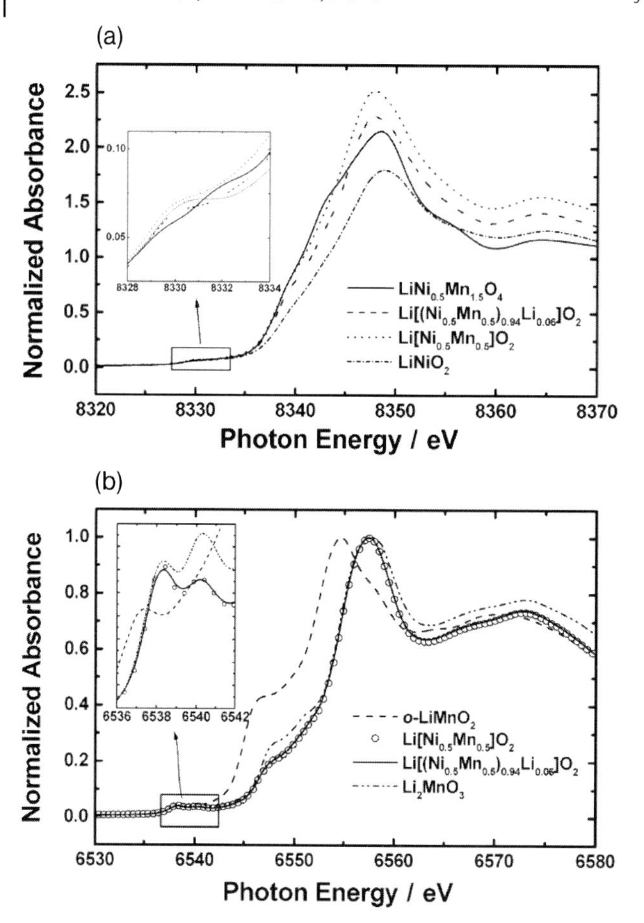

Fig. 3.2 (a) Ni K-edge and (b) Mn K-edge XANES spectra for $Li(Ni_{1/2}Mn_{1/2})O_2$ and $Li[(Ni_{1/2}Mn_{1/2})_{0.94}Li_{0.06}]O_2$.

relative to the $Li(Ni_{1/2}Mn_{1/2})O_2$. This result implies that the oxidation state of Ni in $Li[(Ni_{1/2}Mn_{1/2})_{0.94}Li_{0.06}]O_2$ is greater than that in $Li(Ni_{1/2}Mn_{1/2})O_2$, but lower than that in $LiNiO_2$. For Mn K-edge spectra, the synthesized materials are compared with the orthorhombic $LiMnO_2$ and the rock-salt type Li_2MnO_3, where the oxidation states of Mn are +3 and +4, respectively. The Mn K-edge spectra of both the materials are practically identical to that of Li_2MnO_3, indicating that the Mn ion is tetravalent in both materials.

The oxidation of transition metals would result in changes in the lattice parameters. Figure 3.3 shows variations in the lattice parameters of the $Li[(Ni_{1/2}Mn_{1/2})_{1-x}Li_x]O_2$ ($x = 0{-}0.06$). With an increase in the Li content in the transition metal layer, the a- and c-axis constants and unit-cell volume decrease monotonically without changes in the c/a ratios. As observed by XANES in Figure 3.2, the oxidation state of Ni for $Li[(Ni_{1/2}Mn_{1/2})_{0.94}Li_{0.06}]O_2$ is slightly higher relative to that for $Li(Ni_{1/2}Mn_{1/2})O_2$, which indicates that a larger amount of Ni^{3+} is contained in

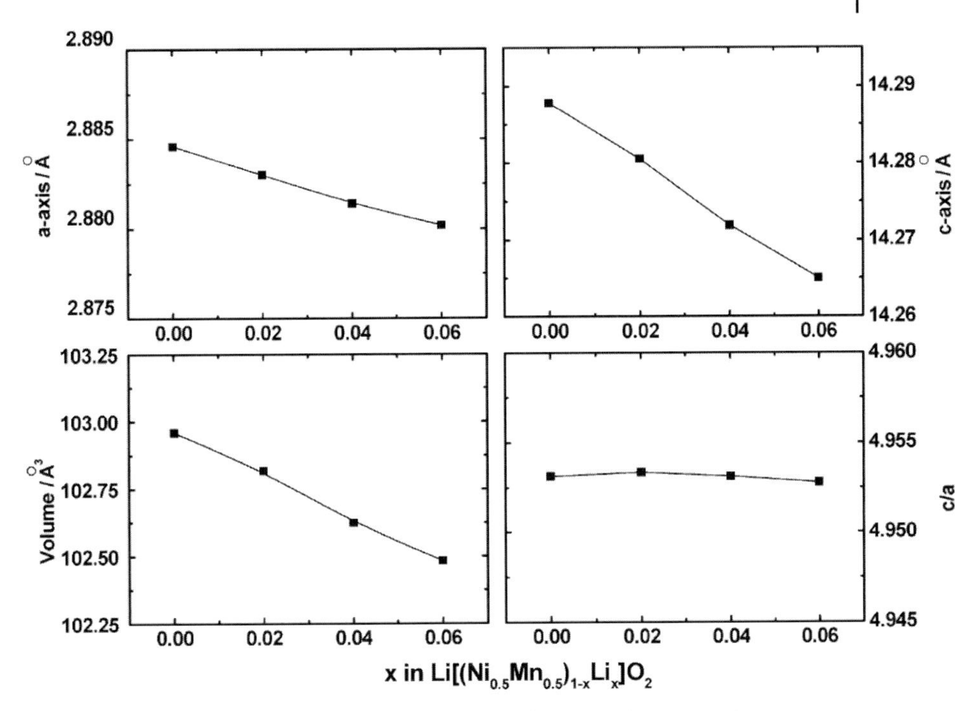

Fig. 3.3 Variations in lattice parameters and c/a ratios of $Li[(Ni_{1/2}Mn_{1/2})_{1-x}Li_x]O_2$.

$Li[(Ni_{1/2}Mn_{1/2})_{0.94}Li_{0.06}]O_2$ compared to $Li(Ni_{1/2}Mn_{1/2})O_2$. When the ionic radii of Ni^{2+} (0.69 Å [8]) and Ni^{3+} (0.56 Å [8]) are taken into account, it is reasonable to think that the formation of the smaller Ni^{3+} and the suppression of cation mixing by excess Li incorporation into the oxide matrix led to the linear decrease in the lattice parameters. Such a linear variation in the lattice parameter is usually observed in solid solution by following Vegard's law, suggesting that excess lithium readily substitutes for the Ni and Mn.

The changes in the structural integrity and oxidation states of the transition metals by the overlithiation affect the electrochemical properties of the materials. Figure 3.4 shows the continuous charge and discharge curves of $Li(Ni_{1/2}Mn_{1/2})O_2$ and $Li[(Ni_{1/2}Mn_{1/2})_{0.94}Li_{0.06}]O_2$. For $Li(Ni_{1/2}Mn_{1/2})O_2$, the capacity fades gradually as the cycle progresses. In addition, the operation voltage drops gradually during cycling On the other hand, $Li[(Ni_{1/2}Mn_{1/2})_{0.94}Li_{0.06}]O_2$ retains its initial capacity well through 50 cycles, and the capacity retention is approximately 95%. The $Li/Li[(Ni_{1/2}Mn_{1/2})_{0.94}Li_{0.06}]O_2$ cell also shows much less of a decrease in the operation voltage upon cycling. Kang *et al.* [16] have reported that the C rate capability improves significantly with increase in lithium excess. The improved rate capability and cyclability of the overlithiated materials could be attributed to the reduced Ni content in the lithium layer and the improved Li^+ ion diffusivity. The thermal stability of the charged electrode with electrolyte is enhanced slightly by the overlithiation [11].

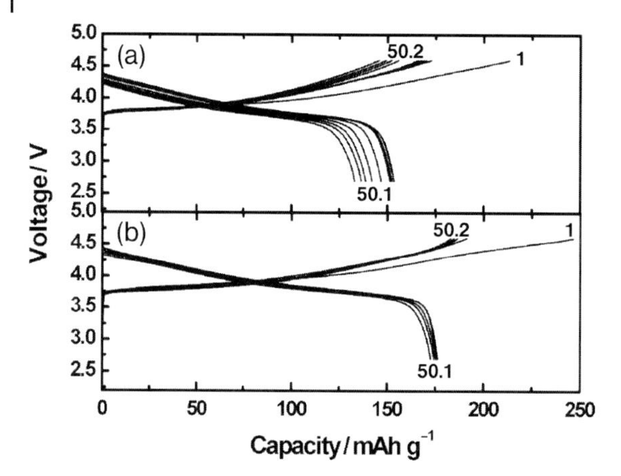

Fig. 3.4 Continuous charge and discharge curves of (a) $Li/Li(Ni_{1/2}Mn_{1/2})O_2$ and (b) $Li/Li[(Ni_{1/2}Mn_{1/2})_{0.94}Li_{0.06}]O_2$ cells. The applied current density across the positive electrode is 20 mA g^{-1} at 25°C.

3.3
$Li_{1+x}(Ni_{1/3}Co_{1/3}Mn_{1/3})_{1-x}O_2$

$Li(Ni_{1/3}Co_{1/3}Mn_{1/3})O_2$ is an attractive material for LIBs owing to its higher reversible capacity and small Co content, and it has been intensively investigated by many researchers and battery developers. On the basis of computational and experimental studies, it is now accepted that $Li(Ni_{1/3}Co_{1/3}Mn_{1/3})O_2$ has Ni^{2+}, Co^{3+}, and Mn^{4+} ions and the intercalation and deintercalation processes of Li^+ ions are accompanied by redox reactions of $Ni^{2+/4+}$ and $Co^{3+/4+}$, whereas Mn^{4+} does not participate in the process [4, 5, 29, 30]. The use of overlithiated $Li_{1+x}(Ni_{1/3}Co_{1/3}Mn_{1/3})_{1-x}O_2$ materials have also been studied to improve electrochemical performances [13, 31–34]. Figure 3.5 shows the X-ray diffraction (XRD) patterns and changes in the lattice parameters of the $Li_{1+x}(Ni_{1/3}Co_{1/3}Mn_{1/3})_{1-x}O_2$ materials synthesized by a spray-drying method. The main intense diffraction peaks could be indexed on the basis of a hexagonal structure, and the hexagonal lattice parameters of $Li_{1+x}(Ni_{1/3}Co_{1/3}Mn_{1/3})_{1-x}O_2$ gradually decrease with increasing x as does $Li_{1+x}(Ni_{1/2}Mn_{1/2})_{1-x}O_2$. X-ray photoelectron spectroscopy (XPS) and XANES studies reveal that one of the reasons for the decrease in the lattice parameters of the material with increasing x is the oxidation of Ni^{2+} ions to Ni^{3+} ions for charge compensation [32, 34]. Figure 3.6 shows the charge–discharge curves of $Li/Li_{1+x}(Ni_{1/3}Co_{1/3}Mn_{1/3})_{1-x}O_2$ cells. All the cells show a similar cycle curve, but the discharge capacity of the cells increases with increase in the value of x when the cells are cycled between 4.5 and 2.8 V versus Li/Li^+, whereas the reversible capacity of $Li/Li_{1+x}(Ni_{1/3}Co_{1/3}Mn_{1/3})_{1-x}O_2$ cells gradually decreases when the cells are cycled between 4.3 and 3.0 V [13]. The overlithiated materials, however, show better capacity retention than the lithium stoichiometric in both the cases. Furthermore, the C rate performance of the $Li/Li_{1+x}(Ni_{1/3}Co_{1/3}Mn_{1/3})_{1-x}O_2$ cell

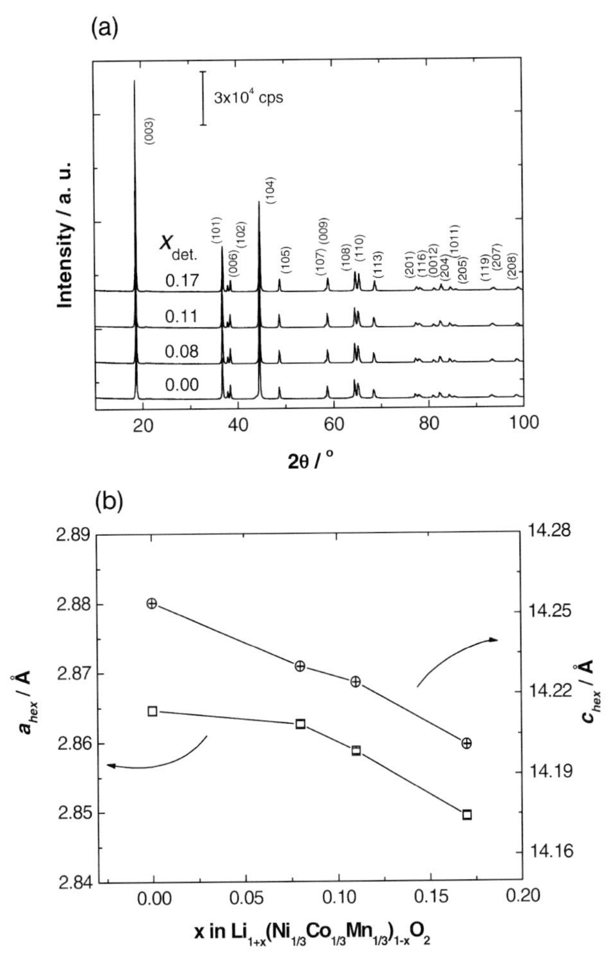

Fig. 3.5 (a) XRD patterns of $Li_{1+x}(Ni_{1/3}Co_{1/3}Mn_{1/3})_{1-x}O_2$ ($x = 0.00$–0.17) materials calcined at 950°C for 24 hours in air and (b) changes in their hexagonal lattice parameters.

improves with increasing x-value. Figure 3.7 shows the C rate performance of $Li/Li_{1+x}(Ni_{1/3}Co_{1/3}Mn_{1/3})_{1-x}O_2$ cells. The rate capability significantly improves with increasing amounts of excess lithium. This improvement in the cell performances could be attributed to increasing structural ordering and electronic conductivity [17, 18].

On the basis of results from first-principle calculations, [29], $Li(Ni_{1/3}Co_{1/3}Mn_{1/3})O_2$ show smaller changes in the lattice parameters than $LiCoO_2$ and $LiNiO_2$ during lithium extraction. Structural studies [35] reveal that $Li_{1-y}(Ni_{1/3}Co_{1/3}Mn_{1/3})O_2$ maintains the initial O3 phase (trigonal, $R3m$; ABCABC oxygen stacking sequence perpendicular to c_{hex} axis) up to 0.70–0.75 mol of Li^+ extraction and an additional

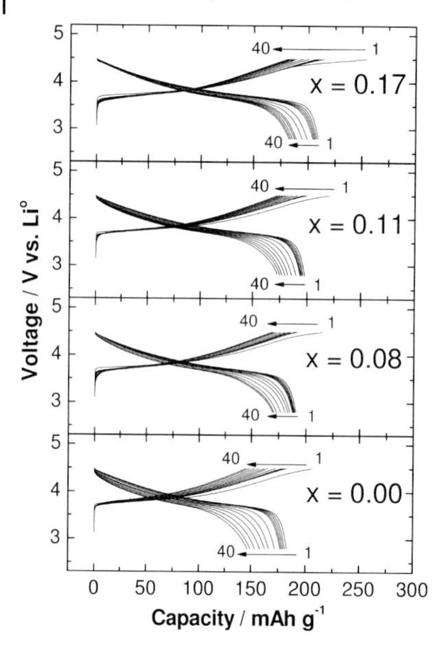

Fig. 3.6 Continuous charge and discharge curves of $Li_{1+x}(Ni_{1/3}Co_{1/3}Mn_{1/3})_{1-x}O_2$ ($x = 0.00$–0.17) materials cycled between 4.5 and 2.8 V versus Li/Li^+ with a current density of $30\,mA\,g^{-1}$ in 1.0 M $LiPF_6$/ethylene carbonate (EC)–diethylene carbonate (DEC) (1 : 2 in volume) at 25 °C.

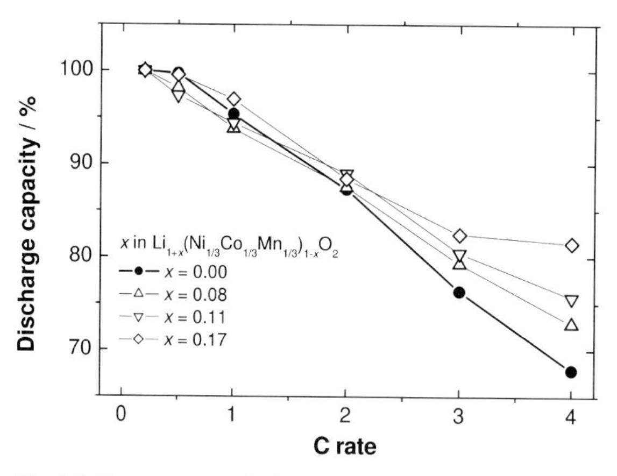

Fig. 3.7 C rates versus discharge capacities of $Li_{1+x}(Ni_{1/3}Co_{1/3}Mn_{1/3})_{1-x}O_2$ ($x = 0.00$–0.17) materials.

Li^+ extraction results in the appearance of the O1 phase (trigonal, P-3 m1; ABAB oxygen stacking sequence). Structural modification of $Li_{1+x}(Ni_{1/3}Co_{1/3}Mn_{1/3})_{1-x}O_2$ is due to the overlithiation. The overlithiated materials retain the pristine structure and show smaller volume change than the lithium stoichiometric one. Figure 3.8 shows *ex situ* XRD patterns of the $Li_{1+x}(Ni_{1/3}Co_{1/3}Mn_{1/3})_{1-x}O_2$ materials at the

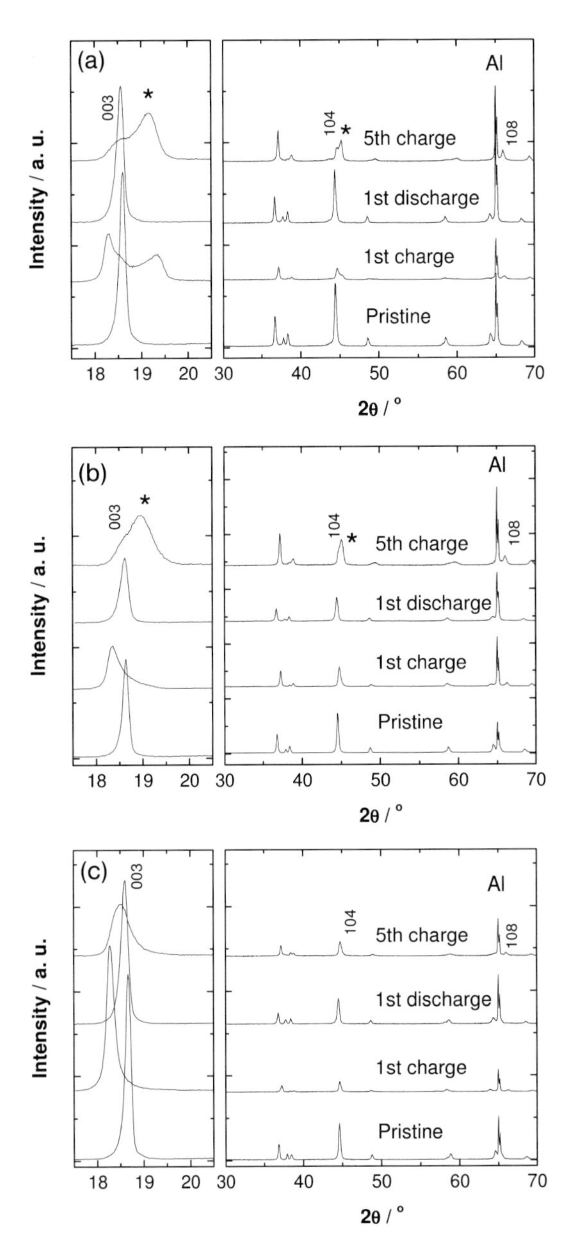

Fig. 3.8 *Ex situ* XRD patterns of the pristine, the first charged and discharged, and fifth charged states for $Li_{1+x}(Ni_{1/3}Co_{1/3}Mn_{1/3})_{1-x}O_2$ materials with $x =$ (a) 0.00, (b) 0.11, and (c) 0.17. The materials were cycled using CC–CV (constant current–constant voltage) mode with about 1/100 C rate until the capacity reached to the theoretical value of 1. Al in the XRD patterns indicates an aluminum foil current collector, which was used as an internal standard for calibration. The new peaks are marked by the asterisk.

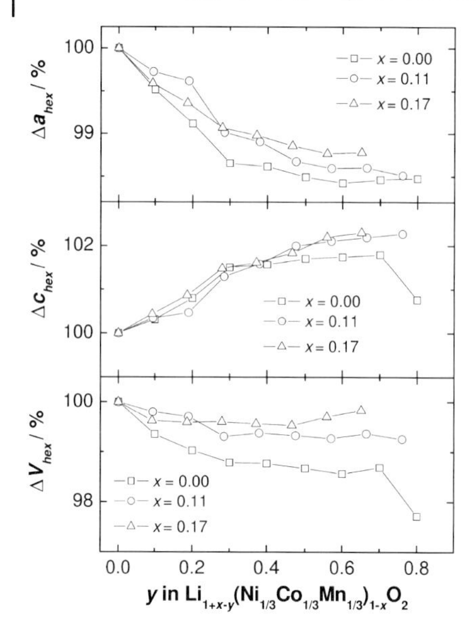

Fig. 3.9 The variations in the hexagonal lattice parameters (a_{hex} and c_{hex}) and unit-cell volume for $Li_{1+x-y}(Ni_{1/3}Co_{1/3}Mn_{1/3})_{1-x}O_2$ ($x = 0.00$, 0.11, and 0.17) during the initial charge. The y values in $Li_{1+x-y}(Ni_{1/3}Co_{1/3}Mn_{1/3})_{1-x}O_2$ material was estimated based on the charge capacities, and a possible error of the y value is in the range -0.02 to $+0.02$.

fully charged and discharged states in comparison with pristine materials. The lithium stoichiometric material shows a significant (003) peak broadening and a new peak relating with O1 phase emerges around 19.32° in 2θ ($d \sim 4.59$ Å) at the first fully charged state, and the new peak disappears at the first fully discharged state, whereas the overlithiated materials show no new peak during the first cycle. At the fifth charged state, the new peak is intensified for the lithium stoichiometric material. A similar peak is observed at the fifth charge for the overlithiated material with $x = 0.11$, whereas the overlithiated material with $x = 0.17$ shows no new peak. Figure 3.9 shows the changes in the lattice parameters of the materials during the first electrochemical lithium extraction. The overlithiated materials show smaller changes in lattice parameters and unit-cell volume than those of the stoichiometric one. Besides, the volume decreases with increasing x-value.

3.4 Other $Li_{1+x}(Ni_zCo_{1-2z}Mn_z)_{1-x}O_2$ Materials

Recently, other overlithiated $Li_{1+x}(Ni_zCo_{1-2z}Mn_z)_{1-x}O_2$ materials have also been reported. Tran *et al.* [15] have reported that $Li_{1+x}(Ni_{0.425}Co_{0.15}Mn_{0.425})_{1-x}O_2$ ($x =$ 0.00–0.12) materials having an $R3m$ structure show a decrease in the Ni occupancy in the Li layer with increasing x-value. Redox titrations and XPS spectra of the transition metals show the oxidation of Ni^{2+} ions into Ni^{3+} ions with increasing

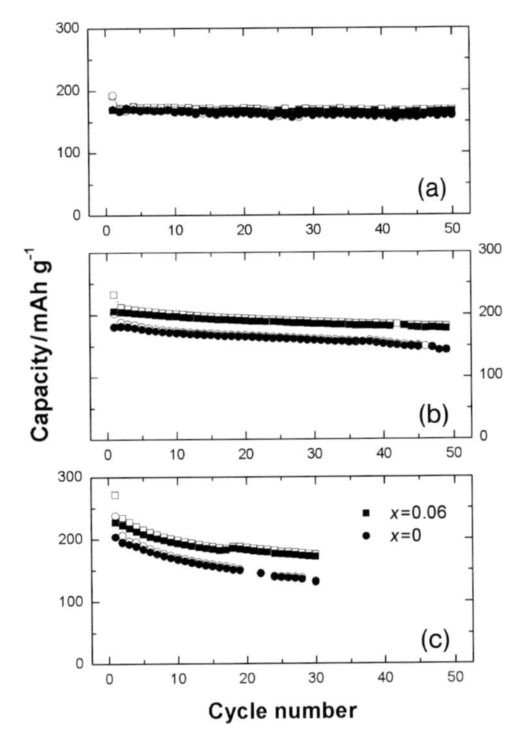

Fig. 3.10 Capacity versus cycle number for $Li/Li_{1+x}(Ni_{0.3}Co_{0.4}Mn_{0.3})_{1-x}O_2$ ($x = 0$ and 0.06) cells cycled in the voltage windows of (a) 4.3–2.8, (b) 4.5–2.8, and (c) 4.7–2.0 V versus Li/Li^+. The open and closed symbols represent charge and discharge capacities, respectively.

overlithiation for the materials; the manganese and cobalt ions are at the tetravalent and trivalent states, respectively. This material also shows a decrease in the Ni occupancy in the Li layer with increasing overlithiation as confirmed by the magnetic measurements and structural refinement results [15]. The electronic conductivities of these materials increase with overlithiation, which is attributed to a hopping between Ni^{2+} and Ni^{3+} ions [15]. However, they found that the thermal stability of the charged electrode with electrolyte is degraded by the overlithiation [15], which is in contrast to the result with $Li_{1+x}(Ni_{1/2}Mn_{1/2})_{1-x}O_2$ [11].

$Li_{1+x}(Ni_{0.4}Co_{0.3}Mn_{0.4})_{1-x}O_2$ materials also show improved electrochemical properties with overlithiation [36]. Figure 3.10 shows the capacity retention characteristics of $Li_{1+x}(Ni_{0.4}Co_{0.3}Mn_{0.4})_{1-x}O_2$ ($x = 0.00$ and 0.06) materials under different operating voltage windows. The overlithiated material shows better cycling behavior than the stoichiometric one in all the operating voltage windows, and conspicuous overlithiation effect on the cycling behavior is observed when the materials are cycled with a higher voltage. Shizuka *et al.* [17] have reported that the C rate performances of $Li_{1+x}(Ni_zCo_{1-2z}Mn_z)_{1-x}O_2$ materials ($x = 0.00$–0.20

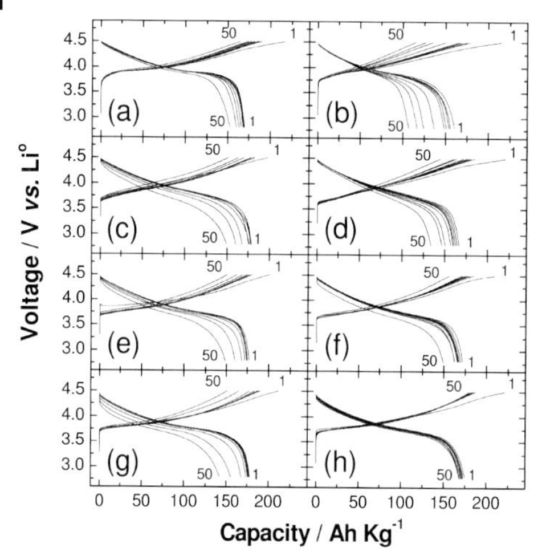

Fig. 3.11 Continuous charge and discharge curves of $Li/Li_{1+x}(Ni_zCo_{1-2z}Mn_z)_{1-x}O_2$ cells with (a) $z = 0.1$ and $x = 0.00$, (b) $z = 0.1$ and $x = 0.05$, (c) $z = 0.2$ and $x = 0.00$, (d) $z = 0.2$ and $x = 0.07$, (e) $z = 0.3$ and $x = 0.00$, (f) $z = 0.3$ and $x = 0.07$, (g) $z = 0.4$ and $x = 0.00$, and (f) $z = 0.4$ and $x = 0.05$ cycled between 4.5 and 2.8 V versus Li/Li^+ with a current density of 30 mA g^{-1} in 1.0 M $LiPF_6$/ethylene carbonate (EC)–dimethyl carbonate (DMC) (3 : 7 in volume) at 25 °C.

and $z = 0.33$–0.475) deteriorate with decreasing Co amount but are improved with increasing overlithiation, relating to both the cation mixing and the electrical conductivity associated with composition variations.

Recently, we have investigated the effects of overlithiation on the cell performances of the $Li_{1+x}(Ni_zCo_{1-2z}Mn_z)_{1-x}O_2$ materials with different Co/(Ni + Mn) ratios. Figure 3.11 shows the continuous charge–discharge curves of $Li/Li_{1+x}(Ni_zCo_{1-2z}Mn_z)_{1-x}O_2$ material ($z = 0.10$–0.40) cells. It is found that the capacity retention ability of the material with a Co/(Ni+Mn) ratio <1 ($z = 0.10$ and 0.20) is improved by overlithiation, whereas it deteriorates when the Co/(Ni+Mn) ratio is >1.

3.5
Conclusion

The overlithiation of $Li_{1+x}(Ni_zCo_{1-2z}Mn_z)_{1-x}O_2$ causes (i) the oxidation of Ni ions for charge compensation, (ii) decreasing Ni amount in the Li layer, and (iii) increasing electronic conductivity caused by the electronic hopping between Ni^{2+} and Ni^{3+} ions. It seems that overlithiation of $Li_{1+x}(Ni_zCo_{1-2z}Mn_z)_{1-x}O_2$ materials improves the structural integrity and electrochemical properties but the effects are more pronounced in materials with higher Ni and Mn content. On the basis of the literature available, it is suggested that overlithiation of $Li_{1+x}(Ni_zCo_{1-2z}Mn_z)_{1-x}O_2$ materials ($0.3 \leq z \leq 0.5$) would be an effective way to improve their electrochemical performances. These materials can be adopted as new positive electrode materials of LIBs for high power consumption electronics or electric vehicles after adjusting the excess lithium amount to obtain the best battery performances.

References

1 Lu, Z., MacNeil, D.D., and Dahn, J.R. (2001) *Electrochem. Solid-State Lett.*, **4**, A200.
2 Ohzuku, T. and Makimura, Y. (2001) *Chem. Lett.*, **30**, 642.
3 Ohzuku, T. and Makimura, Y. (2001) *Chem. Lett.*, **30**, 744.
4 Hwang, B. J., Tsai, Y.W., Carlier, D., and Ceder, G. (2003) *Chem. Mater.*, **15**, 3676.
5 Kim, J.-M. and Chung, H.-T. (2004) *Electrochim. Acta*, **49**, 937.
6 Sun, Y., Ouyang, C., Wang, Z., Huang, X., and Chen, L. (2004) *J. Electrochem. Soc.*, **151**, A504.
7 Jiang, J., Eberman, K.W., Krause, L.J., and Dahn, J.R. (2005) *J.Electrochem. Soc.*, **152**, A566.
8 Shannon, R. D. (1976) *Acta Crystallogr. Sect. A: Cryst. Phys., Diffr., Theor. Gen. Crystallogr.*, **A32**, 751.
9 Kang, K.S., Meng, Y.S., Greger, J., Grey, C.P., and Ceder, G. (2006) *Science*, **311**, 977.
10 Myung, S.-T., Komaba, S., Hosoya, K., Hirosaki, N., Miura, Y., and Kumagai, N. (2005) *Chem. Mater.*, **17**, 2427.
11 Myung, S.-T., Komaba, S., Kurihara, K., Hosoya, K., Kumagai, N., Sun, Y.-K., Nakai, I., Yonemura, M., and Kamiyama, T. (2006) *Chem. Mater.*, **18**, 1658.
12 Stanley Whittingham, M. (2004) *Chem. Rev.*, **104**, 4271.
13 Marinov Todorov, Y. and Numata, K. (2004) *Electrochim. Acta*, **50**, 495.
14 Zhang, L., Wang, X., Muta, T., Li, D., Noguchi, H., Yoshio, M., Ma, R., Takada, K., and Sasaki, T. (2006) *J. Power Sources*, **162**, 629.
15 Tran, N., Croguennec, L., Labrugere, C., Jordy, C., Bieesan, Ph., and Delmas, C. (2006) *J. Electrochem. Soc.*, **153**, A261.
16 Kang, S., Park, S., Johnson, C., and Amine, K. (2007) *J.Electrochem. Soc.*, **154**, A268.
17 Shizuka, K., Kobayashi, T., Okahara, K., Okamoto, K., Kanzaki, S., and Kanno, R. (2005) *J. Power Sources*, **146**, 589.
18 Okamoto, K., Shizuka, K., Akai, T., Tamaki, Y., Okahara, K., and Nomura, M. (2006) *J. Electrochem. Soc.*, **153**, A1120.
19 Gummow, R.J., de Kock, A., and Thackeray, M.M. (1994) *Solid State Ionics*, **69**, 59.
20 Amatucci, G.G., Schmutz, C.N., Blyr, A., Sigala, C., Gozdz, A.S., Larcher, D., and Tarascon, J.M. (1997) *J. Power Sources*, **69**, 11.
21 Saitoh, M., Sano, M., Fujita, M., Sakata, M., Takata, M., and Nishibori, E. (2004) *J. Electrochem. Soc.*, **151**, A17.
22 Levasseur, S., Ménétrier, M., Suard, E., and Delmas, C. (2000) *Solid State Ionics*, **128**, 11.
23 Kang, S.-H., Kim, J., Stoll, M.E., Abraham, D., Sun, Y.K., and Amine, K. (2003) *J. Power Sources*, **112**, 41.
24 Reed, J. and Ceder, G. (2002) *Electrochem. Solid State Lett.*, **5**, A145.
25 Johnson, C. S., Kim, J.-S., Kropf, A.J., Kahaian, A.J., Vaughey, J.T., Fransson, L.M.L., Edstrom, K., and Thackeray, M.M. (2003) *Chem. Mater.*, **15**, 2313.
26 Yoon, W.-S., Grey, C.P., Balasubramanian, M., Yang, X.-Q., and McBreen, J. (2003) *Chem. Mater.*, **15**, 3161.
27 Saiful Islam, M., Davies, R.A., and Gale, J.D. (2003) *Chem. Mater.*, **15**, 4280.
28 Shaju, K.M., Subba Rao, G.V., and Chowdari, B.V.R. (2003) *Electrochim. Acta*, **48**, 1505.
29 Koyama, Y., Yabuuchi, N., Tanaka, I., Adachi, H., and Ohzuku, T. (2004) *J.Electrochem. Soc.*, **151**, A1545.
30 Yabuuchi, N., Koyama, Y., Nakayama, N., and Ohzuku, T. (2005) *J. Electrochem. Soc.*, **152**, A1434.

31 Choi, J. and Manthiram, A. (2004) *Electrochem. Solid State Lett.*, **7**, A365.

32 Kim, J.-M. and Chung, H.-T. (2004) *Electrochim. Acta*, **49**, 3573.

33 Kim, J.-M., Kumagai, N., and Chung, H.-T. (2006) *Electrochem. Solid State Lett.*, **9**, A494.

34 Kim, J.-M., Kumagai, N., Kadoma, Y., and Yashiro, H. (2007) *J.Power Sources*, **174**, 473.

35 Yin, S.-C., Rho, Y.-H., Swainson, I., and Nazar, L.F. (2006) *Chem. Mater.*, **18**, 1901.

36 Kim, J.-M., Kumagai, N., and Komaba, S. (2006) *Electrochim. Acta*, **52**, 1483.

4
Iron-Based Rare-Metal-Free Cathodes

Shigeto Okada and Jun-ichi Yamaki

4.1
Introduction

Ever since the commercialization of lithium-ion batteries that use a layered rocksalt cathode such as $LiCoO_2$ [1], batteries using first-generation 4-V transition-metal oxide cathodes have been widely applied in various portable electronic products such as cellular phones, notebook PCs, and camcorders, owing to the strong demand for downsizing in these markets. Unfortunately, all 4-V-class rechargeable cathodes, that is, $LiCoO_2$, $LiNiO_2$, and $LiMn_2O_4$, have the essential problems of cost and adverse environmental impact, because these cathodes commonly include rare metals as the redox center. Furthermore, the unusual valence state of Co^{4+} or Ni^{4+} on fully charged $LiCoO_2$ or $LiNiO_2$ and Jahn–Teller unstable Mn^{3+} ($3\,d^4$:$t^3{}_{2g\uparrow}e^1{}_{g\uparrow}$) high-spin state on discharged $LiMn_2O_4$ are factors of concern with regard to the thermal and chemical stability of these cathodes. Actually, even almost 20 years after the commercialization of the Li-ion battery, new recall reports calling into question the reliability of Li-ion batteries are often heard. Finding solutions to the problems of economical efficiency and safety is of paramount importance, especially for the use of lithium-ion batteries in electric vehicles (EV). Transition metals are key elements of cathode-active material, but most of them are rare metals, with the exception of iron and titanium. Figure 4.1 shows element content in the Earth's crust, namely, the Clarke numbers. The inclined zigzag line indicates the unstable elements with odd atomic number, and as the figure shows, heavy metals are rare in the earth. If we use $LiCoO_2$ as the cathode for EV, we cannot produce more than 1×10^6 vehicles per year, because cobalt production per year worldwide is only 5×10^4 t, and one EV needs 50 kg cobalt. At present, 25% of the annual production of cobalt in the world is consumed in Japan and 70% of this is used in Li-ion batteries every year. The cobalt resources in the Earth's crust will be exhausted during this century, if cobalt-based cathode is selected for use in the large-scale Li-ion batteries of EV.

To solve these intrinsic problems, we have turned our attention, in this chapter, to rare-metal-free cathodes with low cost and low environmental impact.

Lithium Ion Rechargeable Batteries. Edited by Kazunori Ozawa

ISBN: 978-3-527-31983-1

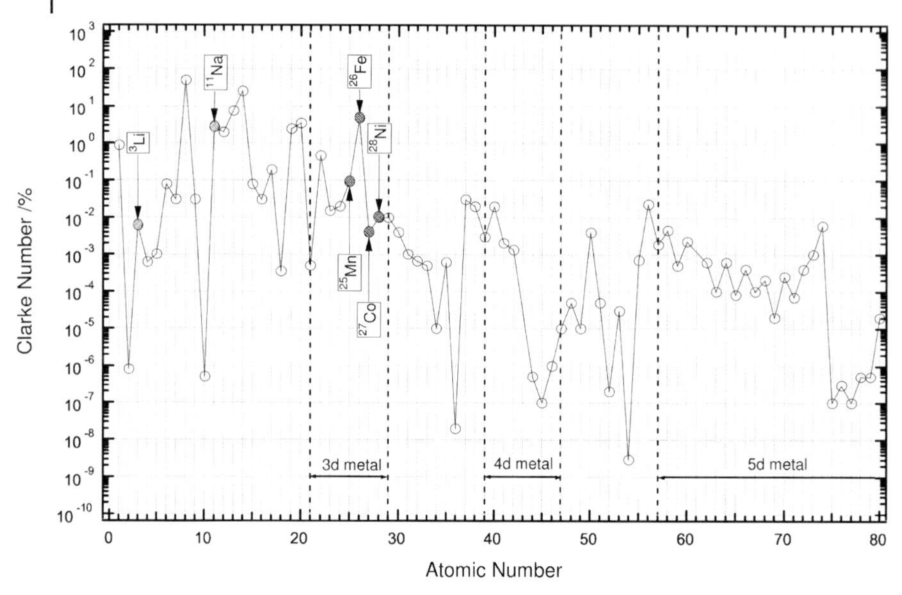

Fig. 4.1 Clarke numbers.

4.2
2D Layered Rocksalt-Type Oxide Cathode

AMO_2 compounds such as $LiCoO_2$ and $LiNiO_2$ have layered rocksalt structure as the stable phase, and they show good cathode utilization, because the layered rocksalt structure has two-dimensional lithium diffusion layers in the matrix. From an environmental point of view, iron is the most attractive redox couple, but, unfortunately, we cannot obtain stable layered rocksalt $LiFeO_2$. To obtain the layered rocksalt structure, we have to avoid cation mixing between A and M in AMO_2. However, similar sized cations tend to mix with each other, leading to a disordered rocksalt structure. This means that the stable structure can be predicted by the cation radius ratio of A versus M, as shown in Figure 4.2. Generally, redox ions with a smaller radius make the layer character more stable. The empirical criterion for the layered rocksalt structure is a cation radius ratio $r_A/r_M > 1.23$. In the case of a lithium system, $LiFeO_2$ and $LiMnO_2$ are located on the dotted border between the layered and disordered structures in Figure 4.2. On the other hand, in the case of a sodium system, all 3d metals bring the layered rocksalt structure to a stable phase, because of the larger sodium ion. This criterion works not only in the binary oxide, but also in the ternary and quaternary oxides, as shown in Table 4.1 [2–7].

These sodium systems included layered rocksalt oxides with an R_3m space group that can be easily obtained by the conventional method of solid-state reaction in air. The first discharge profiles in sodium cells are shown in Figure 4.3. Although the capacity of $NaFeO_2$ is restricted to 80 mAh g^{-1}, the flat discharge profile is similar to that of isostructural $LiCoO_2$. In addition, it is worth noting that

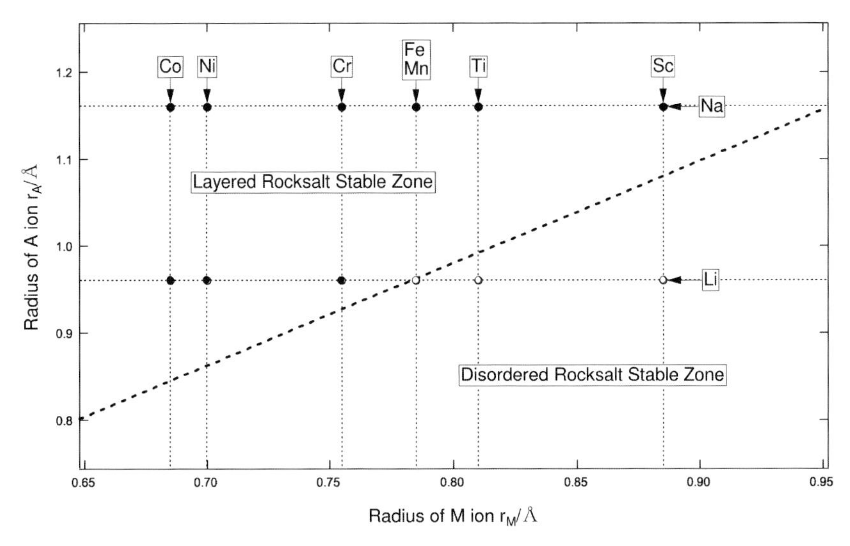

Fig. 4.2 Structure field map for AMO_2.

Na/$NaFeO_2$ is a rare example of the rare-metal-free battery systems combined with an abundant sodium anode and a ferric oxide cathode as shown in Figure 4.1. The 3.3-V discharge plateau against the sodium anode indicates that layered rocksalt $LiFeO_2$ would exhibit a 4-V flat plateau versus Li^+/Li, if it could be synthesized. In the case of the ternary systems, $NaNi_{0.5}Fe_{0.5}O_2$ and $NaNi_{0.5}Ti_{0.5}O_2$, the flatness of the 3.3-V plateau was lost. However, they showed a larger capacity of more than 100 mAh g^{-1}. In particular, it is interesting that the capacity of $NaNi_{0.5}Ti_{0.5}O_2$ in a sodium cell is larger than that of $LiTi_{0.5}Ni_{0.5}O_2$ in a lithium cell, as shown in Table 4.1. It was found that the two-dimensional A-cation layer in α-$NaFeO_2$ type cathodes have a sufficient diffusion field not only for lithium, but also for larger sodium intercalants.

In the initial $NaFeO_2$, Fe^{3+} high-spin state (IS = 0.37 mm s^{-1}, QS = 0.46 mm s^{-1}) was confirmed by ^{57}Fe Mössbauer spectroscopy. On the other hand, a trace of Fe^{4+} (IS = 0.05 mm s^{-1}, QS = 0.55 mm s^{-1}) could be observed in charged $NaFeO_2$ cathode pellets as shown in Figure 4.4. On the other hand, it was confirmed by ^{57}Fe Mössbauer spectroscopy that the iron trivalent state of $NaNi_{0.5}Fe_{0.5}O_2$ did not change on charge/discharge cycling. Instead of Fe, the reversible valence changes of Ni in $NaNi_{0.5}Fe_{0.5}O_2$ and $NaNi_{0.5}Ti_{0.5}O_2$ were also observed by X-ray photoelectron spectroscopy.

4.3
3D NASICON- Type Sulfate Cathode

NASICON (Na Super Ionic Conductor)-type structures can be broken down into fundamental groups of two MO_6 octahedra separated by three XO_4 tetrahedra with which they share common corner oxygens (Figure 4.5). Hexagonal NASICON

Table 4.1 Typical layered rocksalt cathodes.

Cathode	$r(A^+)/r(M^{3+})$	Structure	Voltage	Capacity ($A\,h\,kg^{-1}$)	Ref.
$LiNi_{0.5}Ti_{0.5}O_2$	1.22	Disordered rocksalt (α-$LiFeO_2$ type)	3.8 V vs Li/Li^+	40	[2]
		Layered rocksalt by ion exchange	3.8 V vs Li/Li^+	80	[3]
$Li[Li_{1/3}Fe_{1/3}Mn_{1/3}]O_2$	1.21	Layered rocksalt by hydrothermal	3.8 V vs Li/Li^+	68	[4]
$LiNi_{0.5}Mn_{0.5}O_2$	1.28	Layered rocksalt (α-$NaFeO_2$ type)	4.0 V vs Li/Li^+	200	[5]
$Li[Co_{1/3}Ni_{1/3}Mn_{1/3}]O_2$	1.37	Layered rocksalt (α-$NaFeO_2$ type)	4.0 V vs Li/Li^+	200	[6]
$NaNi_{0.5}Ti_{0.5}O_2$	1.47	Layered rocksalt (α-$NaFeO_2$ type)	3.3 V vs Na/Na^+	100	[7]
$NaFeO_2$	1.48	Layered rocksalt (α-$NaFeO_2$ type)	3.3 V vs Na/Na^+	80	[7]
$NaFe_{0.5}Ni_{0.5}O_2$	1.56	Layered rocksalt (α-$NaFeO_2$ type)	3.3 V vs Na/Na^+	100	[7]

has a total of four possible lithium sites per formula unit. One of them is a sixfold coordinated 6b site between MO_6 octahedra, and the other three are 10-fold coordinated 18e sites located between these ribbons along the *c*-axis. It is not realistic to expect high electronic conductivity in this structure, because all the metal octahedra are isolated. However, there are no shared edges or shared faces in the matrix, so all these large lithium sites are connected to one another. Thus, although it is a three-dimensional framework, it is open and flexible.

The high ionic conductivity of NASICON was reported first by Goodenough, 20 years ago [8]. Since then, researchers have found that NASICON is a suitable and attractive framework not only for solid electrolyte but also for cathode-active material on the basis of the study of NASICON cathodes such as $LiTi_2(PO_4)_3$ [9] and $Li_3Fe_2(PO_4)_3$ [10].

In the NASICON matrix, a transition metal and a countercation share a common oxygen in an M–O–X linkage. As the electronegativity of the countercation, X, becomes closer to that of oxygen (3.44) the bonding in an XO_4 polyanion becomes covalent. On the other hand, Fe–O bonding becomes ionic by the inductive effect [11], making the Fe^{3+}/Fe^{2+} redox level lower and the cell voltage higher. There is a linear relationship between the electronegativity of X and the discharge voltage

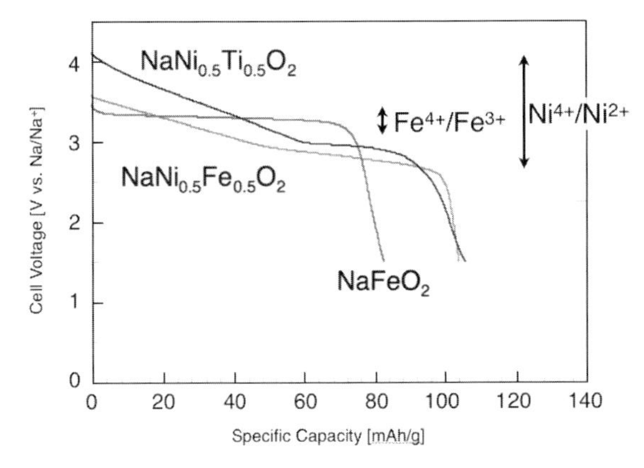

Fig. 4.3 The discharge profiles of $NaFeO_2$, $NaNi_{0.5}Fe_{0.5}O_2$, and $NaNi_{0.5}Ti_{0.5}O_2$ cathodes against a sodium metal anode at a rate of $0.2\,mA\ cm^{-2}$.

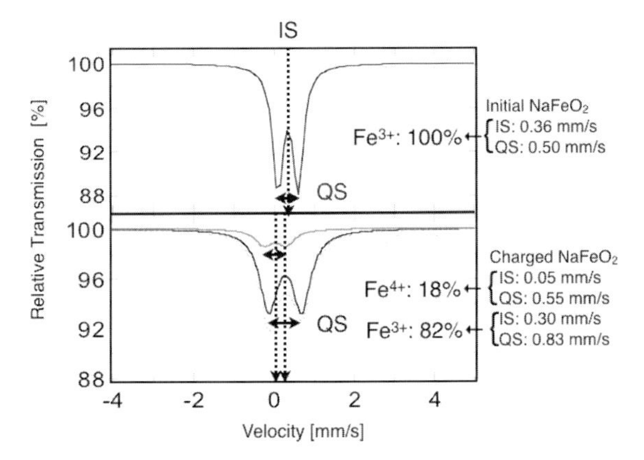

Fig. 4.4 ^{57}Fe Mössbauer spectroscopy of initial and 3.6-V charged $NaFeO_2$ cathodes.

in NASICON cathodes, as shown in Figure 4.6. Actually, ferric sulfate with the highest electronegative hetero atom in polyanion provides the highest discharge voltage so far reported for an iron cathode.

Although ferric sulfate has two crystal forms, hexagonal [12] and distorted monoclinic [11], as shown in Figure 4.5, both of them have the same basic structural unit in the matrix and show a similar 3.6-V plateau on the cycle profiles. The lantern units in the hexagonal phase are stacked in parallel and they are stacked alternately in the monoclinic phase. The 3.6-V region corresponds to the reduction from ferric ions to ferrous ions. The theoretical capacity of two lithium insertions reaches $134\,mAh\,g^{-1}$. At both end points of the two-phase region, the voltage changes rapidly. It is easy to control the charge/discharge depth by monitoring the cell voltage even in the rocking-chair system. It is both interesting and important that the system realizes low hysteresis, low irreversibility, and high discharge voltage without an unusual valence state such as Fe^{4+}. This voltage is compatible

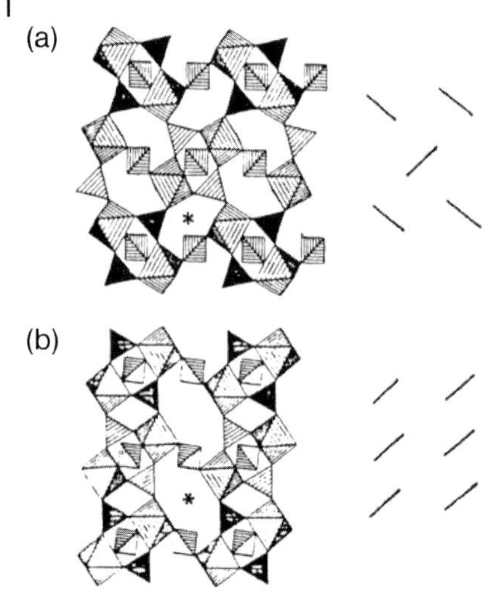

Fig. 4.5 NASICON structure of (a) monoclinic P1/n and (b) hexagonal $R\bar{3}c$ [22]. Asterisk denotes bottleneck for ionic transport.

with commercial 3.6-V Li-ion batteries. In addition, it is possible to charge the system with less than 4 V, due to the flat profile and the small charge overvoltage.

4.4 3D Olivine-Type Phosphate Cathode

Among various polyanionic cathodes, phosphates provide the largest rare-metal-free cathode-active material group, because condensate salts such as pyrophosphate $(P_2O_7)^{4-}$, tripolyphosphate $(P_3O_{10})^{5-}$, or polymetaphosphate $(P_nO_{3n})^{n-}$ can easily produce iron phosphates such as $LiFePO_4$, $LiFeP_2O_7$, and $Fe_4(P_2O_7)_3$. These iron phosphates show similar discharge plateaus at 3 V [13]. In particular, olivine $LiFePO_4$ has the highest theoretical energy density (560 mWh g^{-1}) among the iron-based polyanionic cathodes. The discharged $LiFePO_4$ of the triphylite structure and the fully charged $FePO_4$ phase of the heterosite structure have the same space group, *Pnma*. $LiFePO_4$ shows a 3.3-V flat discharge profile corresponding to the two-phase reaction. In the first paper concerning the olivine cathode by Goodenough [14], the capacity was restricted to 120 mAh g^{-1} in spite of the low rate of 0.05 mA cm^{-2}, caused by the low Li diffusivity of the interface between the two phases, the restricted Li diffusion direction along the *b*-axis, and the low electrical conductivity through the long metal atom distance in the olivine matrix.

The poor rate capability, which is the largest weakness of $LiFePO_4$, is being gradually resolved by carbon nanocoating using organic precursors [15], a carbothermal reduction process [16], minimizing the particle size [17], and substitutional doping to Fe [18]. According to the patent of Hydro-Québec, the addition of carbon precursors such as polypropylene and cheaper saccharides on calcinating

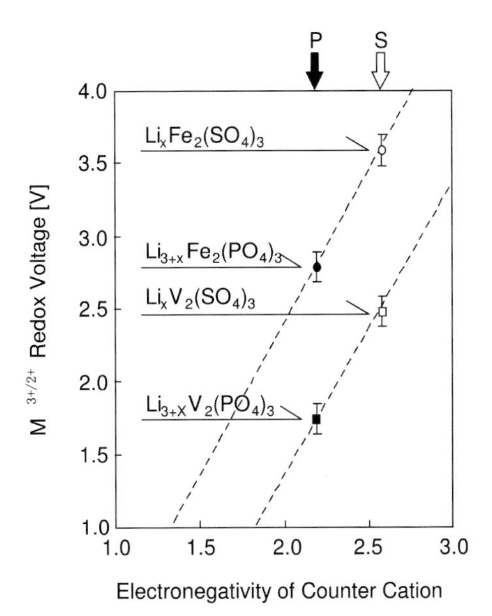

Fig. 4.6 Relationship between the electronegativity of X and the discharge voltage in a NASICON cathode.

$LiFePO_4$ raw material at 700 °C in Ar apparently improves the electric contact through carbon coating on $LiFePO_4$ particles and prevents oxidation from Fe^{2+} to Fe^{3+}. As a consequence, it showed a high capacity, close to the theoretical limit of 170 mAh g^{-1} at operating temperature of 80 °C. In addition, Nazar proposes that reduced conductive iron phosphide Fe_2P on the surface of the $LiFePO_4$ is apparently effective in reducing the contact resistance between $LiFePO_4$ particles [16]. On the other hand, Yamada succeeded in increasing the capacity up to 160 mAh g^{-1} at room temperature at a rate of 0.12 mA cm^{-2} by low-temperature synthesis at less than 600 °C to suppress particle growth [17]. Moreover, Chung has reported that the bulk electronic conductivity can be enhanced from 10^{-9} to 10^{-1} S cm^{-1} by 1 atom % substitutional doping of various metal ions to Fe [18]. This means that the improved conductivity of doped $LiFePO_4$ is larger than the 10^{-5} S cm^{-1} of $LiMn_2O_4$ and the 10^{-3} S cm^{-1} of $LiCoO_2$. According to the home page (*http://a123systems.textdriven.com*) of A123 founded by Chung, the rate capability of their battery with an olivine cathode shows 95% utilization at 17 C rate at room temperature. It is actually the best data among Li-ion batteries on the market so far.

Following the improvement of the conductivity, various synthesis approaches such as hydrothermal synthesis [19], microwave synthesis [20], and a high-temperature quick melting method [21] were developed to reduce the production cost, as shown in Table 4.2. The hatched zone in Figure 4.7 shows the glass-phase area in the Li–Fe–P ternary system. This figure shows that all the known iron phosphate cathodes such as $LiFePO_4$ and $Fe_4(P_2O_7)_3$ can be made amorphous by using a quenching method. General expectations for an amorphous

Table 4.2 Typical synthesis conditions of $LiFePO_4$.

Starting materials			Heating condition	Ref.
Fe source	**Li source**	**P source**		
$FeC_2O_4 \cdot 2H_2O$	$LiOH \cdot H_2O$	$(NH_4)_2HPO_4$	800°C, 6 h in N_2	[23]
$Fe_3(PO_4)_2 \cdot 8H_2O$	Li_3PO_4 with PP(3 w/o)		350°C, 3 h - >700°C, 7 h in Ar	[15]
$(CH_3COO)_2Fe$	CH_3COOLi	$NH_4H_2PO_4$	350°C, 5 h - >700°C, 10 h with 15 w/o sol–gel carbon in N_2	[16]
$FeC_2O_4 \cdot 2H_2O$	Li_2CO_3	$(NH_4)_2HPO_4$	320°C, 12 h - >800°C, 24 h with 12 w/o sugar in Ar	[24]
$FeC_2O_4 \cdot 2H_2O$	Li_2CO_3	$NH_4H_2PO_4$	600–850°C with 1 atm% dopant in Ar	[18]
$(CH_3COO)_2Fe$	Li_2CO_3	$NH_4H_2PO_4$	320°C, 10 h - >550°C, 24 h in N_2	[17]
$FeSO_4$	LiOH	H_3PO_4	Hydrothermal synthesis at 120°C 5 h in Teflon reactor	[19]
$(NH_4)_2Fe(SO_4)_2 \cdot 2H_2O$	LiOH	H_3PO_4	A few minutes of microwave heating with 5 w/o carbon black in air	[20]
Fe_2O_3	LiH_2PO_4		Carbothermal reduction at 750°C, 8 h with carbon in Ar	[25]
FeO	$LiOH \cdot H_2O$	P_2O_5	High-temperature quick melting synthesis at 1500°C in Ar	[21]

cathode are that it will have quick synthesis, employ inexpensive but inactive raw materials such as iron oxide by heating above the melting temperature, have a large bottleneck framework because of its corner-sharing glass matrix, have continuous composition control, and so on. As an example, some reported properties of iron phosphate cathodes and anodes are plotted in Figure 4.8. From the standpoint of cell voltage, the phosphate-rich phase is attractive. On the other hand, when considering the specific capacity, the iron-rich phase is more attractive. Under this

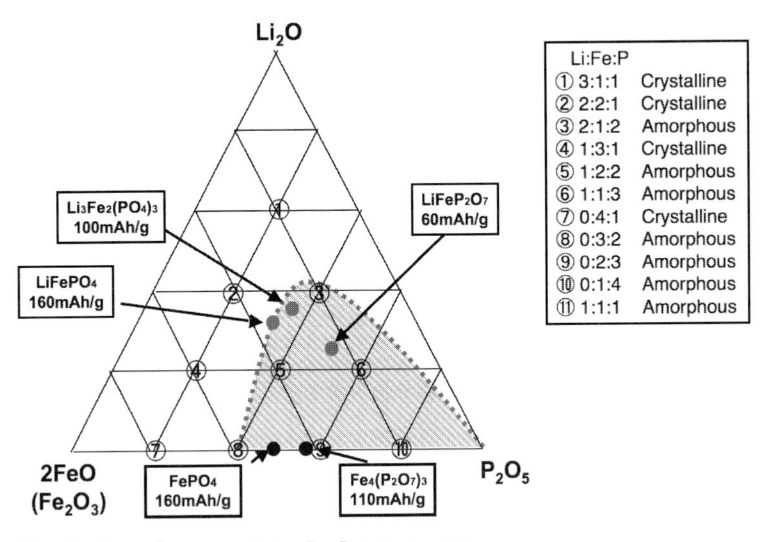

Fig. 4.7 Glass-phase area in Li–Fe–P ternary system.

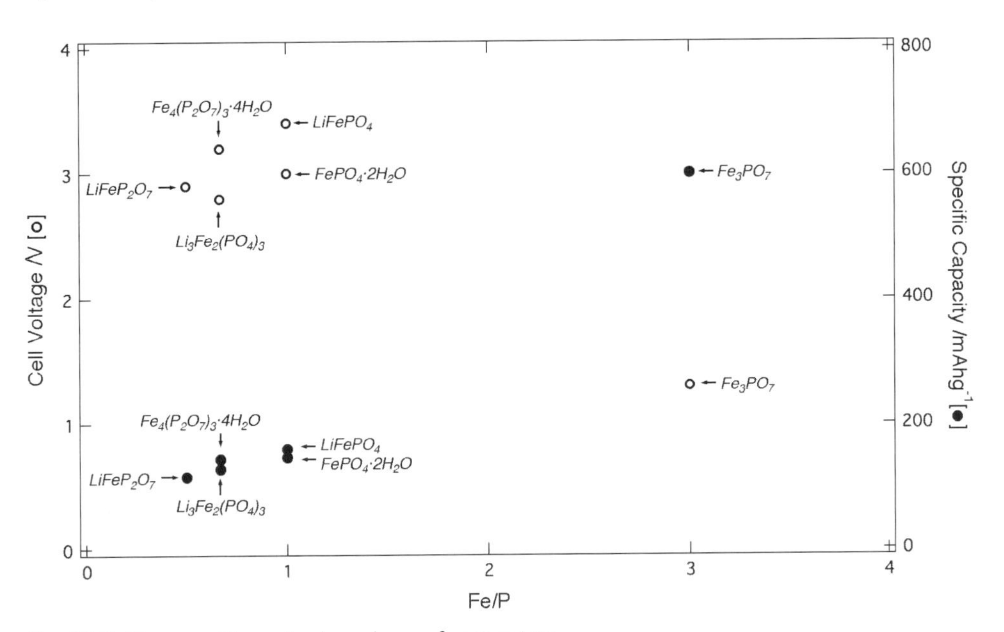

Fig. 4.8 Fe/P composition ratio dependence of various iron phosphate cathode/anode performances.

trade-off situation, we expect that unknown well-balanced iron phosphates may be hidden in the large blank area between $FePO_4$ and Fe_3PO_7. To find new iron phosphate in this unexplored vast area, amorphosizing may be useful.

Another contrastive approach is low-temperature synthesis of $FePO_4$ from the liquid phase. In contrast with $LiFePO_4$, $FePO_4$ can be synthesized in air. In

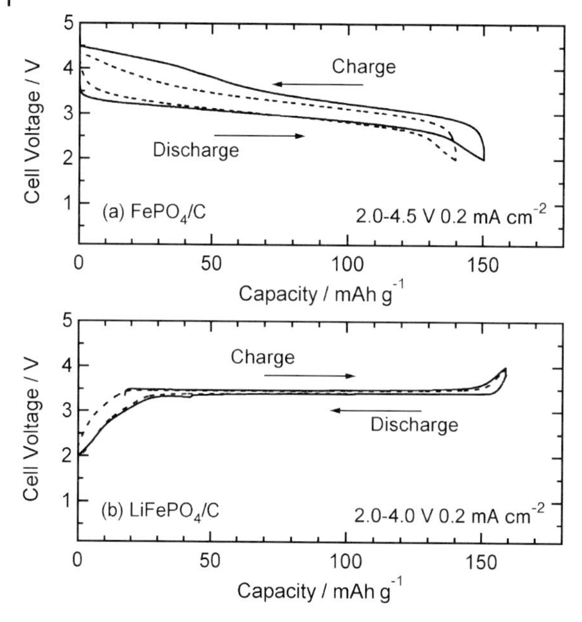

Fig. 4.9 The cyclability for (a) amorphous $FePO_4$/C composite, and (b) olivine-type $LiFePO_4$/C.

addition, inexpensive starting materials such as P_2O_5 and metallic iron powder can be used. The P_2O_5 and metallic iron powders were reacted in water at room temperature. The precursor solution was mixed for 24 hours by a planetary ball mill at 200 rpm at room temperature. The crystalline phase annealed at 650 °C was identified as trigonal $FePO_4$ with a *P*321 space group. On the other hand, the annealed sample below crystallization temperature showed amorphous diffuse scattering in the X-ray diffraction (XRD) profile. Both amorphous and trigonal $FePO_4$ showed a monotonous discharge slope without a discharge plateau as shown in Figure 4.9(a). Although the rechargeable capacity was 140 mAh g^{-1}, the $FePO_4$ obtained without Li cannot be used as the cathode against a carbon anode in an Li-ion battery. However, Li predoping to $FePO_4$ is possible by chemical lithiation using LiI acetonitrile solution in Ar atmosphere. After annealing, the XRD profile of the lithiated $FePO_4$ showed an olivine phase and the 3.3-V discharge plateau characteristic of an olivine cathode was recognized, as shown in Figure 4.9(b) [26].

4.5
3D Calcite-Type Borate Cathode

The mean discharge voltage of a borate cathode is lower than that of polyanionic cathodes, because the inductive effect caused by boron with low electronegativity is weaker than that of sulfur, or phosphorus (Figure 4.10). However, ferric borates

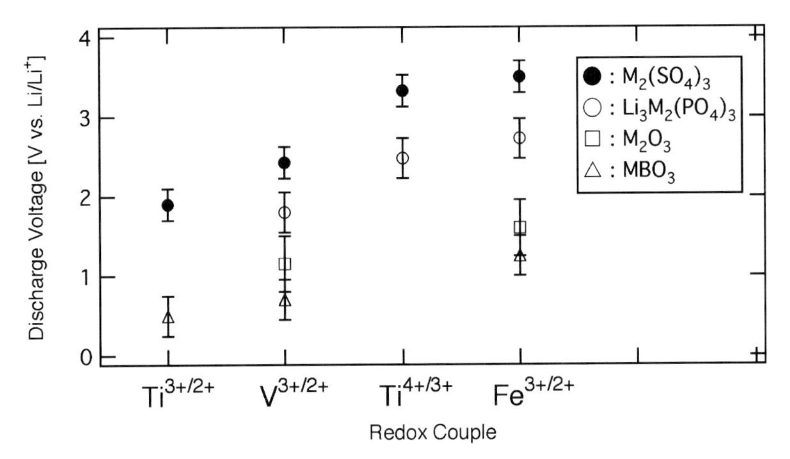

Fig. 4.10 Redox potential map of various polyanionic cathodes.

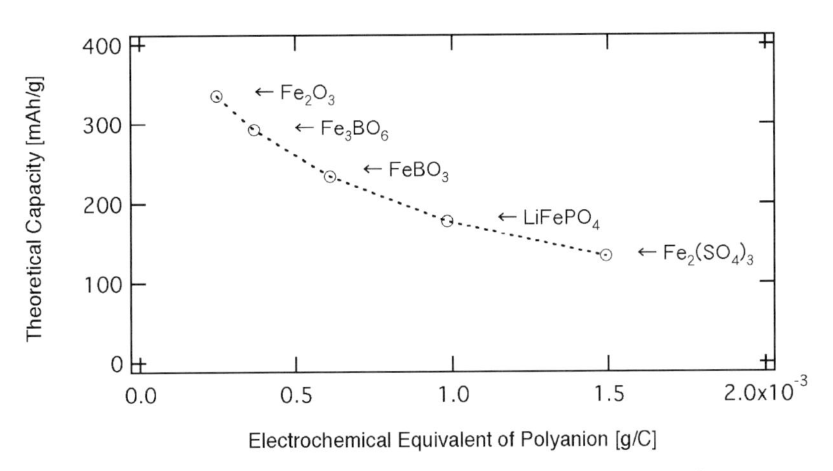

Fig. 4.11 Theoretical capacities of various iron-based cathode-active materials.

having the lightest weight borate polyanions are attractive in terms of their specific capacity in various iron-based polyanionic cathodes or anode materials (Figure 4.11).

The theoretical capacity for calcite $FeBO_3$ estimated by the Fe^{3+}/Fe^{2+} redox reaction are 234 mAh g^{-1} (856 mAh cc^{-1}). The volumetric capacity of $FeBO_3$ is equal to that of a graphite anode (855 mAh cc^{-1}). The mean voltage on lithium intercalation into ferric borates is 1.5 V, unfortunately, too high for anodic use in lithium-ion batteries. However, the mean charge/discharge voltage of $FeBO_3$ can be tuned by V^{3+} or Ti^{3+} substitution in $FeBO_3$ on the basis of the analogy of the 3.6-V NASICON cathode, $Fe_2(SO_4)_3$ and 2.6-V NASICON cathode, $V_2(SO_4)_3$ (Figure 4.6) [27].

4.6
3D Perovskite-Type Fluoride Cathode

The expectation of Fe^{3+}/Fe^{2+} as the cheapest redox couple and fluorine as the anion with the highest electronegativity and the smallest electrochemical equivalent seems to be quite natural. The first report of the FeF_3 cathode performances goes back to more than 10 years ago [28]. To date there have been few reports about perovskite-type oxide cathodes, because it is difficult to satisfy the tolerance factor (t) of perovskite ABO_3 with monovalent Li A cation and pentavalent transition metal B cation.

$$t = \frac{\{r_A + r_O\}}{2^{1/2}\{r_B + r_O\}} \sim 1 \tag{4.1}$$

In contrast to perovskite-type oxide, there are many candidates for trivalent transition metal B cation for perovskite-type fluoride ABF_3. However, the FeF_3 perovskite fluoride cathode had two defects: (i) their solubility in polar solvents such as propylene carbonate and ethylene carbonate, because of the ionic compound, and (ii) the limitation in their use for Li-ion batteries with carbon anodes, because the initial composition does not include any Li. The former defect was addressed recently: a reversible capacity larger than that of olivine-type $LiFePO_4$ has been obtained in carbon-coated FeF_3 as shown in Figure 4.12 [29, 30]. For the latter defect, Na predoping to FeF_3 by planetary ball milling through the reaction formula FeF_2+ NaF − >$NaFeF_3$ was carried out successfully and the electrochemical activity of the mechanochemically doped sodium has been confirmed by the charge/discharge cycle test of the predoped $NaFeF_3$/Na cell [31].

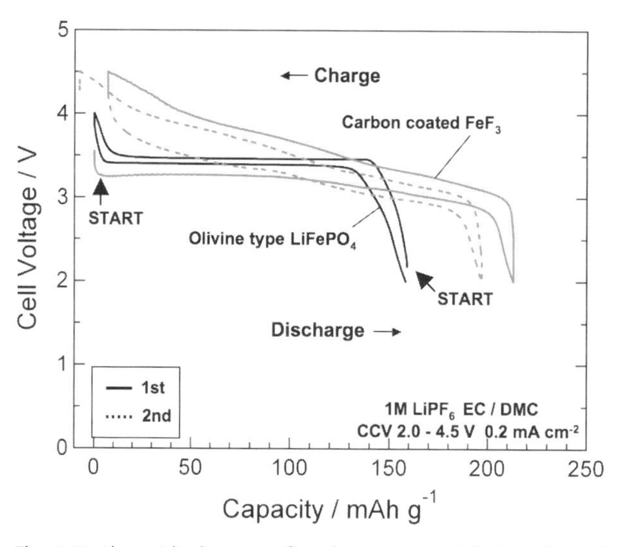

Fig. 4.12 Charge/discharge profile of carbon-coated FeF_3 and $LiFePO_4$.

4.7 Summary

With plug-in type electric vehicles entering the market, second-generation Li-ion batteries are required. The new battery should have the capacity of being charged quickly and well below the freezing point or well above the boiling point of water. The cathode of the battery must be inexpensive and stable in electrolytes under various external temperatures.

The basic concepts that need to be realized the second-generation cathode, as outlined in this chapter, are as follows:

1. a rare-metal-free low-cost and low-environmental-impact cathode;
2. a corner-sharing framework for the large bottleneck of Li or Na diffusion;
3. polyanion or fluoride for a high discharge voltage cathode;
4. polyanion or antioxide for reduction of the exothermic heat at elevated temperatures.

We hope that the above guiding principles will help lead to development of an inexpensive, environment-friendly second-generation Li-ion battery in the near future.

References

1 Nagaura, T. and Tozawa, K. (1990) *Prog. Batteries Sol. Cells*, **209**, 9.

2 Noguchi, H., Muta, T., and Decheng, L. (2004) Abstract of the 45th Battery Symposium, Vol. 3B09, Japan, p. 248.

3 Tsuda, M., Arai, H., Takahashi, M., Ohtsuka, H., Sakurai, Y., Sumitomo, K., and Kageshima, H. (2005) *J. Power Sources*, **144**, 183.

4 Tabuchi, M., Nakashima, A., Shigemura, H., Ado, K., Kobayashi, H., Sakaebe, H., Kageyama, H., Nakamura, T., Kohzaki, M., Hirano, A., and Kanno, R. (2002) *J. Electrochem. Soc.*, **149**, A509.

5 Makimura, Y. and Ozuku, T. (2003) *J. Power Sources*, **119**, 156.

6 Yabuuchi, N. and Ozuku, T. (2003) *J. Power Sources*, **119**, 171.

7 Okada, S., Takahashi, Y., Kiyabu, T., Doi, T., Yamaki, J., and Nishida, T. (2004) Abstract of the 210th Meeting of the Electrochemical Society, Hawaii, USA, p. 584.

8 Goodenough, J.B., Hong, H.Y.-P., and Kafalas, J.A. (1976) *Mater. Res. Bull.*, **11**, 203.

9 Delmas, C., Nadiri, A., and Soubeyroux, J.L. (1988) *Solid State Ionics*, **28–30**, 419.

10 Nanjundaswamy, K.S., Padhi, A.K., Goodenough, J.B., Okada, S., Ohtsuka, H., Arai, H., and Yamaki, J. (1996) *Solid State Ionics*, **92**, 1.

11 Manthiram, A. and Goodenough, J.B. (1987) *J. Solid State Chem.*, **71**, 349.

12 Okada, S., Ohtsuka, H., Arai, H., and Ichimura, M. (1993) Proceedings of the Symposium on New Sealed Rechargeable Batteries and Super Capacitors, Philadelphia, USA, Vol. 93–23, p. 431.

13 Padhi, A.K., Nanjundaswamy, K.S., Masquelier, C., Okada, S., and Goodenough, J.B. (1997) *J. Electrochem. Soc.*, **144**, 1609.

14 Padhi, A.K., Nanjundaswamy, K.S., and Goodenough, J.B. (1997) *J. Electrochem. Soc.*, **144**, 1188.

15 Ravet, N., Chouinard, Y., Magnan, J.F., Besner, S., Gauthier, M., and Armand, M. (2001) *J. Power Sources*, **97–98**, 503.

16 Huang, H., Yin, S.-C., and Nazar, L.F. (2001) *Electrochem. Solid-State Lett.*, **4**, A170.

17 Yamada, A., Chung, S.C., and Hinokuma, K. (2001) *J. Electrochem. Soc.*, **148**, A224.

18 Chung, S.-Y., Bloking, J.T., and Chiang, Y.-M. (2002) *Nat. Mater.*, **1**, 123.

19 Yang, S., Zavalij, P.Y., and Whittingham, M.S. (2001) *Electrochem. Commun.*, **3**, 505.

20 Park, K.S., Son, J.T., Chung, H.T., Kim, S.J., Lee, C.H., and Kim, H.G. (2003) *Electrochem. Commun.*, **5**, 839.

21 Okada, S., Okazaki, Y., Yamamoto, T., Yamaki, J., and Nishida, T. (2004) Meeting Abstract of 2004 ECS Joint International Meeting, Hawaii, Abstract No. 584.

22 Masquelier, C., Padhi, A.K., Nanjundaswamy, K.S., Okada, S., and Goodenough, J.B. (1996) Proceedings of the 37th Power Sources Conference, Philadelphia, USA, p. 184.

23 Paques-Ledent, M.Th. (1974) *Ind. Chim. Belg.*, **39**, 845.

24 Chen, Z. and Dahn, J.R. (2002) *J. Electrochem. Soc.*, **149**, A1184.

25 Barker, J., Saidi, M.Y., and Swoyer, J.L. (2003) *Electrochem. Solid-State Lett.*, **6**, A53.

26 Shiratsuchi, T., Okada, S., Yamaki, J., Yamashita, S., and Nishida, T. (2007) *J. Power Sources*, **173**, 979.

27 Okada, S., Tonuma, T., Uebo, Y., and Yamaki, J. (2003) *J. Power Sources*, **119–121**, 621.

28 Arai, H., Okada, S., Sakurai, Y., and Yamaki, J. (1997) *J. Power Sources*, **68**, 716.

29 Badway, F., Pereira, N., Cosandey, F., and Amatucci, G.G. (2003) *J. Electrochem. Soc.*, **150**, A1209.

30 Nishijima, M. *et al.* (2006) Abstract of the 47th Battery Symposium, Vol. 3F02, Japan, p. 560.

31 Gocheva, I.D., Nishijima, N., Doi, T., Okada, S., and Yamaki, J. (2007) Abstract of the 74th Meeting of Electrochemical Society of Japan, Japan, p. 1H27.

5
Thermodynamics of Electrode Materials for Lithium-Ion Batteries

Rachid Yazami

5.1
Introduction

As a high energy density storage and conversion system, lithium-ion batteries require the negative electrode material (or the anode) and the positive electrode material (or the cathode) to store and release large amounts of lithium ions during charge and discharge cycles. The lithium exchange between anode and cathode occurs owing to ion transport within the electrolyte. Such an electrode reaction involves changes in the lithium-ion composition in each electrode, which in turn induces changes in the electrode material characteristics and properties including the crystal and electronic structures, and hence the chemical potential of lithium ions. As long cycle life is also an important battery requirement, structural changes within each electrode should be as benign as possible to allow for a high lithium-ion storage capability and fast electrode kinetics. In fact, from the viewpoint of electrode material, cycle life strongly depends on the initial characteristics of the active material such as high crystallinity and low level of impurities. Cycle life also relates to the thermodynamics parameters such as the lithium stoichiometry, which relates to the state of charge (SOC) or state of discharge (SOD), the electrode potential, and the kinetics parameters including the charge and discharge current rate (or C−rate) and charge and discharge voltage limits, temperature, and pressure.

The total energy stored in and released by a cell during charge and discharge is controlled by the thermodynamics of the active electrode processes. Basically electrode reactions at the anode (AN) and cathode (CA) consist of the exchange of lithium ions and electrons with the electrolyte and an external circuit, respectively. Electrode reactions during discharge can schematically be simplified as

$$\mathrm{Li}_x\mathrm{AN} \xrightarrow{\text{discharge}} \mathrm{AN} + x\mathrm{Li}^+(\text{electrolyte}) + xe^-(\text{external circuit}) \tag{5.1}$$

and

$$\mathrm{CA} + y\mathrm{Li}^+(\text{electrolyte}) + ye^-(\text{external circuit}) \xrightarrow{\text{discharge}} \mathrm{Li}_y\mathrm{CA} \tag{5.2}$$

Lithium Ion Rechargeable Batteries. Edited by Kazunori Ozawa

ISBN: 978-3-527-31983-1

The electrode reactions (5.1) and (5.2) can be investigated using half cells or full cells. In a half cell usually metallic lithium is used as the counterelectrode versus the working electrode (AN or CA). It also serves as the reference electrode.

In a full cell, both (AN) and (CA) are used as the working electrodes. In this case, the electrons produced at the cathode are those consumed at the anode, which means $x = y$, in Equations 5.1 and 5.2. This is true provided no side reaction takes place at either one or both electrodes. Accordingly, the full lithium-ion cell discharge reaction is given by

$$\mathrm{Li}_x\mathrm{AN} + \mathrm{CA} \xrightarrow{\text{discharge}} \mathrm{AN} + \mathrm{Li}_x\mathrm{CA} \tag{5.3}$$

At each stage of the cell reaction, lithium ion and electron fluxes can be reversed and thermodynamic equilibrium can be reached. Such an equilibrium is defined, among others, by an open-circuit voltage (OCV), hereafter denoted by $E_0(x)$. $E_0(x)$ grows from differences in the lithium chemical potential (ionic and electronic) between the anode and the cathode. In fact the cell OCV is a function of composition "x", pressure "P", and temperature "T": $E_0(x, P, T)$. In most cases, the pressure is the normal one or close enough to normal. Therefore, only composition and temperature need to be considered as the key parameters.

The free energy of the full-cell reaction 5.3, $\Delta G(x, T)$, relates to $E_0(x, T)$, according to Equation (5.4):

$$\Delta G(x, T) = -nFE_0(x, T) \tag{5.4}$$

where n is the charge number carried by the exchanged ion (here $n = 1$ for Li^+) and F is the Faraday constant. Therefore direct measurement of $E_0(x, T)$ enables the free energy $\Delta G(x, T)$ to be determined and vice versa.

Moreover, the free energy relates to the heat of reactions ΔH and to entropy ΔS following the Equation (5.5):

$$\Delta G(x, T) = \Delta H(x, T) - T\Delta S(x, T) \tag{5.5}$$

The temperature dependence of the heat and the entropy functions are small and can be neglected within a small temperature range. Equation (5.5) reduces to

$$\Delta G_0(x, T) = \Delta H(x) - T\Delta S(x) \tag{5.6}$$

Combining Equations (5.4) and (5.6) yields

$$\Delta S(x) = \mathrm{F}\left(\left.\frac{\partial E_0(x, T)}{\partial T}\right|_x\right) \tag{5.7}$$

$$\Delta H(x) = \mathrm{F}\left(-E_0(x, T) + T\left.\frac{\partial E_0(x, T)}{\partial T}\right|_x\right) \tag{5.8}$$

where the symbol $\left.\frac{\partial E_0(x,T)}{\partial T}\right|_x$ denotes the temperature slope of $E_0(x, T)$, which should be measured at constant composition x. To this end, it is important to avoid side reactions taking place during the measurement of $E_0(x, T)$ as they affect the composition. Self-discharge is one of the obvious side reactions and this can be minimized if $E_0(x, T)$ is measured at temperatures below ambient, when possible. Also the electrochemical cell used for the experimental measurement should be isolated from any electrical current or chemical leaks and the OCV measurement system should have a large internal resistance.

Combining Equations (5.7) and (5.8) enables the change in the entropy and enthalpy of the cell reaction to be determined. The total change in entropy and in enthalpy between two electrode compositions x_1 and x_2, $\overline{\Delta S}|_{x_1}^{x_2}$, and $\overline{\Delta H}|_{x_1}^{x_2}$, respectively can be obtained from integration of the $\Delta S(x)$ and $\Delta H(x)$ curves:

$$\overline{\Delta S}\Big|_{x_1}^{x_2} = \mathrm{F}\left(\int_{x_1}^{x_2} \left.\frac{\partial E_0(x,T)}{\partial T}\right|_x \mathrm{d}x\right) \tag{5.9}$$

$$\overline{\Delta H}\Big|_{x_1}^{x_2} = \mathrm{F}\int_{x_x}^{x_2}\left(-E_0(x,T) + T\left.\frac{\partial E_0(x,T)}{\partial T}\right|_x \mathrm{d}x\right) \tag{5.10}$$

Normalizing the electrode composition to the full compositional range Δx_{max} of reversible electrode processes ($y = \frac{x}{\Delta x_{\mathrm{max}}}$, $0 < y < 1$) gives the total changes in entropy and enthalpy on a full-cell operation range:

$$\overline{\Delta S}\Big|_0^1 = \mathrm{F}\left(\int_0^1 \left.\frac{\partial E_0(y,T)}{\partial T}\right|_y \mathrm{d}y\right) \tag{5.11}$$

$$\overline{\Delta H}\Big|_0^1 = \mathrm{F}\left(\int_0^1 (-E_0(y,T) + T\left.\frac{\partial E_0(y,T)}{\partial T}\right|_y \mathrm{d}y\right) \tag{5.12}$$

The amount of heat a cell is susceptible to release is an important data for predicting its thermal behavior and assessing its safety. In fact, undesirable phenomena within the cell, including thermal runaway and incidence of fire, are triggered by the heat generated owing to reactions at one or both electrodes. To tackle this sensitive issue, extensive studies have been devoted to the thermal behavior of various lithium-ion cells with the aim of understanding the basic processes involved in the heat generation and, to some extent, controlling their kinetics [1–8]. Thermodynamic studies on full cells have also been carried out to determine the entropy and/or the enthalpy component of free energy [9–16]. Computer modeling and simulation have been extensively used and have proved very helpful in understanding and predicting the thermal behavior of lithium-ion cells [17–22]. Other theoretical studies based on first-principles methods are being increasingly applied in the field of lithium-ion batteries, in particular, for computing thermodynamic functions associated with lithiation and delithiation processes [23–36]. Specific thermodynamics studies of anodes [37–46], cathodes [47–65], and electrolytes [66, 67] for lithium-ion batteries have been the focus of recent studies to assign contributions of each of the active materials to the thermal behavior of the cell. It has been recognized that composition-induced phase transformations occurring in electrode materials bear a thermodynamics signature, particularly in the OCV profile curve [68–73]. Cation ordering [73] and cation mixing [74] are among typical phenomena in lithium transition metal oxide based cathode materials that affect their thermodynamics.

The recent emergence of nanostructure materials as promising alternative materials for anode and cathode application has also generated interest in the thermodynamics aspects of these materials. It has been theorized that particle size may be considered as an independent parameter in the thermodynamics as are composition and temperature. In fact, when the particle size falls to the nanometer scale, many of the electrode material characteristics and properties become increasingly governed by the surface to bulk atomic ratio and by the spatial extent of interphase domains and grain boundaries [75–79].

Although the temperature dependence of OCV $\left.\frac{\partial E_0(x,T)}{\partial T}\right|_x$ (occasionally referred to as the *entropy term*) was measured on electrode materials and reported in the literature [57, 65, 80, 81], there was no attempt to correlate the results with structure defects in the starting material and to find the origin and the scale of the "entropy term" changes with the electrode composition, cycle, and thermal history.

The study, we carried out over the last five years on the thermodynamics of anode and cathode materials was driven by the need for filling the gap in the fundamental understanding of electrode processes behind changes in the entropy- and the enthalpy-state functions in the course of lithium intercalation and deintercalation. It was theorized that thermodynamics studies can be used as a new tool in the characterization of electrode materials, complementary to conventional physical–chemical techniques such as those based on matter–wave scattering including diffraction (electron, neutron, and X ray) and physical spectrometry (XAS, NMR, FTIR, Raman, EPR, XPS, EELS, etc.) Most of theses techniques require heavy and expensive equipment and can be destructive. This is in contrast to the method we introduce here. We will show that thermodynamics data can be acquired quite conveniently using an electrochemical cell and a nondestructive measurement technique. The new method applies to half cells together with full cells with focus on electrode reaction thermodynamics of a specific material for anode or cathode application or on a full system consisting of two working electrodes. Owing to space constraints, this chapter will deal only with the entropy function because it reveals fine details on the lithiation/delithiation processes, in particular, on the phase transitions that take place in the electrode material. The enthalpy function is complementary to the entropy one and relates to the heat generation within the cell, an important parameter in the cell's cycle life and thermal behavior. The materials covered in this study are carbonaceous anode materials heat-treated at different temperatures and transition-metal-based cathode materials such as $LiCoO_2$, and

$LiMn_2O_4$. Other cathode materials such as $Li(MnCoNi)_{1/3}O_2$ and $LiFePO_4$ were also investigated, but, however, are not discussed here.

5.2
Experimental

5.2.1
The ETMS

Our early laboratory setup for the $E_0(x, T)$ measurements was mostly manual and required real-time control of the cell temperature and the electrode composition. We then set up an automatic laboratory system that we called the *ETMS* (electrochemical thermodynamics measurement system), the schematic diagram of which is displayed in Figure 5.1(a).

The ETMS consisted of the following functional components:

1. A battery cycling system: The battery cycling system applies changes in the electrode composition x under either a galvanostatic mode (constant current) or a potentiostatic mode (voltage steps).

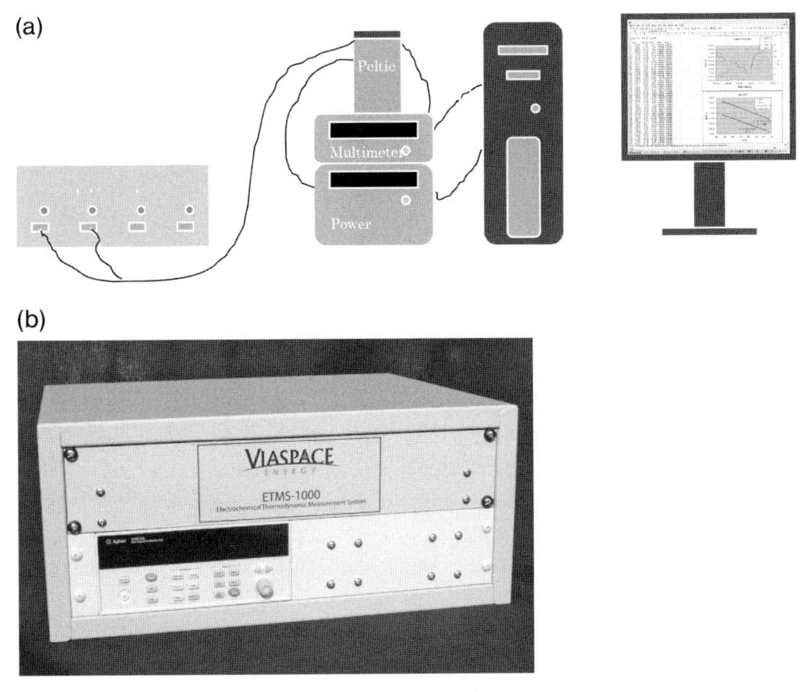

Fig. 5.1 (a) Overview of the automated electrochemical thermodynamic measurement system (ETMS). (b) Picture of the ETMS-1000 system developed by Viaspace for automatic thermodynamics data acquisition.

As a constant or a variable current $i(t)$ passes across the cell, the corresponding change in the electrode composition Δx is obtained by Faraday's Law:

$$\Delta x = \frac{1}{mQ_{\text{th}}} \int_0^t i(t)\mathrm{d}t, \tag{5.13}$$

In Equation (5.13), m is the total mass (usually in grams) of the active material in the electrode and Q_{th} is the specific (mass) capacity (usually in milliampere hours per gram) of the electrode active material. The latter is given by

$$Q_{\text{th}} = \frac{n\text{F}}{3.6M} \tag{5.14}$$

In Equation (5.14) n is the number of electrons per electrode mole, F is the Faraday constant (in C), M is the molar mass of the anode or the cathode material in Equations (5.1) and (5.2) (in grams) and 3.6 is a unit-conversion constant. Accordingly, in Equation (5.13) $i(t)$ should be expressed in milliamperes and time t in hours.
In fact, in most electrode materials Q_{th} is achieved experimentally such as by very low charge and discharge rate, within a voltage window where fully reversible electrode processes take place with minimal side reactions such as electrolyte decomposition and irreversible electrode transformations.

2. A temperature-controlled cell holder: Peltier plaques are used to set the temperature in the cell holder. The temperature is varied stepwise with ΔT of about 5 °C increments between ambient and 0 °C. Cooling the cells from the ambient temperatures was preferred to heating them to minimize the self-discharge phenomenon, which is thermally activated and may affect the electrode composition.
3. A high accuracy OCV measurement system: The system has a high internal resistance to avoid discharging the cell in the course of OCV versus T measurement. Because the electrode equilibration at set temperature may require a long period of rest time, in practice, it is convenient to set an upper limit for the absolute value of the OCV time dependence $\left(\left|\frac{\partial E_0(x,T)}{\partial t}\right|_{x,T}\right)$ below about 0.1–1 mV h^{-1} for ending the OCV measurement and triggering the next step.
An OCV drift may be observed during measurement, caused by the electronic circuit fluctuations or owing to a small self-discharge. Such a drift is easily erased by the program.
4. A computer controlled power system: The power system sets the cell temperature by adjusting the amount of electric power in the Peltier plaques.
5. A computer system: The computer system runs the program that monitors the battery cycler and the electric power in the Peltier plaques. It also acquires and processes the data and converts them to thermodynamics functions and graphic presentations according to Equations (5.7) and (5.8).

Viaspace, Inc. (Pasadena, California) have recently developed and commercialized a fully integrated ETMS (BA-1000) shown in Figure 5.1(b). External cell holders with temperature control were built up for both coin cells and cylindrical cells, including the 18650 size.

All thermodynamics data presented here were obtained on our in-house-built ETMS except for that presented in the last section on the effect of electrode cycling where the BA-1000 was used.

5.2.2 Electrochemical Cells: Construction and Formation Cycles

In our study, we used 2016 coin cells for running thermodynamics tests because of their convenience and easy preparation in an argon glove box. At least two cells were run for each electrode material. Our half cells used metallic lithium as the counter and the reference electrode, a microporous polymer separator wet with the electrolyte, and a working electrode.

A typical working electrode consists of a mixture of the following components:

1. an active electrode material such as a carbonaceous material for anode studies or a lithium transition metal oxide ($LiCoO_2$, $LiMn_2O_4$) for cathode materials studies;
2. a conductive additive usually consisting of acetylene black, natural or synthetic graphite or carbon nanofibers; and
3. a polymer binder such as polyvinylidene fluoride (PVDF).

The electrode mixture is thoroughly stirred in an organic solvent; the slurry is then spread on a polytetrafluoroethylene (Teflon) plate. The uniformly thick film of about 100 μm so obtained is dried in air and then in vacuum for several hours. Discs ~17 mm in diameter are cut from the film, further dried overnight in vacuum at 80 °C, and introduced into the dry box for final cell assembly.

The electrolyte composition is 1 M $LiPF_6$ solution in an equal volume mixture of ethylene carbonate (EC) and dimethyl carbonate (DMC) for anode materials studies and 1 M $LiClO_4$ solution in propylene carbonate (PC) for the cathode ones. Coin cells are then assembled in dry argon atmosphere and transferred into the cell holder of ETMS for immediate electrochemical and thermal investigation. The cells are first cycled under C/10 charge and discharge rate for a few cycles until a stable capacity is reached. The cells are then either fully charged (delithiated) to 1.5 V for carbon anodes or fully discharge (lithiated) to 3 V for cathodes studies. At 1.5 V, the lithium content in the carbon anode is considered close to zero (Li_0C_6) and at 3 V the lithium content in the cathode is considered close to unity (Li_1CoO_2, $Li_1Mn_2O_4$,). This sets the initial state of discharge of the cells.

5.2.3 Thermodynamics Data Acquisition

The cells are first either discharged (anode) or charged (cathode) typically under C/100 to C/20 rate for a period of time long enough to achieve a fixed increment in the lithium composition (usually in the order of $\Delta x = 0.05$). The cell is then rested until acceptable constant OCV is reached at the ambient temperature. The temperature is then decreased stepwise (ΔT of about 5 °C per step) until a new stable OCV is reached. When data acquisition at the lowest set temperature is achieved, the cell is then brought to the ambient temperatures and a new composition increment Δx is applied. This protocol is repeated until a target lowest (anode) or highest (cathode) voltage is reached, and then the current is reversed to incrementally vary the electrode composition in the opposite direction.

This protocol allows the thermodynamics data to be acquired during a full charge/discharge cycle and to observe whether a hysteresis appeared.

5.3 Results

5.3.1 Carbonaceous Anode Materials

We carried out a systematic study on the effects of the heat treatment temperature (HTT) of a carbon material precursor on the thermodynamics of lithiation and delithiation. Carbonaceous samples were provided by Superior Graphite, Co. using the same petroleum pre-coke precursor HTT below 500 °C. The pre-coke was further heated to different temperatures between 700 and 2600 °C.

Figure 5.2(a) shows the X-ray diffraction (XRD) patterns of cokes heat treated either below 500 °C (no HTT) or HTT between 900 and 2600 °C. A sketch of the evolution of the carbonaceous structure upon heat treatment is shown in Figure 5.2(b) [82]. Using the Scherrer equation, the crystal size along the c-axis (L_c) and the a-axis (L_a) was determined. L_c and L_e are good indicators of the degree of graphitization G (3D ordering) of the carbonaceous material. The L_c and L_a curve profiles depicted in Figure 5.3 are typical of graphitizing (soft) carbons and are in good agreement with the literature [83]. Table 5.1 summarizes the characteristic of the coke materials as derived from XRD analysis. Cokes HTT ≤ 1700 °C have two diffraction peaks in the 002 area: a sharp peak corresponding to a graphitic phase and a broader peak corresponding to highly disordered carbons. The table gives their respective degree of graphitization G together with L_c values.

Raman scatting is also a powerful tool for characterizing the rate of carbonization and graphitization process in soft carbons [84]. Typically, two Raman-active modes are observed in carbonaceous materials. The first one is the G-mode at approximately 1590 cm^{-1} that relates to vibrational modes within well-ordered polyaromatic carbon hexagons in the graphene layer. The second active mode is the D-mode at approximately 1360 cm^{-1} and is usually associated with lattice modes of carbons present at the layers edges or in smaller polyaromatic domains. In disordered carbons, the D-mode has stronger intensity I_D relative to the G-mode (I_G). In fact the I_G/I_D intensity ratio is used as a metric for the

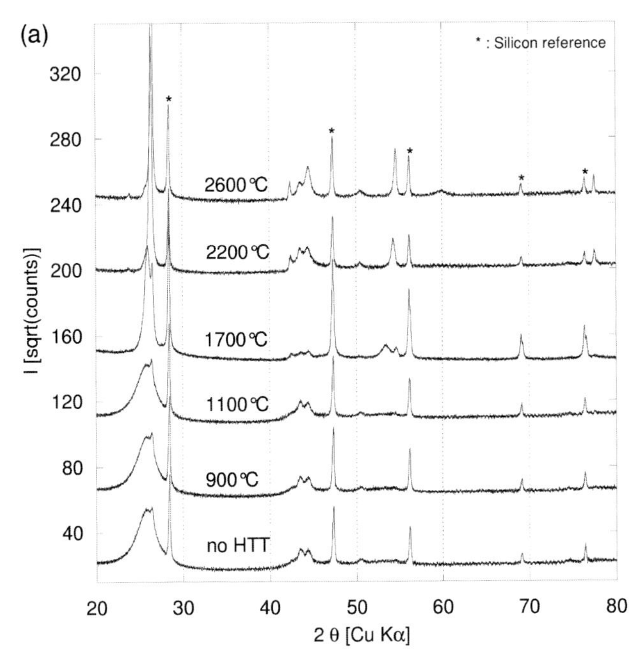

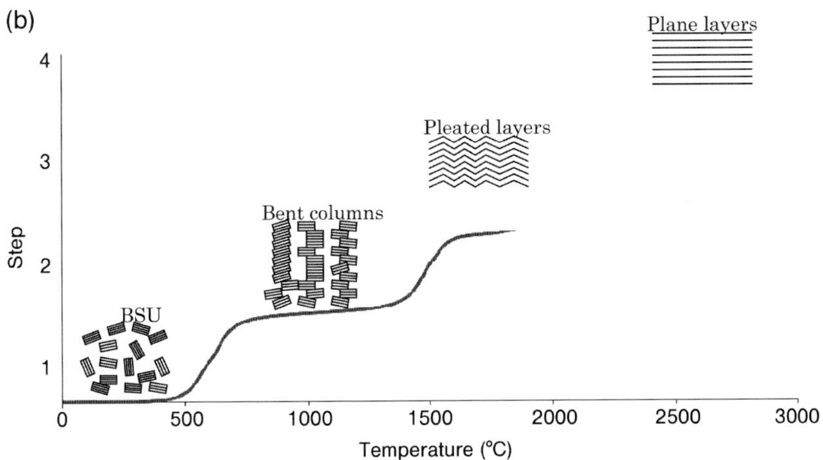

Fig. 5.2 (a) XRD pattern of the coke samples HTT at different temperature with an internal silicon reference (labeled*). (b) Structural evolution of a graphitic carbon as a function of temperature (BSU = basic structural units) (After 82).

degree of graphitization, in a way similar to the L_c and L_a obtained from XRD analysis [85].

Figure 5.4 shows the evolution of the Raman spectra with the HTT. The results are in good agreement with the literature [85], that is, the D-mode intensity decreases and the G-mode intensity increases with the HTT.

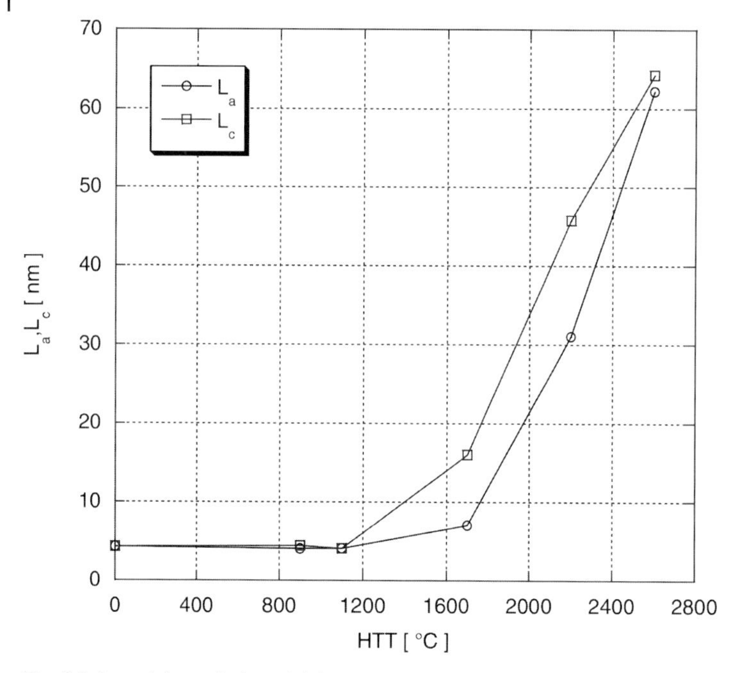

Fig. 5.3 L_a and L_c evolution with heat treatment temperature based on Raman spectrometry (L_a) and on XRD (L_c).

Table 5.1 Crystallite size in the *c* direction based on the 002 peak broadening, L_c, and graphitization degree, *G*.

Sample	Full width at half maximum – FWHM (2θ)	d-spacing (Å)	002 peak angle (2θ)	G(%)	L_c(Å)
Coke no HTT	1.97	3.461	25.72	0	43
	0.37	3.372	26.41	82	277
Coke HTT 900°C	1.93	3.461	25.72	0	44
	0.35	3.372	26.41	82	297
Coke HTT 1100°C	2.05	3.461	25.72	0	41
	0.33	3.372	26.41	82	320
Coke HTT 1700°C	0.58	3.428	25.97	30	160
	0.23	3.359	26.51	94	523
Coke HTT 2200°C	0.25	3.377	26.37	77	458
Coke HTT 2600°C	0.20	3.361	26.50	92	643

As is shown in the next section, we found that the thermodynamics of the lithiation reaction of carbonaceous anode materials strongly depends on their degree of graphitization.

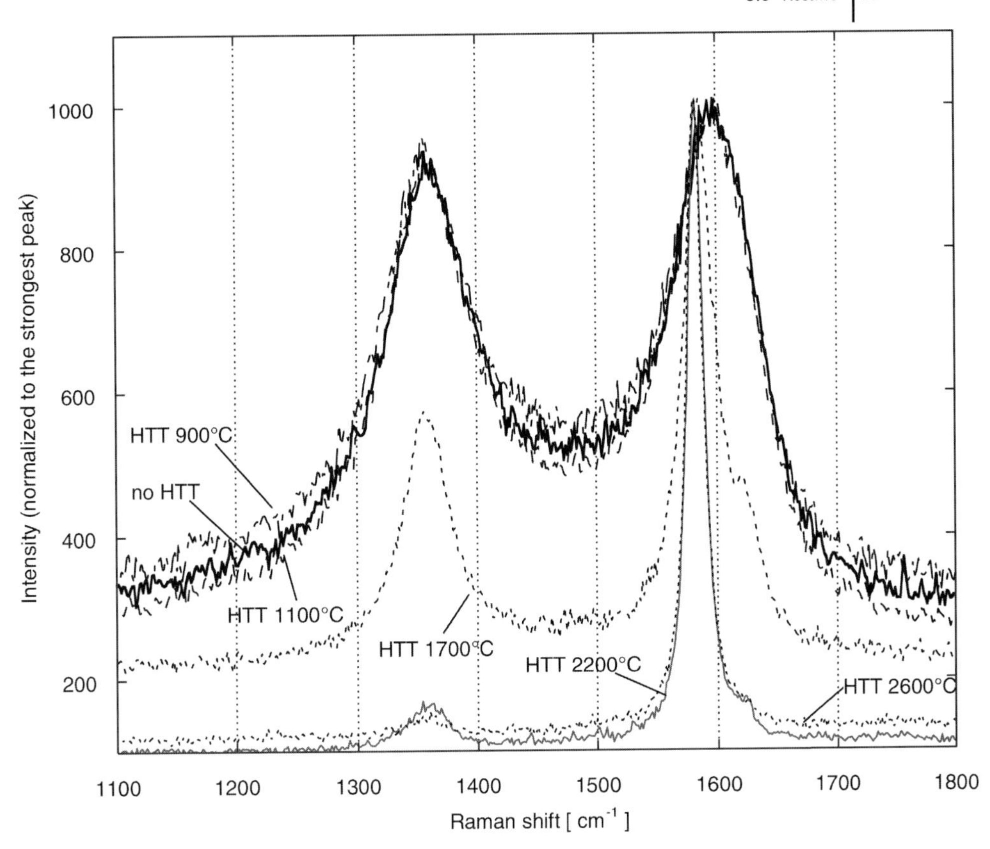

Fig. 5.4 Raman spectra of coke materials used in this study.

5.3.1.1 Pre-coke (HTT < 500 C)

Lithium intercalation into carbonaceous materials is usually schematized as

$$6C + xLi^{+} + xe^{-} \longleftrightarrow Li_xC_6, \quad 0 \leq x \leq 1 \tag{5.15}$$

Figure 5.5 shows the $E_0(x)$ and the $\Delta S(x)$ curves of the pre-coke heat-treated below 500 °C. Whereas the OCV curve monotonously decreased with x, a result typical of disordered carbons, the $\Delta S(x)$ shows more pronounced changes in the slope value and sign indicative of different steps in the lithiation mechanism. The $\Delta S(x)$ curve in Figure 5.5 can be divided into seven different composition areas:

- Area-I ($0 < x < 0.073$): where a sharp decrease in entropy is observed until a minimum is reached at $x \sim 0.073$;
- Area-II ($0.073 < x < 0.162$): where $\Delta S(x)$ increases;
- Area-III ($0.162 < x < 0.384$): where the entropy decreased almost linearly until $x = 0.384$;
- Area-IV ($0.384 < x < 0.420$): where the $\Delta S(x)$ slope becomes more negative and then makes a minimum at $x \sim 0.420$;

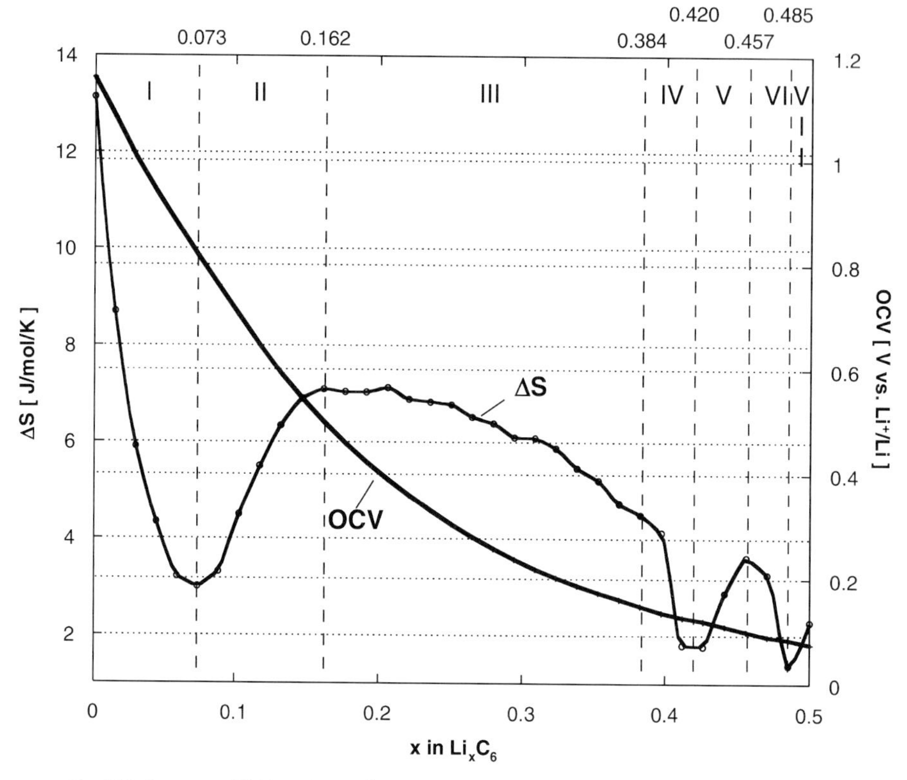

Fig. 5.5 Entropy of lithium intercalation into coke with no HTT and corresponding OCV during charge.

- Area-V ($0.420 < x < 0.457$): where $\Delta S(x)$ changes in sign and goes through a maximum at $x = 0.457$;
- Area-VI ($0.457 < x < 0.485$): where $\Delta S(x)$ makes another minimum at $x \sim 0.48$;
- Area-VII ($0.485 < x < 0.50$): where ΔS increases again.

Variations in the entropy value and slope at relatively well-defined compositions are the signature of dramatic changes in the energetics of lithiation. In fact, lithium ions progressively populate carbon sites and domains depending upon their energy; more energetically favored sites are first populated followed by less energetic ones. This site segregation induces phase transitions such as those from insulator to semiconductor and from semiconductor to metallic. Lithiation may also induce crystallographic reorganization within specific carbon domains. In such cases, a sharp decrease in entropy in Area-I may be associated with lithium cathodic deposition on carbon sites present at the surfaces of disordered domains. It is most likely that lithium atoms arrange on the surface to form a short-range order, which induces a sharp decrease in entropy. Area-II appears as a smooth transition area between Areas-I and III. In the latter, the entropy decreases uniformly in a composition range Δx of about 0.25, which accounts for half of the total lithium uptake in the electrode. Where Area-I relates to carbon atoms present on the grains surface, Area-III should be associated with lithium intercalation into the bulk of disordered carbon domains. The uniform slope suggests a solid-solution behavior in which disordered domains are progressively populated with lithium. As these domains become saturated with lithium, a sharp decrease in entropy is observed as we enter Area-IV. The latter should be associated with lithium intercalation in smaller and better crystallized domains close to graphite. Additional lithiation leads to Areas-VI and VII that may relate to staging phenomena in the graphitic domains. Stage number in graphite intercalation compounds refers to the number of graphene layers between adjacent intercalate layers. For instance, a Stage II lithium compound has a stacking sequence of LiGGLiGG. Accordingly, Areas-II and IV should be associated with transition areas during the lithiation of different carbonaceous domains. The coexistence of crystallographically disordered and ordered carbon domains in the pre-coke is evidenced by XRD and Raman scattering measurements. In the pre-coke with HTT $< 500\,^{\circ}$C, a sharp 002 peak of graphite is superposed on a broader 002 peak of disordered carbon in the XRD pattern of Figure 5.2 and a relatively strong G-mode is observed in the corresponding Raman spectra in Figure 5.4.

Although our proposed mechanism of entropy changes in the pre-coke-based anode needs to be supported by further independent characterizations of the corresponding Li_xC_6 materials, it is clear that the entropy curve shows well-defined steps in the electrode process, a feature absent in the OCV curve. XRD and Raman studies do not give direct evidence of multidomain presence in the pre-coke. However, mathematical simulations of the XRD and Raman results in the light of the entropy ones may lead to a similar conclusion, that is, the pre-coke material consists of multicarbon domain structure, including highly active surface states.

It is worth noting that the $\Delta S(x)$ results allow for a direct titration of the relative amounts of different carbon domains according to the energetics of interaction with lithium. In fact, the high resolution achieved in the lithium composition where transitions occur can hardly be matched by XRD, Raman, or any single physical spectrometry technique.

5.3.1.2 **Cokes HTT 900–1700°C**

As shown in Figure 5.6, a similar "seven domains" feature described in the previous section is observed in cokes HTT at 900 and 1100 °C with, however, slight differences in the composition thresholds of successive transitions. This result correlates well with the X-ray and Raman results, which showed no significant changes in the diffraction patterns or Raman spectra in the temperature range 900–1100 °C.

In contrast to the coke HTT 900 and 1100 °C, sizable change in the entropy profile is observed in the coke HTT at 1700 °C as depicted in Figure 5.7. Here again the OCV curve shows a monotonous decay similar to the one observed in cokes

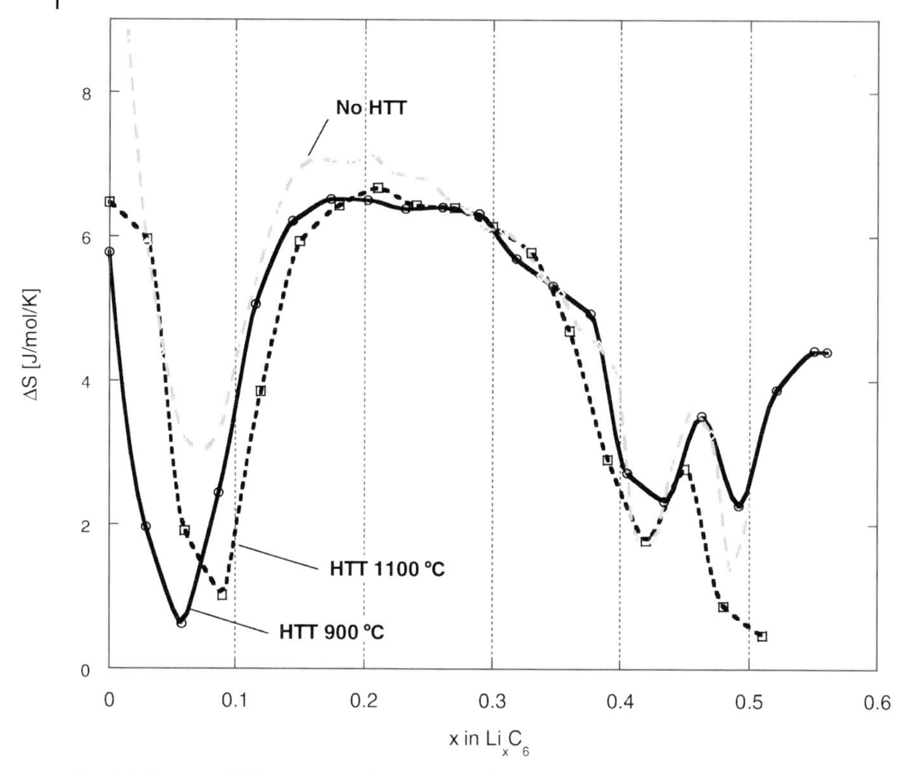

Fig. 5.6 Entropy of lithium intercalation into cokes HTT 900 and 1100 °C during charge. The coke with no HTT is shown for comparison.

HTT at lower temperatures. However, the entropy curve features well-defined maxima and minima and, surprisingly, an increase in the early stage of lithiation unlike in the previous cokes where it sharply decreased. This suggests domain I associated with surface states has disappeared. Such a disappearance correlates with a decrease in the total lithium uptake, which translates to a lower discharge capacity ($\Delta x_{max} \sim 0.55$ at HTT = 1100 °C versus $\Delta x_{max} \sim 0.45$ at HTT = 1700 °C). Also at 1700 °C, L_a and L_c feature a sharp increase because of enhanced crystal ordering of the carbon structure as sketched in Figure 5.2(b). Also smoother entropy in Area-V is indicative of larger graphitic domains in the coke HTT 1700 °C.

5.3.1.3 **Cokes HTT 2200 and 2600°C**

The entropy curve of the coke HTT 2200 °C is shown in Figure 5.8. It resembles that in Figure 5.7 of the coke HTT 1700 °C as Area-II appeared with sharp increase in entropy up to $x = 0.0582$. Areas III and IV are less distinguishable (only Area-III is noted here) and extend to $x = 0.421$ to cover $\Delta x = 0.3628$ range. Transition Area-V covers $\Delta x = 0.489 - 0.421 = 0.068$; then Area-VI with a smooth dome shape covers $\Delta x = 0.361$ up to $x = 0.850$. It is interesting to note the extent of

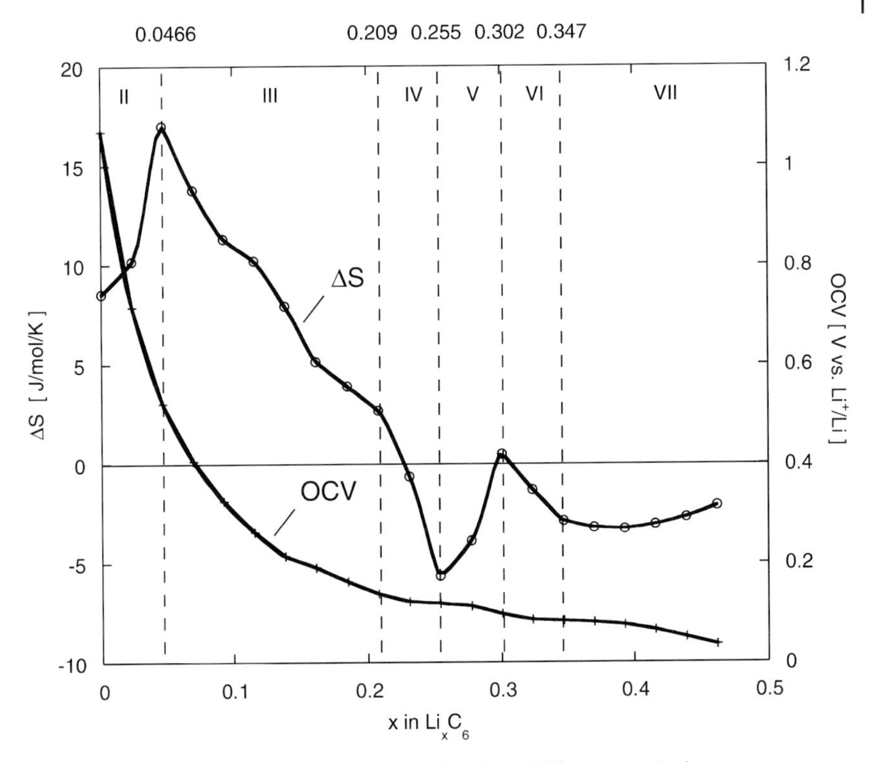

Fig. 5.7 Entropy of lithiation and OCV of a coke HTT 1700 °C during discharge.

Areas III and VI are in the same order of magnitude of $\Delta x \sim 0.36$, which supports the staging model, that is, Area-III is where Stage II forms from pure graphite and Area-VI is where Stage II converts to Stage I.

A sharp transition in Area-IV is indicative of Stage II to Stage I phase transition as is shown in the next section. It is interesting to note the lithium composition covered in Areas-III and V is in the same order of magnitude ($\Delta x \sim 0.2$), which supports the staging nature of the phase transition. In fact, an equal amount of lithium is needed to make Stage II from pure graphite and Stage I from Stage II. However, owing to the persistence of nongraphitic domains in the coke HTT 2200 °C, the amount of lithium involved in the stage transitions is lower than in purely graphitic material as is discussed in the next sections ($\Delta x \sim 0.2$ versus 0.5, respectively).

A dramatic change in the entropy profile is observed in the coke HTT 2600 °C compared to the coke materials discussed earlier. In addition to Area-I, Area-II vanished at the expense of Area-III (see Figure 5.9), and the slope of entropy in Areas-IV and V is smoother. The total lithium intercalation capacity increased to $\Delta x_{max} \sim 0.72$. The slope of the entropy curve in Area-III is smoother indicating a well-defined transition in the solid-solution domain. This may be the signature of stage formation in the graphitic domains.

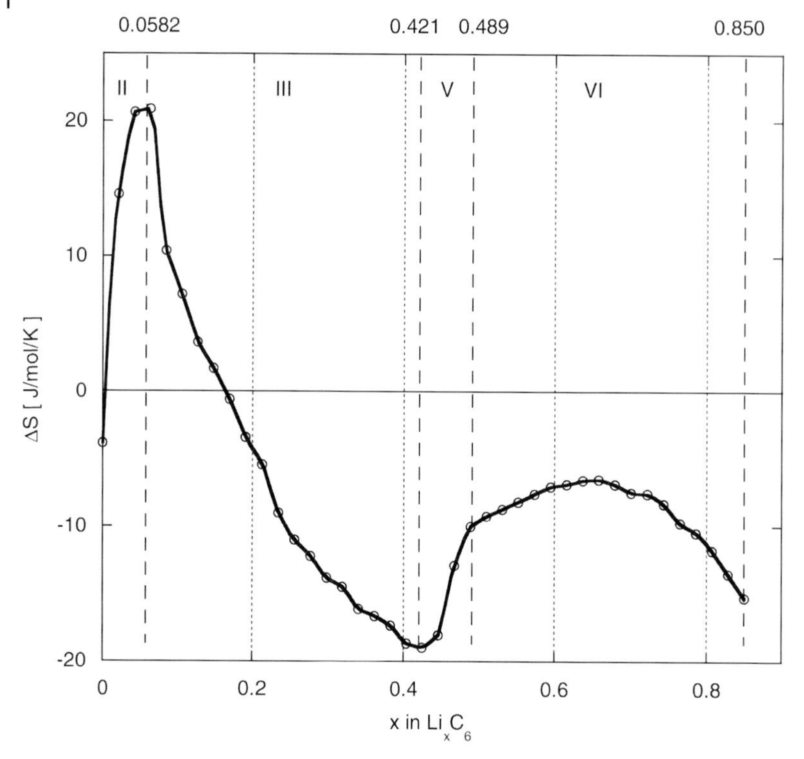

Fig. 5.8 Entropy of lithiation during discharge of a coke HTT 2600 °C.

5.3.1.4 Natural Graphite

The OCV curve in Figure 5.9 of natural graphite shows six well-separated composition domains corresponding to the step-by-step ordering of the lithium ions into the graphite structure. The sharp decrease in OCV and entropy at the early stages of the lithiation process in Area-I corresponds to the solid-solution behavior. Specifically, it corresponds to the formation of so-called dilute Stage I intercalation compound, in which lithium ions randomly occupy the van der Waals space between the graphene layers (i.e., the gas lattice model). Dilute Stage I occurs in a narrow composition range of $\Delta x \sim 0.0495$.

With increased in-plane lithium concentration, the alkali cations tend to condense to form a liquid-like phase up to $x = 0.125$ of Area-II. This is followed by a solid-solution behavior in the composition range $0.125 < x < 0.25$ that translates to 3D ordering as higher intercalation stages VIII, IV, and III form (Area-III). Noteworthy is a discrepancy in the upper composition thresholds at which staging take place between the OCV ($x = 0.19$) and the entropy ($x = 0.25$) curves. This highlights differences in sensitivity between the entropy and the free energy functions, the latter being constant between two stages, whereas the former can vary owing to configurational entropy of the lithium within the interstitial sites.

Slightly sloping OCV and entropy curves are found in Area-V usually associated with transition between liquid (disordered) Stage II to solid (ordered) Stage II.

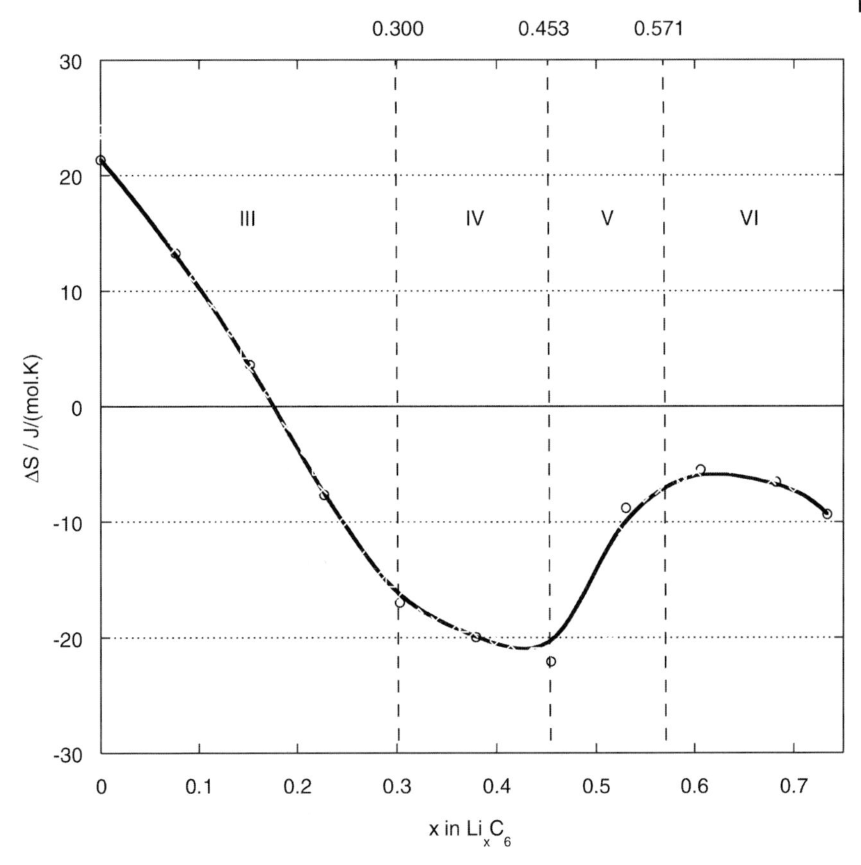

Fig. 5.9 Entropy of lithiation during discharge for the coke material HTT 2600 °C.

According to the entropy curve, this area covers a composition range $Li_{0.25}C_6$ (or LiC_{24}) to $Li_{0.5}C_6$ (or LiC_{12}). Liquid Stage II has so far been associated with $Li_{0.33}C_6$ (LiC_{18}), and not with $Li_{0.25}C_6$. Both OCV and entropy curves do not show any anomaly at $x = 0.33$, which questions the LiC_{18} composition for liquid Stage II [86]. Sloping OCV and entropy curves suggests that the Stage II conversion follows a solid-solution behavior rather than a two-phase one.

A sharp increase in entropy occurs at $x \sim 0.5$ (Figure 5.10a and magnified Figure 5.10b), which correlates very well with the change in the voltage profile as the Stage II to I phase transition takes place. A very interesting difference, however, is a sloping entropy curve in the composition range $0.55 < x < 0.88$, whereas the OCV curve is much flatter. Constant OCV is usually associated with a two-phase system behavior (here $Li_{0.5}C_6$ and Li_1C_6, Stage II and I, respectively). However, the beginning and the end of the stage transition ($0.5 < x < 0.55$ and $0.88 < x < 1$, respectively) show rather a single-phase behavior with slopping OCV. Accordingly, the two-phase model applies only in the composition range $0.55 < x < 0.88$, and not in the full $0.5 < x < 1$ range as commonly accepted in the literature [86] and

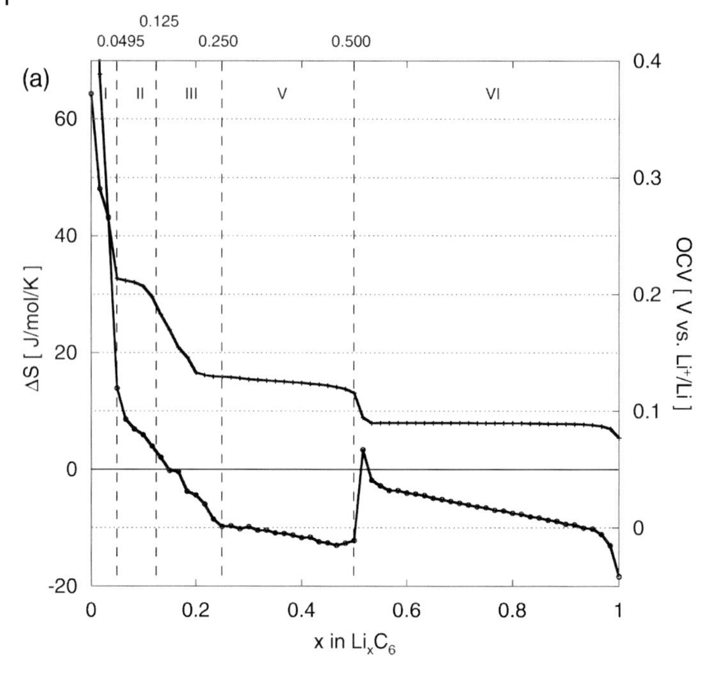

Fig. 5.10 (a) Entropy and OCV of lithiation during discharge of natural graphite.

references therein]. In a recent coupled experimental and theoretical study, we found that intermediary phases form in the composition range $0.5 < x < 1$. These new phases are thermodynamically more favored compared to a mixture of pure Stage II and Stage I compounds [87].

5.3.1.5 **Entropy and Degree of Graphitization**

Since the entropy function describes the thermodynamics behavior of a carbonaceous material much better than does the free energy function (or the OCV) and distinguishes materials very well according to their rate of crystallization (or graphitization degree G in the case of soft carbons), it is tempting to inversely use the entropy function as a tool to determine G. In fact, G is usually calculated from XRD using an empiric equation:

$$G = \frac{3.461 - d_{002}}{3.461 - 3.352} \tag{5.16}$$

where d_{002} is the interlayer spacing (inangstroms) of the carbonaceous material. The values 3.461 and 3.352 Å are those of d_{002} in fully disordered ($G = 0$) and highly graphitized carbons ($G = 1$), respectively.

Our thermodynamics approach for determining G is based on the assumption that a carbonaceous material having an average degree of graphitization G consists of a mixture of interconnected domains of highly graphitized carbon in the relative amount of G and disordered carbons in the amount of $1 - G$. When used as an

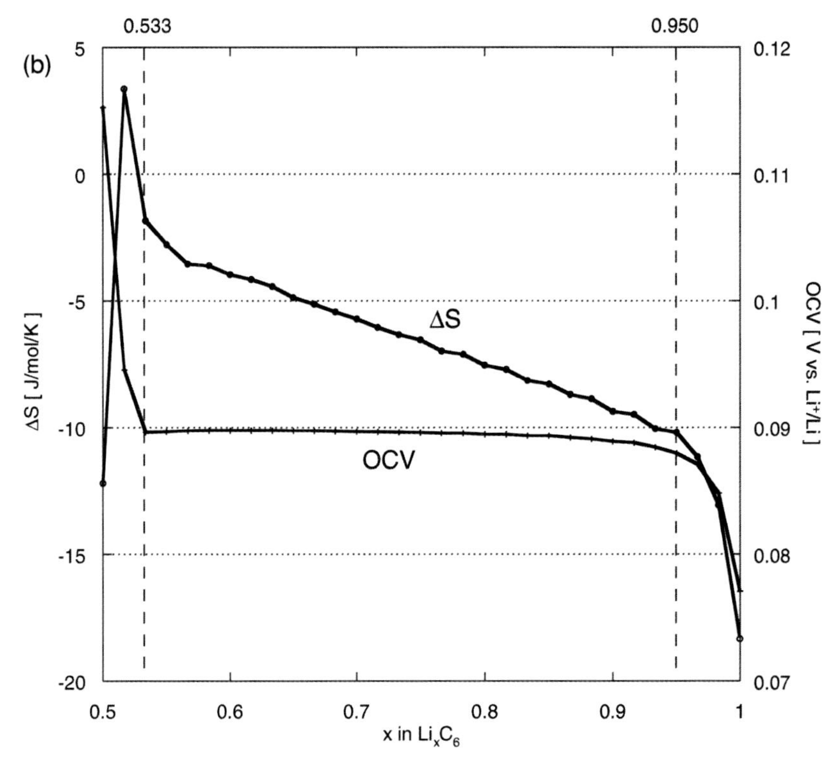

Fig. 5.10 (b) Partial entropy and OCV curves of lithiation of natural graphite, $0.5 < x < 1$.

electrode material for lithium intercalation, at any composition 'x' in Li_xC_6, lithium will be distributed between ordered and disordered domains. This distribution is not homogeneous, however, owing to differences in intercalation site energy in both types of carbon domains. As electron conductivity is high in carbonaceous materials, the Li_xC_6 electrode should be equipotential at any composition x at equilibrium. Therefore, at $E_0(x)$ the total entropy variation $\Delta S_T(E_0)$ is the sum of partial entropies coming from the graphitic domains $\Delta S_G(E_0)$ and from disordered domains $\Delta S_D(E_0)$, weighted by their molar fractions G and $1 - G$ respectively:

$$\Delta S_T(E_0) = G\Delta S_G(E_0) + (1 - G)\Delta S_D(E_0) \tag{5.17}$$

To check the validity of Equation (5.17), we applied it to the coke HTT 1700 °C. Figure 5.11(a) and (b) shows the $\Delta S_G(E_0)$ and $\Delta S_D(E_0)$, of the coke HTT below 500 °C and graphite HTT 2600 °C, respectively. Figure 5.11(c) shows the $\Delta S_T(E_0)$ of the coke HTT 1700 °C together with best fit, using Equation (5.17). The latter was achieved with $G = 0.32$, which is in very good agreement with that obtained by XRD expressed by Equation (5.15), taking $G = 0.94$ and $G = 0.15$ in the cokes HTT 2600 °C and HTT below 500 °C, respectively. Figure 5.11(d) shows the entropy vs. composition of the coke sample HTT 1700 °C together with a fit curve using $G = 0.32$. The experimental and calculated curves show a good agreement.

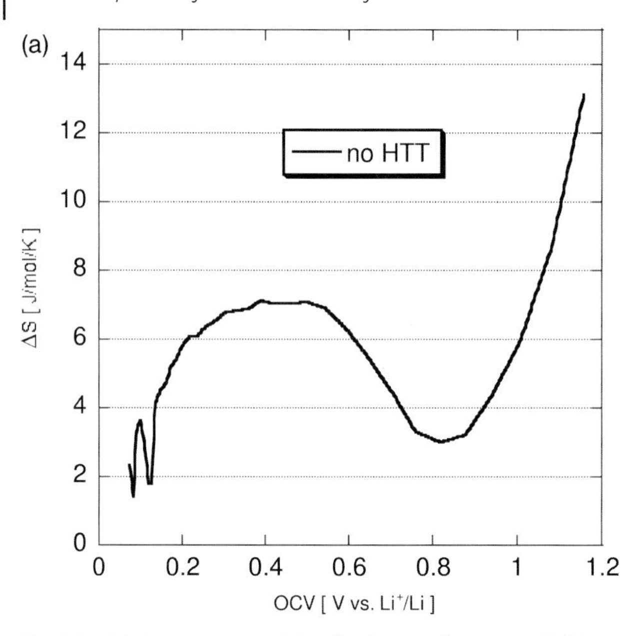

Fig. 5.11 (a) Entropy versus OCV of coke sample HTT $< 500\,^{\circ}$C.

In conclusion, the entropy function can be used to determine the degree of graphitization of a carbonaceous material as do XRD and Raman spectrometry. More generally, Equation (5.17) applies to any anode and cathode materials consisting of a mixture of "i" components, provided the $\Delta S_i(E_0)$ function for each component "i" is known.

A generalized equation can be expressed as

$$\Delta S_{\mathrm{T}}(E_0) = \sum_i x_i \Delta S_i(E_0), \tag{5.18}$$

where $\Delta S_{\mathrm{T}}(E_0)$ is the OCV dependence of the entropy function experimentally achieved in the multicomponent electrode, x_i is the molar fraction of component "i" in the mixture, and $\Delta S_i(E_0)$ is the OCV dependence of the entropy of component "i", provided an easy electron flow between components to insure an equipotential electrode.

5.3.2 Cathode Materials

5.3.2.1 $LiCoO_2$

$LiCoO_2$ has been the standard cathode material in lithium-ion batteries since their commercialization in the early 1990s. Its crystal structure can be described as regular stacking of CoO_2 slabs in which the Co^{3+} cations are sandwiched between two close-packed oxygen layers in CoO_6 octahedra. Lithium ions are also sandwiched between two CoO_2 slabs and occupy octahedral sites. The CoO_2 slabs are stacked

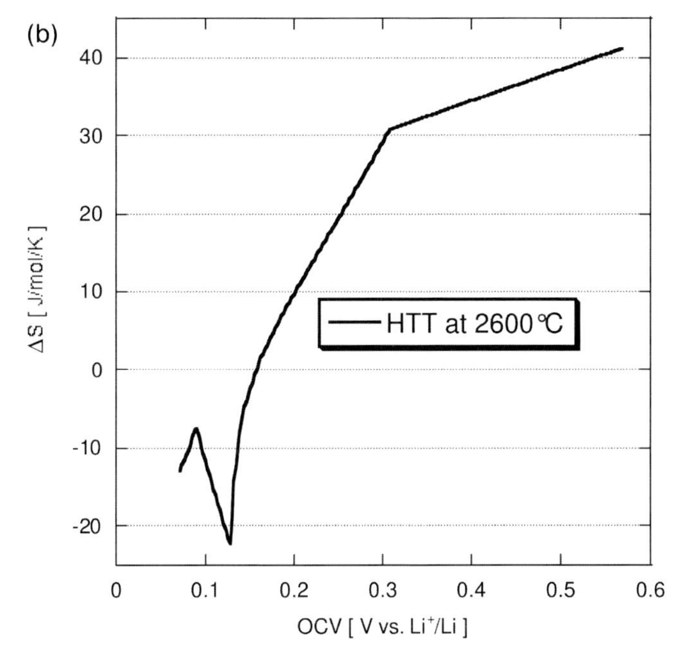

Fig. 5.11 (b) Entropy versus OCV of coke sample HTT 2600 °C.

along the c-axis in the ABC sequence to form the O3 structure. This arrangement yields a rhombohedral symmetry of the $LiCoO_2$ crystal ($R\bar{3}m$ as the space group). Where the Co^{3+} cations are covalently bonded to the oxygen atoms, the Li^+ form weaker ionic bonds with oxygen. Such a difference in binding energy allows for fast 2D diffusion of the lithium ions, whereas the Co^{3+} cations remain basically immobile. As lithium intercalate into and deintercalate from the CoO_2 structure, the latter change in their average partial charge to compensate for lithium cation gains and losses. The charge compensation mechanism can be schematized as follows:

$$Li^+(CoO_2)^- \xrightarrow{\text{discharge}} (Li_{1-y})^{1-y}(CoO_2)^{-(1-y)} + yLi^+ + ye^- \tag{5.19}$$

Equation (5.19) describes the charge mechanism of the positive electrode (commonly called the cathode although, in fact, Equation (5.19) describes an anodic process) in a lithium cell, during which lithium is deintercalated from $LiCoO_2$. Basically up to $y = 1$ lithium can be deintercalated from $LiCoO_2$ to form CoO_2 [88]. However, y exceeding ~ 0.5 has proved to adversely affect the 2D structure of the CoO_2 slabs, hence deteriorating the battery cycle life [89].

The phase diagram of the Li_xCoO_2 ($x = 1 - y$ in Equation (5.19)) system has been studied by in situ XRD [90]. These experimental results were supported by theoretical calculations as well [91]. In fact, as lithium is extracted from $LiCoO_2$, hexagonal phases form as single- or two-phase systems [92]. At $x \sim 0.5$, a monoclinic phase of distorted LiO_6 octahedra forms and may be the precursor of less stable hexagonal phases as the lithium composition decreases. Among

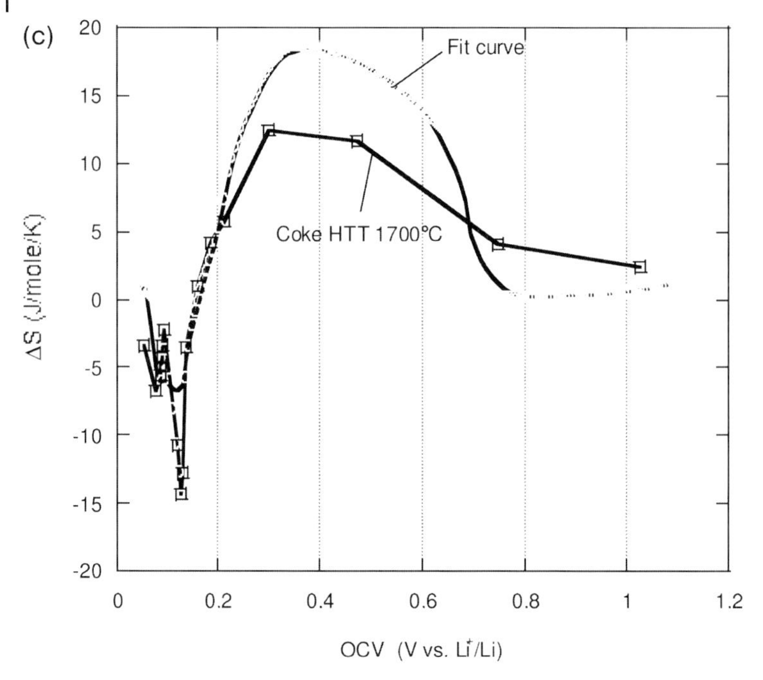

Fig. 5.11 (c) Entropy versus OCV of coke sample HTT 1700 °C together with a fitting curve.

proposed trigonal phases for $x < 0.5$ are the H1-3 and the O1 phases of Stage II and dilute Stage- I-like structures, respectively [90].

Our thermodynamics study on the Li_xCoO_2 system was limited to the composition range $0.5 < x < 1$. Electrochemical measurements were performed in an $Li/LiClO_4$ 1 M PC/Li_xCoO_2 coin shaped half cell. Figure 5.12 is the OCV(x) trace during discharge. The OCV curve decreases monotonously at the beginning ($0.5 < x < 0.75$), makes a semiplateau at $0.75 < x < 0.95$, and then declines again at the end of discharge ($0.95 < x < 1$). This profile suggests a solid-solution behavior where the OCV declines and a two-phase behavior where it makes a semiplateau. The nonobvious information in the OCV curve is signature of the hexagonal to the monoclinic phase transition at $x \sim 0.5$. This lack of clear evidence from the OCV curve indicates that the phase transition involves a very small change in the free energy. This feature strongly contrasts with that achieved owing to the entropy study as evidenced in Figure 5.13. In fact, the $\Delta S(x)$ curve shows dramatic features, in which six areas can be distinguished as follows:

- Area-I ($0.49 < x < 0.511$): $\Delta S(x)$ sharply decreases and makes a minimum;
- Area-II ($0.511 < x < 0.551$): $\Delta S(x)$ sharply increases and makes a maximum;
- Area-III ($0.551 < x < 0.581$): $\Delta S(x)$ decreases sharply again;
- Area-IV ($0.581 < x < 0.823$): $\Delta S(x)$ decreases uniformly;
- Area-V ($0.823 < x < 0.953$): where $\Delta S(x)$ makes a sloping plateau;
- Area-VI ($0.95 < x < 1$): where $\Delta S(x)$ sharply increases and makes a short semiplateau.

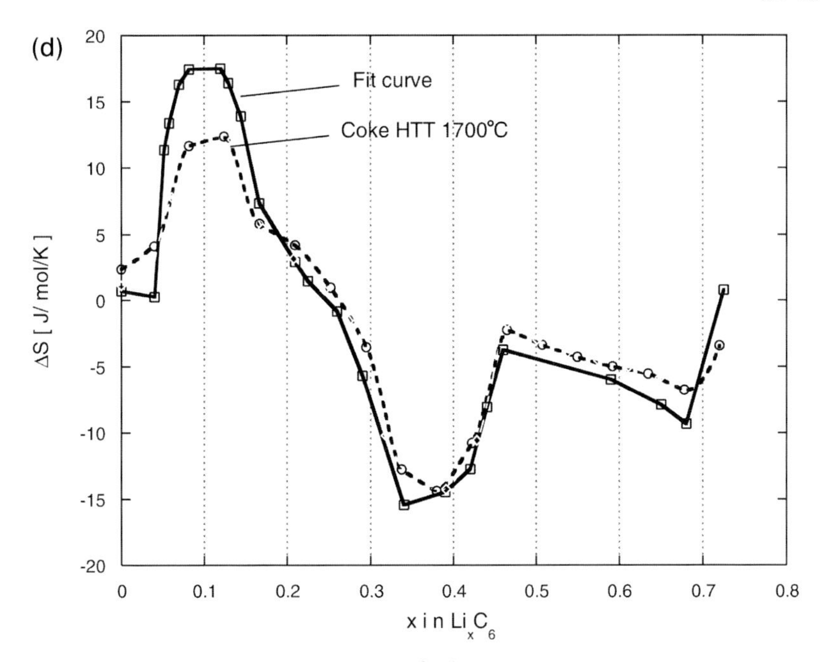

Fig. 5.11 (d) Entropy versus composition of coke HTT 1700 °C and a fitting curve corresponding to G = 0.32 in Equation 17.

Sharp decreases and increases in the $\Delta S(x)$ function occur at phase transitions boundaries as discussed in the previous section on carbonaceous anode materials. Accordingly, a decrease in Area-I and III should be associated with the H1-3 to the monoclinic and to the monoclinic to a hexagonal phase transitions, respectively. The latter hexagonal phase covers Area-IV with a typical solid-solution behavior. A second hexagonal phase forms at $x = 0.823$. Therefore, the almost flat entropy in the composition range $0.823 < x < 0.953$ should be associated with a two-phase system (first and second hexagonal phases). An increase in entropy above $x = 0.953$ should signal a transition between the second hexagonal phase and a new Li-rich phase such as spinel $Li_2Co_2O_4$.

In conclusion, the entropy study allows phase transition boundary in the Li_xCoO_2 system to be defined with a much higher accuracy than by in situ techniques such as XRD and X-ray absorption spectrometry. The reason may lie in the fact that most in situ methods are applied under a dynamic lithiation regime although at slow rates, which may not allow for the system to reach an equilibrium state. Our entropy measurements are performed at quasi-equilibrium, and therefore, they describe the phase transitions with better accuracy. Despite being taken at equilibrium, OCV measurements are less sensitive to phase transitions than the entropy ones. This is particularly true for the monoclinic phase transition, which, in addition to a large change in entropy versus OCV, has better defined transition boundaries. It is our argument that crystal imperfection or chemical impurity present in the starting $LiCoO_2$ material will affect both the OCV and the entropy curves, more

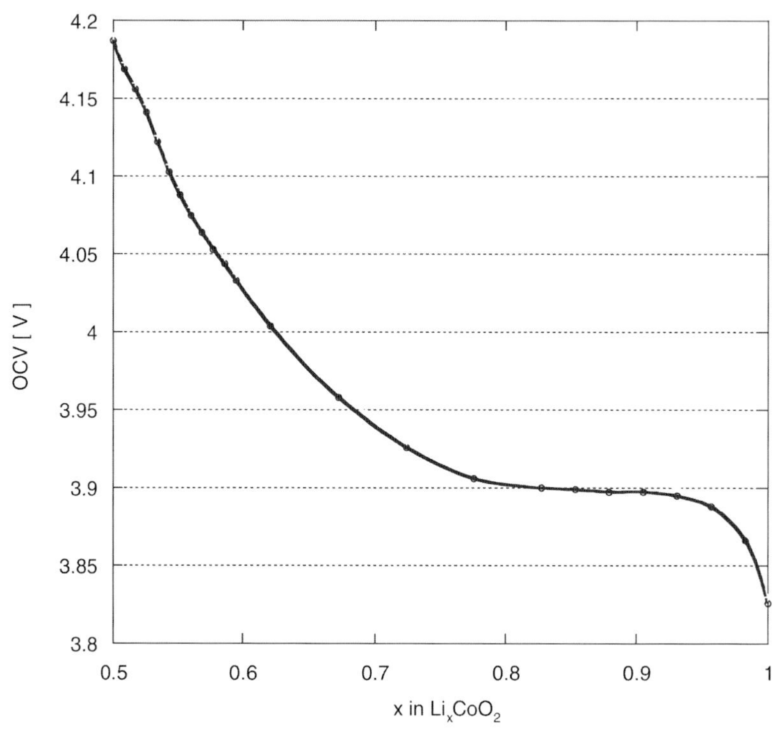

Fig. 5.12 Open circuit voltage (OCV) as a function of lithium concentration in Li_yCoO_2.

significantly for the latter, thus allowing to distinguish between materials for their long-term electrochemical performances such as cycle life and thermal aging.

5.3.2.2 $LiMn_2O_4$

$LiMn_2O_4$ with cubic symmetry (spinel structure) is another very important cathode material that has been widely studied as an alternative to more expensive and less environmentally benign $LiCoO_2$ material. Among the important attractive features of $LiMn_2O_4$ are higher discharge voltage and faster charge and discharge kinetics. However, owing to inherent chemical and crystal structure instability involving Mn dissolution, especially at high temperatures, and the Jahn–Teller distortion, $LiMn_2O_4$ has not enjoyed the same commercial success as $LiCoO_2$ despite a lower cost [93–95]. Therefore, tremendous research and R&D efforts have been directed toward enhanced cycle life and thermal stability, in particular, through cation substitution (including Li^+) [96, 97], anion substitution [98], and surface coating [99]. Very promissing results have been achieved, which may enable the use of $LiMn_2O_4$ in commercial batteries in the near future.

The redox process in the $Li_xMn_2O_4$ system is quite well understood as it involves mostly the Mn^{IV}/Mn^{III} solid state redox couple, which results in a smaller change in the oxygen oxidation state unlike in the Li_xCoO_2 system [100, 101]. Accordingly, the

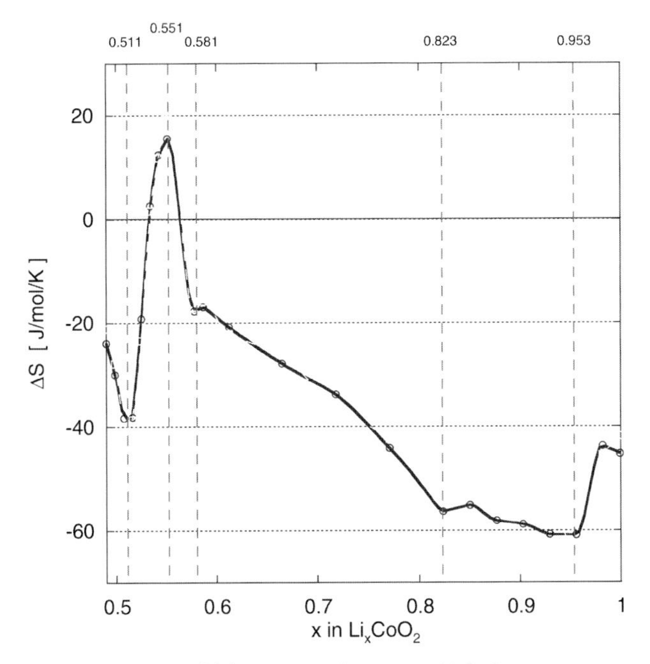

Fig. 5.13 Entropy of lithium intercalation into Li_xCoO_2.

lithium deintercalation mechanism during charge can be schematized as follows:

$$Li^+(Mn^{III}Mn^{IV}O_4)^- \xrightarrow{\text{discharge}} [(Li_{1-y})^{1-y}]\left[\left(\left(Mn^{III}_{\frac{1-y}{2}}Mn^{IV}_{\frac{1+y}{2}}\right)_2 O_4\right)^{-(1-y)}\right] + yLi^+ + ye^-, \quad 0 < y = 1 - x < 1 \qquad (5.20)$$

At low constant rate, the discharge profile of a $Li/Li_xMn_2O_4$ cell in the composition range $0 < x < 1$ typically shows two voltage plateaus at around 4 V of approximately the same length of $\Delta x = 0.5$. These were associated with an order/disorder transition in the spinel phase that occurs at approximately $x = 0.5$ [96].

In our thermodynamics study, we used two manganese spinel materials – a close to stoichoimetric compound ($Li_{1.0}Mn_2O_4$) and a lithium-rich compound ($Li_{1.08}Mn_{1.92}O_4$). Figure 5.14 shows their respective OCV profiles with typical two 4-V plateaus; the one at 4.13 V roughly covered $0.20 < x < 0.55$ and the other averaged at 3.98 V covering $0.55 < x < 1$. A close analysis of the OCV profile shows slight differences between the two spinel materials. The stoichiometric material has a flatter high-voltage plateau, whereas the nonstoichiometric one shows a steeper OCV decrease at the end of lithiation ($x \sim 1$).

The corresponding entropy curves are depicted in Figure 5.15. In the non-stoichiometric material, four composition areas can be distinguished: Area-I ($0.2 < x < 0.518$) where the entropy curve declines monotonously suggesting a

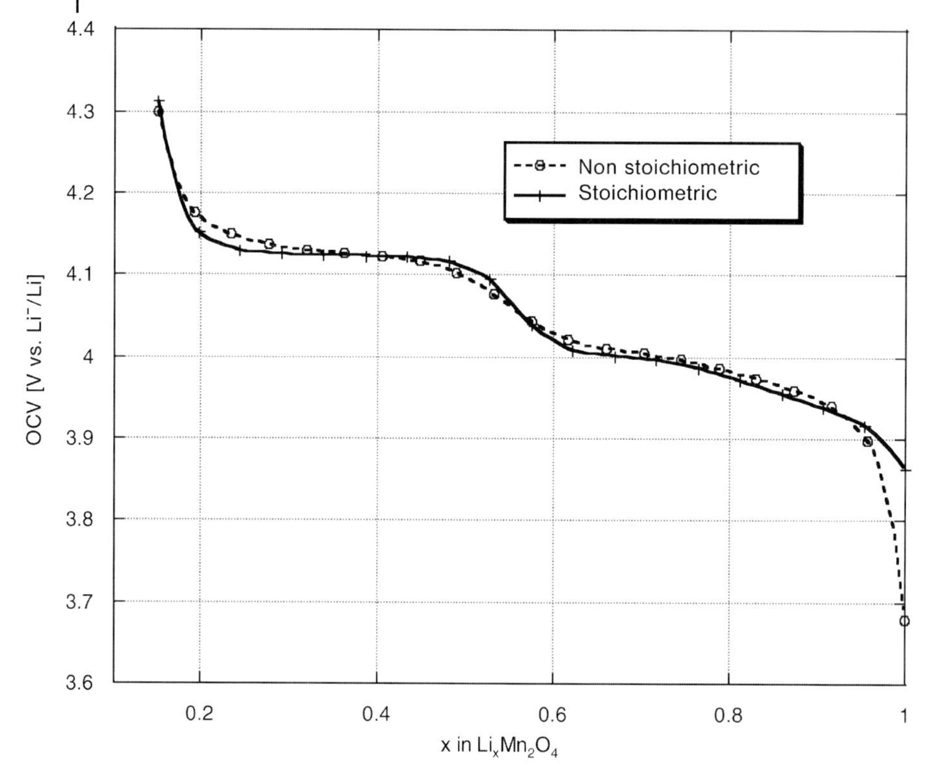

Fig. 5.14 Open-circuit voltage (OCV) as a function of lithium concentration in $Li_xMn_2O_4$ shown for stoichiometric and nonstoichiometric materials.

solid-solution behavior, Area-II ($0.518 < x < 0.678$) where the entropy increases in a transition zone between two solid solutions, Area-III ($0.678 < x < 0.963$) where $\Delta S(x)$ decreases in a second solid-solution system, and Area-IV ($0.963 < x < 1$) where the $\Delta S(x)$ makes a steep increase in a new transition zone probably as the onset to the tetragonal phase formation.

The stoichiometric material, however, shows a different thermodynamics path: Area-I ($0.2 < x < 0.339$) where $\Delta S(x)$ varies a little a signature of two-phase system behavior, Area-II ($0.339 < x < 0.518$) where the entropy shows a negative slope indicative of a solid-solution behavior, Area-III ($0.518 < x < 0.641$) where $\Delta S(x)$ increases in a transition zone, and Area-IV ($0.641 < x < 1$) were the entropy decreases as we enter a second solid solution. Unlike for the nonstoichiometric material, there is no onset of the cubic to tetragonal phase transition at $x < 1$ in the stoichiometric one. Differences in the thermodynamics behavior of the two spinel materials show the lithium excess in the nonstoichimetric material and account for an earlier cubic to tetragonal phase transition compared to the stoichiometric one. In fact, in the latter, we found that the OCV makes a steep drop to 2.9 V at $x = 1$, where the tetragonal phase starts to form. The delay in the tetragonal phase

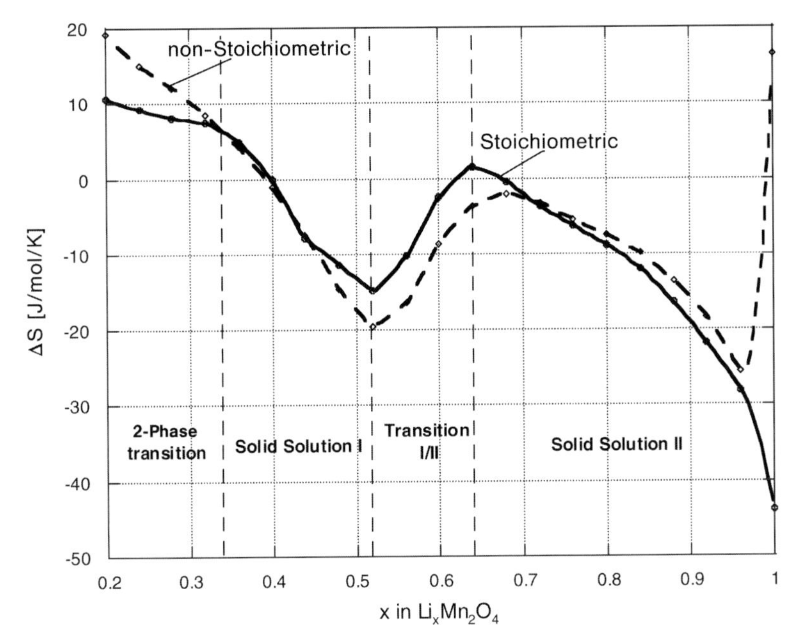

Fig. 5.15 Entropy as a function of lithium concentration in $Li_xM_2O_4$ shown for two materials: one stoichiometric and the other nonstoichiometric.

formation may play an important role in the cycle life as the tetragonal phase is known to be less cyclable.

5.3.2.3 **Effect of Cycling on Thermodynamics:**

Lithium-ion 4-Ah cells with graphitic carbon anode and $LiNi_{1/3}Co_{1/3}Mn_{1/3}O_2$ cathode were prepared courtesy of ENAX, Co. (Yonezawa Labs., Japan). Cells were cycled 100% DOC–DOD for three formation cycles and others for 500 cycles, respectively denoted fresh and cycled cells. The capacity loss after 500 cycles was about 10.5%. Fully discharged cells were then opened in a dry atmosphere. Electrodes were retrieved, washed, and dried in ambient temperatures. New lithium half cells were made from the anode and cathode for thermodynamics study as described in the previous sections. Here the BA-1000 system (Figure 5.1b) was used.

The entropy and enthalpy versus SOC curves of the half cells using fresh and cycled $LiNi_{1/3}Co_{1/3}Mn_{1/3}O_2$ cathodes are displayed in Figures 5.16 and 5.17, respectively. Where the entropy profile of the fresh cell features dramatic changes in slope, that of the cycled cell shows much smoother slopes except at the early stage of charging where both curves have steep negative slopes. As discussed previously, changes in the entropy slope occur at phase transitions boundaries. Phase transitions in $LiNi_{1/3}Co_{1/3}Mn_{1/3}O_2$, the origin of which is beyond the scope of this chapter, still need to be clarified, as they are dramatically affected by the cycling. Our recent transmission electron microscopy study on the $LiNi_{1/3}Co_{1/3}Mn_{1/3}O_2$

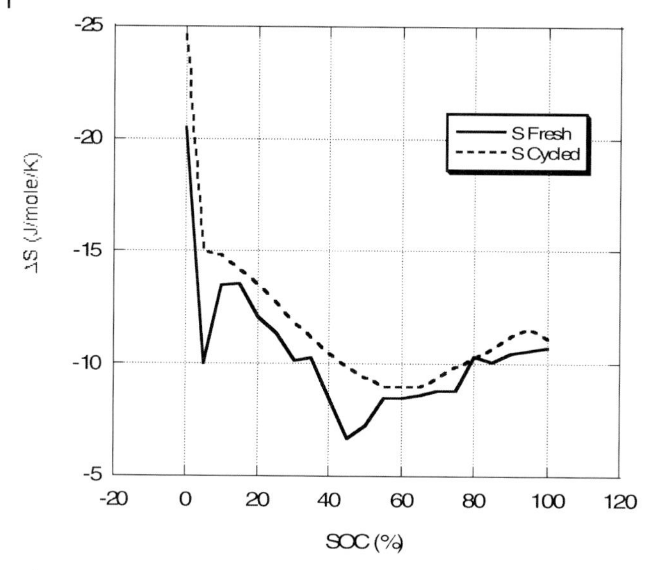

Fig. 5.16 Entropy profile of fresh and cycled $LiNi_{1/3}Co_{1/3}Mn_{1/3}O_2$ cathodes.

cathode before and after thermal aging showed dramatic changes in the crystal structure, which we attribute to transition metal (TM) cation rearrangement and to Li–TM cation mixing 102. In fact, the $\sqrt{3} \times \sqrt{3}$TM superlattice that occurs in fresh $LiNi_{1/3}Co_{1/3}Mn_{1/3}O_2$ with rhombohedral symmetry is thermodynamically instable and converts to more disordered (mixed TM) structure and/or to cubic spinel upon thermal aging. It is our argument that prolonged cycling may lead to similar crystal structure phenomena, which then bear a thermodynamics signature.

It is interesting, however, to see that the enthalpy curve of Figure 5.17 does not show any significant difference between fresh and cycled cathodes. This suggests that the thermochemistry behind lithium intercalation and deintercalation is not very sensitive to the cation ordering within the TM layer or to 10.5% decrease in cycle capacity.

This example clearly illustrates the higher sensitivity of the entropy function to the local TM ordering in $LiNi_{1/3}Co_{1/3}Mn_{1/3}O_2$, unlike the enthalpy and free energy functions.

5.4 Conclusion

A new method for characterization of electrode materials based on thermodynamics measurements has been developed and applied to anode and cathode materials for lithium-ion batteries. An automatic system (ETMS) has been set up (Figure 5.1(b)) and it allows for a full control of the cell state of charge and

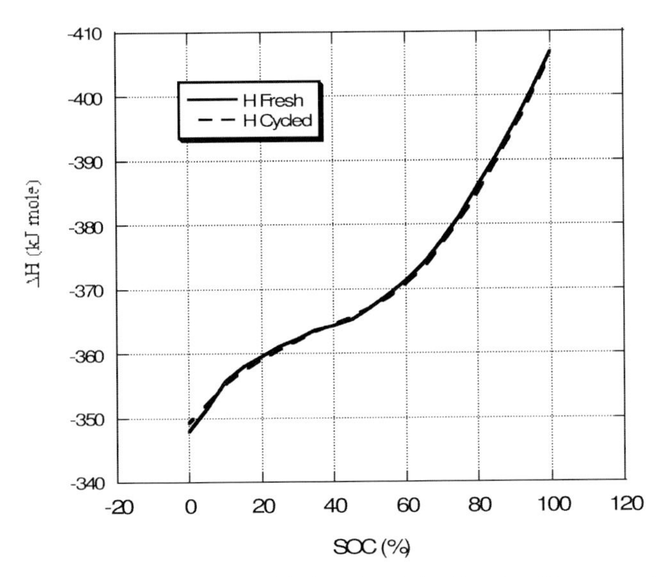

Fig. 5.17 Enthalpy profile of fresh and cycled $LiNi_{1/3}Co_{1/3}Mn_{1/3}O_2$ cathodes.

discharge, temperature, and full data collection and conversion to thermodynamics state functions.

The main features of this new thermodynamics characterization methodology, in particular, the entropy measurement (or entropymetry) can be summarized here:

- high sensitivity to phase transitions taking place in electrode material;
- high sensitivity to departure from stoichiometry and presence of defects and impurities in the electrode material;
- high resolution in composition boundaries of phase transitions; method can be used as a titration technique for each phase present;
- high energy and voltage resolution of phase transitions;
- high sensitivity to local disorder and degree of crystallization;
- nondestructive nature of the method;
- allows the heat of electrode reaction to be accurately determined, therefore can be used as calorimetry;
- applies to half cells and to full cells;
- applies to multicomponent electrode material and distinguishes among them;
- versatile method that should be applicable to any battery chemistry, including rechargeable alkaline and lead-acid batteries.

Our thermodynamics results were obtained in lithium half-cell configuration. Therefore, only processes involved in one working electrode were characterized. In a full-cell configuration, the OCV results from difference in OCV^+ of the positive and OCV^- of the negative electrode:

$$OCV = OCV^+ - OCV^-$$

The temperature dependence follows and so does the cell entropy:

$$\Delta S(\text{cell}) = \Delta S(\text{positive}) - \Delta S(\text{negative})$$

Therefore, any change in the entropy function of the positive and the negative electrodes should translate to that of the full cell. The same principle applies to the heat of cell reaction. Owing to the high sensitivity of the entropy and enthalpy functions to preexisting or induced structural disorders, it is expected that the functions will bear the signature of any disorder source such as cycle number, overcharge and overdischarge, and exposure to high and possibly low temperatures.

Acknowledgments

The work presented in this chapter was performed at the CNRS-CALTECH International Associated Laboratory codirected by Prof. Brent Fultz and by the author. It was part of the PhD research project of Dr Yvan Reynier. The author acknowledges scientific and experimental contributions of Prof. Fultz and Dr Reynier to this work.

The author thanks ENAX, Co., Japan for providing the cathode materials used in this study.

References

1 Kumaresan, K., Sikha, G., and White, R.E. (2008) *J. Electrochem. Soc.*, **155**, A164–A171.

2 Lu, W., Belharouak, I., Liu, J., and Amine, K. (2007) *J. Power Sources*, **174**, 673–77.

3 Okamoto, E., Nakamura, M., Akasaka, Y., Inoue, Y., Abe, Y., Chinzei, T., Saito, I., Isoyama, T., Mochizuki, S., Imachi, K., and Mitamura, Y. (2007) "Analysis of heat generation of lithium ion rechargeable batteries used in implantable battery systems for driving undulation pump ventricular assist device." *Artif. Organs*, **31**(7), 538–41.

4 Lu, W., Belharouak, I., Park, S.H., Sun, Y.K., and Amine, K. (2007) "Isothermal calorimetry investigation of Li1+xMn2-yAlzO4 spinel." *Electrochim. Acta*, **52**(19), 5837–42.

5 Lu, W., Belharouak, I., Vissers, D., and Amine, K. (2006) "In situ thermal study of Li1+x[Ni1/3Co1/3Mn1/3](1-x)O-2 using isothermal micro-calorimetric techniques." *J. Electrochem. Soc.*, **153**(11), A2147–A2151.

6 Ohshima, T., Nakayama, M., Fukuda, K., Araki, T., and Onda, K. (2006) "Thermal behavior of small lithium-ion secondary battery during rapid charge and discharge cycles." *Electr. Eng. Japan*, **157**(3), 17–25.

7 Wang, M.J. and Navrotsky, A. (2005) "LiMO2 (M = Mn, Fe, and Co): Energetics, polymorphism and phase transformation." *J. Solid State Chem.*, **178**(4), 1230–40.

8 Rao, L. and Newman, J. (1997) "Heat-generation rate and general energy balance for insertion battery systems." *J. Electrochem. Soc.*, **144**(8), 2697–704.

9 Huang, Q., Yan, M.M., and Jiang, Z.Y. (2006) "Thermal study on single electrodes in lithium-ion battery." *J. Power Sources*, **156**(2), 541–46.

10 Takano, K., Saito, Y., Kanari, K., Nozaki, K., Kato, K., Negishi, A., and Kato, T. (2002) "Entropy change in lithium ion cells on charge and discharge." *J. Appl. Electrochem.*, **32**(3), 251–58.

11 Funahashi, A., Kida, Y., Yanagida, K., Nohma, T., and Yonezu, I. (2002) "Thermal simulation of large-scale lithium secondary batteries using a graphite-coke hybrid carbon negative electrode and lini0.7Co0.3O2 positive electrode." *J. Power Sources*, **104**(2), 248–52.

12 Selman, J.R., Al Hallaj, S., Uchida, I., and Hirano, Y. (2001) "Cooperative research on safety fundamentals of lithium batteries." *J. Power Sources*, **97-8**, 726–32.

13 Al Hallaj, S., Prakash, J., and Selman, J.R. (2000) "Characterization of commercial Li-ion batteries using electrochemical-calorimetric measurements." *J. Power Sources*, **87**(1–2), 186–94.

14 Sandhu, S.S. and Fellner, J.P. (1999) "Thermodynamic equations for a model lithium-ion cell." *Electrochim. Acta*, **45**(6), 969–76.

15 Hong, J.S., Maleki, H., Al Hallaj, S., Redey, L., and Selman, J.R. (1998) "Electrochemical-calorimetric studies of lithium-ion cells." *J. Electrochem. Soc.*, **145**(5), 1489–501.

16 Saito, Y., Kanari, K., and Takano, K. (1997) "Thermal studies of a lithium-ion battery." *J. Power Sources*, **68**(2), 451–54.

17 Doi, T., Fukudome, H., Okada, S., and Yamaki, J.I. (2007) "Computer simulation of a porous positive electrode for lithium batteries." *J. Power Sources*, **174**(2), 779–83.

18 Thomas, K.E. and Newman, J. (2003) "Thermal modeling of porous insertion electrodes." *J. Electrochem. Soc.*, **150**(2), A176–A192.

19 Kalikmanov, V.I. and de Leeuw, S.W. (2002) "Role of elasticity forces in thermodynamics of intercalation compounds: Self-consistent mean-field theory and Monte Carlo simulations." *J. Chem. Phys.*, **116**(7), 3083–89.

20 Kim, S.W. and Pyun, S.I. (2001) "Thermodynamic and kinetic approaches to lithium intercalation into a Li1-delta Mn2O4 electrode using Monte Carlo simulation." *Electrochim. Acta*, **46**(7), 987–97.

21 Botte, G.G., Subramanian, V.R., and White, R.E. (2000) "Mathematical modeling of secondary lithium batteries." *Electrochim. Acta*, **45**(15–16), 2595–609.

22 Kuko, T. and Hibino, M. (1998) "Theoretical dependence of the free energy and chemical potential upon composition in intercalation systems with repulsive interaction between guest ions." *Electrochem. Acta*, **43**(7), 781–89.

23 Shi, S., Ouyang, C., Lei, M., and Tang, W. (2007) "Effect of Mg-doping on the structural and electronic properties of LiCoO2: A first-principles investigation." *J. Power Sources*, **171**(2), 908–12.

24 Kuhn, A., Diaz-Carrasco, P., de Dompablo, M.E.A.Y., and Garcia-Alvarado, F. (2007) "On the synthesis of ramsdellite LiTiMO4 (M = Ti, V, Cr, Mn, Fe): An experimental and computational study of the spinel-ramsdellite transformation." *Eur. J. Inorg. Chem.*, **21**, 3375–84.

25 Wang, L., Maxisch, T., and Ceder, G. (2007) "A first-principles approach to studying the thermal stability of oxide cathode materials." *Chem. Mater.*, **19**(3), 543–52.

26 Zhou, F., Maxisch, T., and Ceder, G. (2006) "Configurational electronic entropy and the phase diagram of mixed-valence oxides: The case of lixfepo4." *Phys. Rev. Lett.*, **97**(15), 155704.

27 Wagemaker, M., Van Der Ven, A., Morgan, D., Ceder, G., Mulder, F.M., and Kearley, G.J. (2005) "Thermodynamics of spinel LixTiO2 from first principles." *Chem. Phys.*, **317**(2–3), 130–36.

28 Koudriachova, M.V., Harrison, N.M., and de Leeuw, S.W. (2004) "First principles predictions for

intercalation behavior." *Solid State Ionics*, **175**(1–4), 829–34.

29 Chen, Z.W., Xuan, C., Ying, Z., and Yong, Y. (2003) "First principle investigation of positive electrode material for lithium ion batteries." *Rare Metal Mater. Eng.*, **32**(9), 693–98.

30 Carlier, D., Van der Ven, A., Delmas, C., and Ceder, G. (2003) "First-principles investigation of phase stability in the O-2-LiCoO2 system." *Chem. Mater.*, **15**(13), 2651–60.

31 Shi, S.Q., Wang, D.S., Meng, S., Chen, L.Q., and Huang, X.J. (2003) "First-principles studies of cation-doped spine LiMn2O4 for lithium ion batteries." *Phys. Rev. B* **67**(11).

32 Ceder, G. and Van der Ven, A. (1999) "Phase diagrams of lithium transition metal oxides: investigations from first principles." *Electrochim. Acta*, **45**(1–2), 131–50.

33 Benco, L., Barras, J.L., Atanasov, M., Daul, C., and Deiss, E. (1999) "First principles calculation of electrode material for lithium intercalation batteries: TiS2 and LiTi2S4 cubic spinel structures." *J. Solid State Chem.*, **145**(2), 503–10.

34 Ceder, G., Kohan, A.F., Aydinol, M.K., Tepesch, P.D., and Van der Ven, A. (1998) "Thermodynamics of oxides with substitutional disorder: A microscopic model and evaluation of important energy contributions." *J. Am. Ceram. Soc.*, **81**(3), 517–25.

35 Deiss, E., Wokaun, A., Barras, J.L., Daul, C., and Dufek, P. (1997) "Average voltage, energy density, and specific energy of lithium-ion batteries - Calculation based on first principles." *J. Electrochem. Soc.*, **144**(11), 3877–81.

36 Aydinol, M.K., Kohan, A.F., Ceder, G., Cho, K., and Joannopoulos, J. (1997) "Ab initio study of lithium intercalation in metal oxides and metal dichalcogenides." *Phys. Rev. B*, **56**(3), 1354–65.

37 Hallstedt, B. and Kim, O. (2007) "Thermodynamic assessment of the Al-Li system." *Int. J. Mater. Res.*, **98**(10), 961–69.

38 Paddon, C.A., Jones, S.E.W., Bhatti, F.L., Donohoe, T.J., and Compton, R.G. (2007) "Kinetics and thermodynamics of the Li/Li+ couple in tetrahydrofuran at low temperatures (195–295 K)." *J. Phys. Org. Chem.*, **20**(9), 677–84.

39 Reynier, Y.F., Yazami, R., and Fultz, B. (2004) "Thermodynamics of lithium intercalation into graphites and disordered carbons." *J. Electrochem. Soc.*, **151**(3), A422–A426.

40 Reynier, Y., Yazami, R., and Fultz, B. (2003) "The entropy and enthalpy of lithium intercalation into graphite." *J. Power Sources*, **119**, 850–55.

41 Lee, H.H., Wan, C.C., and Wang, Y.Y. (2003) "Identity and thermodynamics of lithium intercalated in graphite." *J. Power Sources*, **114**(2), 285–91.

42 Ol'shanskaya, L.N. and Astaf'eva, E.N. (2002) "Thermodynamics of lithium intercalates in carbonized fabric." *Russian J. Appl. Chem.*, **75**(5), 740–44.

43 Al Hallaj, S., Venkatachalapathy, R., Prakash, J., and Selman, J.R. (2000) "Entropy changes due to structural transformation in the graphite anode and phase change of the LiCoO2 cathode." *J. Electrochem. Soc.*, **147**(7), 2432–36.

44 Gong, J.B. and Wu, H.Q. (2000) "Electrochemical intercalation of lithium species into disordered carbon prepared by the heat-treatment of poly (p-phenylene) at 650 degrees C for anode in lithium-ion battery." *Electrochim. Acta*, **45**(11), 1753–62.

45 Yamaki, J., Egashira, M., and Okada, S. (2000) "Potential and thermodynamics of graphite anodes in Li-ion cells." *J. Electrochem. Soc.*, **147**(2), 460–65.

46 Huggins, R.A. (1999) "Lithium alloy negative electrodes." *J. Power Sources*, **82**, 13–19.

47 Attidekou, P.S., Garcia-Alvarado, F., Connor, P.A., and Irvine, J.T.S. (2007) "Thermodynamic aspects of the reaction of lithium with SnP2O7

based positive electrodes." *J. Electrochem. Soc.*, **154**(3), A217–A220.

48 Takahashi, Y., Kijima, N., Dokko, K., Nishizawa, M., Uchida, I., and Akimoto, J. (2007) "Structure and electron density analysis of electrochemically and chemically delithiated LiCoO2 single crystals." *J. Solid State Chem.*, **180**(1), 313–21.

49 Lu, W.Q., Yang, H., and Prakash, J. (2006) "Determination of the reversible and irreversible heats of LiNi(0.8)Co(0.2)O(2)/mesocarbon microbead Li-ion cell reactions using isothermal microcalorimetery." *Electrochim. Acta*, **51**(7), 1322–29.

50 Wang, M.J. and Navrotsky, A. (2004) "Enthalpy of formation of LiNiO2, LiCoO2 and their solid solutions LiNi1-xCoxO2." *Solid State Ionics*, **166**(1–2), 167–73.

51 Vicente, C.P., Lloris, J.M., and Tirado, J.L. (2004) "Understanding the voltage profile of Li insertion into LiNi0.5-yFeyMn1.5O4 in Li cells." *Electrochim. Acta*, **49**(12), 1963–67.

52 Kobayashi, H., Arachi, Y., Kageyama, H., and Tatsumi, K. (2004) "Structural determination of Li1-yNi0.5Mn0.5O2 (y=0.5) using a combination of Rietveld analysis and the maximum entropy method." *J. Mater. Chem.*, **14**(1), 40–42.

53 Fujiwara, H., Ueda, Y., Awasthi, A., Krishnamurthy, N., and Garg, S.P. (2003) "Determination of standard free energy of formation for niobium silicides by EMF measurements." *J. Electrochem. Soc.*, **150**(8), J43–J48.

54 Thomas, K.E. and Newman, J. (2003) "Heats of mixing and of entropy in porous insertion electrodes." *J. Power Sources*, **119**, 844–49.

55 Yamaki, J., Egashira, M., and Okada, S. (2001) "Voltage prediction from Coulomb potential created by atoms of spinel LiMn2O4 cathode active material for Li ion cells." *J. Power Sources*, **97-8**, 349–53.

56 Barbato, S. and Gautier, J.L. (2001) "Hollandite cathodes for lithium ion batteries. 2. Thermodynamic and kinetics studies of lithium insertion into BaMmn7O16 (M = Mg, Mn, Fe, Ni)." *Electrochim. Acta*, **46**(18), 2767–76.

57 Thomas, K.E., Bogatu, C., and Newman, J. (2001) "Measurement of the entropy of reaction as a function of state of charge in doped and undoped lithium manganese oxide." *J. Electrochem. Soc.*, **148**(6), A570–A575.

58 Idemoto, Y., Ogawa, S., Uemura, Y., and Koura, N. (2000) "Thermodynamic stability and cathode performance of Li1+xmn2-xo4 as a cathode active material for lithium secondary battery." *J. Ceram. Soc. Jpn.*, **108**(9), 848–53.

59 Idemoto, Y., Ogawa, S., Koura, N., and Udagawa, K. (2000) "Thermodynamic stability and cathode performance of limn2-xmgxo4 as cathode active material for the lithium secondary battery." *Electrochemistry*, **68**(6), 469–73.

60 Kumagai, N., Koishikawa, Y., Komaba, S., and Koshiba, N. (1999) "Thermodynamics and kinetics of lithium intercalation into Nb2O5 electrodes for a 2 V rechargeable lithium battery." *J. Electrochem. Soc.*, **146**(9), 3203–210.

61 Korovin, N.V. (1998) "Electrochemical intercalation into cathodic materials: electrode potentials." *Russian J. Electrochem.*, **34**(7), 669–675.

62 Kumagai, N., Yu, A.S., Kumagai, N., and Yashiro, H. (1997) "Electrochemical intercalation of lithium into hexagonal tungsten trioxide." *Thermochim. Acta*, **299**(1–2), 19–25.

63 Guzman, G., Yebka, B., Livage, J., and Julien, C. (1996) "Lithium intercalation studies in hydrated molybdenum oxides." *Solid State Ionics*, **86-8**, Part 1, 407–13.

64 Kumagai, N., Fujiwara, T., Tanno, K., and Horiba, T. (1993) "Thermodynamic and kinetic-studies of electrochemical lithium insertion into quaternary Li-Mn-V-O spinel as positive materials for rechargeable lithium batteries." *J. Electrochem. Soc.*, **140**(11), 3194–99.

65 Baddour, R., Pereiraramos, J.P., Messina, R., and Perichon, J. (1991)

"A thermodynamic, structural and kinetic-study of the electrochemical lithium intercalation into the xerogel V2O5.1.6 H2O in a propylene carbonate solution." *J. Electroanal. Chem.*, **314**(1–2), 81–101.

66 Joo, J.H. and Bae, Y.C. (2006) "Molecular thermodynamics approach for phase behaviors of solid polymer electrolytes/salt system in lithium secondary battery on the nonrandom mixing effect: Applicability of the group-contribution method." *Polymer*, **47**(20), 7153–59.

67 Joo, J.H., Bae, Y.C., and Sun, Y.K. (2006) "Phase behaviors of solid polymer electrolytes/salt system in lithium secondary battery by group-contribution method: The pressure effect." *Polymer*, **47**(1), 211–17.

68 Limthongkul, P., Jang, Y.I., Dudney, N.J., and Chiang, Y.M. (2003) "Electrochemically-driven solid-state amorphization in lithium-metal anodes." *J. Power Sources*, **119**, 604–9.

69 Yamaki, J., Egashira, M., and Okada, S. (2001) "Thermodynamics and phase separation of lithium intercalation materials used in lithium ion cells." *Electrochemistry*, **69**(9), 664–69.

70 Hong, J.S. and Selman, J.R. (2000) "Relationship between calorimetric and structural characteristics of lithium-ion cells - I. Thermal analysis and phase diagram." *J. Electrochem. Soc.*, **147**(9), 3183–89.

71 Vitins, G. and West, K. (1997) "Lithium intercalation into layered LiMnO2." *J. Electrochem. Soc.*, **144**(8), 2587–92.

72 Kumagai, N., Yu, A.S., and Yashiro, H. (1997) "Thermodynamics and kinetics of electrochemical intercalation of lithium into Li0.50WO3.25 with a hexagonal tungsten bronze structure." *Solid State Ionics*, **98**(3–4), 159–66.

73 Barker, J., West, K., Saidi, Y., Pynenburg, R., Zachauchristiansen, B., and Koksbang, R. (1995) "Kinetics and thermodynamics of the lithium insertion reaction in spinel phase LixMn2O4." *J. Power Sources*, **54**(2), 475–78.

74 Idemoto, Y., Sakaya, T., and Koura, N. (2006) "Dependence of properties, crystal structure and electrode characteristics on Li content for LixCo1/3Ni1/3Mn1/3O2+delta as a cathode active material for Li secondary battery." *Electrochemistry*, **74**(9), 752–57.

75 Maier, J. (2007) *J. Power Sources*, **174**, 569–74.

76 Quintin, M., Devos, O., Delville, M.H., and Campet, G. (2006) "Study of the lithium insertion-deinsertion mechanism in nanocrystalline gamma-Fe2O3 electrodes by means of electrochemical impedance spectroscopy." *Electrochim. Acta*, **51**(28), 6426–34.

77 Garcia-Belmonte, G., Garcia-Canadas, J., and Bisquert, J. (2006) "Correlation between volume change and cell voltage variation with composition for lithium intercalated amorphous films." *J. Phys. Chem. B*, **110**(10), 4514–18.

78 Xu, J.J. and Jain, G. (2003) "Nanocrystalline ferric oxide cathode for rechargeable lithium batteries." *Electrochem. Solid State Lett.*, **6**(9), A190–A193.

79 Schoonman, J. (2003) "Nanoionics." *Solid State Ionics*, **157**(1–4), 319–26.

80 Hill, I.R., Sibbald, A.M., Donepudi, V.S., Adams, W.A., and Donaldson, G.J. (1992) "Microcalorimetric studies on lithium thionyl chloride cells - temperature effects between 25-degrees-C and -40-degrees-C." *J. Power Sources*, **39**(1), 83–94.

81 Gautier, J.L., Meza, E., Silva, E., Lamas, C., and Silva, C. (1997) "Effect of the ZnNiyMn2-yO4 ($0 \leq y \leq 1$) spinel composition on electrochemical lithium insertion." *J. Solid State Electrochem.*, **1**(2), 126–33.

82 Oberlin, A. (1984) "Carbonization and graphitization." *Carbon*, **22**(6), 521–41.

83 Franklin, E.R. (1951) "Crystallite growth in graphitizing and non-graphitizing carbons." *Proc.*

Roy. Soc. London Series A, Math. Phys. Sci., **209**(1097), 196–218.

84 Tuinstra, F. and Koenig, J.L. (1970) "Raman spectrum of graphite." *J. Chem. Phys.*, **53**, 1126.

85 Nikiel, L.W. (1993) "Raman-spectroscopic characterization of graphites - a reevaluation of spectra/structure correlation." *Carbon*, **31**(8), 1313–17.

86 Billaud, D., Henry, F.X., Lelaurain, M., and Willmann, P. (1996) "Revisited structures of dense and dilute stage II lithium-graphite intercalation compounds." *J. Phys. Chem. Solids*, **57**(6–8), 775–81.

87 Filhol, J.-S., Combelles, C., Yazami, R., and Doublet, M.-L. (2008) "Phase diagrams for systems with low free energy variation: a coupled theory/experiments method applied to Li-graphite." *J. Phys. Chem. C*, **112**(10), 3982–88.

88 Amatucci, G.G., Tarascon, J.M., and Klein, L.C. (1996) "CoO2, the end member of the LixCoO2 solid solution." *J. Electrochem. Soc.*, **143**(3), 1114–123.

89 Gabrisch, H., Yazami, R., and Fultz, B. (2004) "Hexagonal to cubic spinel transformation in lithiated cobalt oxide – TEM investigation." *J. Electrochem. Soc.*, **151**(6), A891–A897.

90 Chen, Z.H., Lu, Z.H., and Dahn, J.R. (2002) "Staging phase transitions in LixCoO2." *J. Electrochem. Soc.*, **149**(12), A1604–A1609.

91 Van der Ven, A., Aydinol, M.K., and Ceder, G. (1998) "First-principles evidence for stage ordering in LixCoO2." *J. Electrochem. Soc.*, **145**(6), 2149–55.

92 Ohzuku, T. and Ueda, A. (1994) "Solid-state redox reactions of LiCoO2 (R(3)over-bar-m) for 4 volt secondary lithium cells." *J. Electrochem. Soc.*, **141**(11), 2972–77.

93 Tarascon, J.M. and Guyomard, D. (1991) "Li metal-free rechargeable batteries based on Li1+xMn2O4 cathodes (0 less-than-or-equal-to x less-than-or-equal-to 1) and carbon anodes." *J. Electrochem. Soc.*, **138**(10), 2864–68.

94 Aurbach, D., Levi, M.D., Gamulski, K., Markovsky, B., Salitra, G., Levi, E., Heider, U., Heider, L., and Oesten, R. (1999) "Capacity fading of LixMn2O4 spinel electrodes studied by XRD and electroanalytical techniques." *J. Power Sources*, **81**, 472–79.

95 Shin, Y.J. and Manthiram, A. (2004) "Factors influencing the capacity fade of spinel lithium manganese oxides." *J. Electrochem. Soc.*, **151**(2), A204–A208.

96 Wakihara, M. (2005) "Lithium manganese oxides with spinel structure and their cathode properties for lithium ion battery." *Electrochemistry*, **73**(5), 328–35.

97 Huang, H., Vincent, C.A., and Bruce, P.G. (1999) "Correlating capacity loss of stoichiometric and nonstoichiometric lithium manganese oxide spinel electrodes with their structural integrity." *J. Electrochem. Soc.*, **146**(10), 3649–54.

98 Amatucci, G., Du Pasquier, A., Blyr, A., Zheng, T., and Tarascon, J.M. (1999) "The elevated temperature performance of the LiMn2O4/C system: failure and solutions." *Electrochim. Acta*, **45**(1–2), 255–71.

99 Amatucci, G.G., Blyr, A., Sigala, C., Alfonse, P., and Tarascon, J.M. (1997) "Surface treatments of Li1+xMn2-xO4 spinels for improved elevated temperature performance." *Solid State Ionics*, **104**(1–2), 13–25.

100 Shiraishi, Y., Nakai, I., Kimoto, K., and Matsui, Y. (2001) "EELS analysis of electrochemically deintercalated Li1-xMn2O4 and substituted spinels LiMn1.6 M0.4O4 (M = Co, Cr, Ni)." *J. Power Sources*, **97-8**, 461–64.

101 Graetz, J., Hightower, A., Ahn, C.C., Yazami, R., Rez, P., and Fultz, B. (2002) "Electronic structure of chemically-delithiated $LiCoO_2$ studied by electron energy-loss spectrometry." *J. Phys. Chem. B*, **106**(6), 1286–89.

Further Reading

Gabrisch, H., Yi, T., and Yazami, R. (2008) "Transmission electron microscope studies of $LiNi_{1/3}Mn_{1/3}Co_{1/3}/O_2$ before and after long-term aging at 70 °C." *Electrochem. Solid-State Lett.*, **11**(7), 119–24.

6
Raman Investigation of Cathode Materials for Lithium Batteries

Rita Baddour-Hadjean and Jean-Pierre Pereira-Ramos

6.1
Introduction

The basic process that occurs in a rechargeable lithium battery is the intercalation of lithium ions into different host materials. One fundamental problem with these materials is the loss of capacity during successive electrochemical insertion/deinsertion processes. As it is known that the electrochemical properties of such materials (e.g., specific capacity, reversibility, rate capability, and cycling behavior) are strongly dependent on the structural changes induced by the lithium insertion reaction. Therefore, the establishment of clear relationships between electrochemical and structural data seems to be one of the key issues for better understanding and further improving the electrochemical performances of electrode materials for rechargeable lithium batteries. The experimental techniques carried out to fulfill this requirement allow to provide either long-range [X-ray, neutron, or electron diffraction, etc.) or short-range data (X-ray absorption, nuclear magnetic resonance (NMR), electron paramagnetic resonance (EPR), X-ray photoelectron spectroscopy (XPS), etc.]. Among the local probes, Raman spectroscopy is very appropriate since it is a nondestructive characterization technique that is able to detect structural variations on the atomic level. Indeed unique molecular and crystalline information is then accessible: local disorder, changes in bond lengths, bond angles, coordination, Li dynamics, cation ordering, etc.

However, most of the Raman literature data reported on cathode materials are very scattered and qualitative, mainly dealing with the starting materials. This has strongly limited the use of Raman spectroscopy for the understanding of the structural changes induced by the electrochemical discharge/charge process.

This chapter describes the approach taken by us and other researchers over the past few years to interpret and extract quantitative information from the Raman microprobe analysis of metal oxide-based materials used as positive electrodes. The examples have been selected from an exhaustive analysis to highlight the great potentiality of Raman microspectrometry to obtain relevant and useful information on some aspects of the electrochemical behavior exhibited by these electrode materials.

Lithium Ion Rechargeable Batteries. Edited by Kazunori Ozawa

ISBN: 978-3-527-31983-1

6.2 Raman Microspectrometry: Principle and Instrumentation

6.2.1 Principle

The Raman effect was named after the Indian scientist C. V. Raman, who observed the effect by means of sunlight in 1928. The Raman effect is a phenomenon that results from the interaction of light and matter. When a photon of light interacts with a molecule, it is either absorbed or scattered to a virtual state, as shown in the energy level diagram (Figure 6.1). When scattering of the photon occurs without any change in the atomic coordinates of the molecule, the photon is elastically scattered. It exhibits the same energy (frequency) and, therefore, wavelength, as the incident photon. This process is commonly referred to as *Rayleigh scattering*. However, for a very small fraction of light (approximately 1 in 10^7 photons), atomic, i.e., vibrational motion occurs. The process leading to this inelastic scatter constitutes the Raman effect. In this case, the molecule may either gain energy from or lose energy to the photon (Figure 6.1). If the transfer of energy in the virtual state is from the photon to the molecule, the scattered photon will be lower in energy than the incident photon and the phenomenon is referred to as *Stokes Raman scattering*. Conversely, if the transfer of energy in the virtual state is from the molecule to the photon, the scattered photon will be higher in energy than the incident photon and the phenomenon is referred to as *anti-Stokes Raman scattering*.

A Raman spectrum is a plot of the intensity of the Raman-scattered radiation as a function of its frequency difference from the incident radiation (Figure 6.2). This difference is called the *Raman shift*. Figure 6.2 demonstrates the symmetry of Stokes and anti-Stokes bands. Raman spectra are often plotted as a function of intensity versus the Stokes-shifted frequencies in wavenumbers (cm^{-1}). The Raman shift corresponds to the vibrational energy levels of the molecule or the crystal. Local environment, symmetry of the crystal, atomic mass, bond order, H-bonding, molecular substituents, atomic environment, molecular geometry, structural disorder, strains, all these parameters affect the vibrational force constants, which, in turn, dictate the vibrational energy. Hence Raman spectroscopy allows to study intramolecular vibrations, crystal-lattice vibrations, and other motions of extended solids.

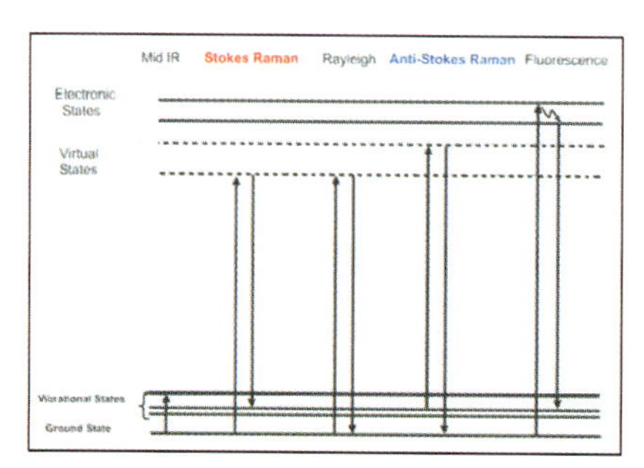

Fig. 6.1 Energy-level diagram for different processes.

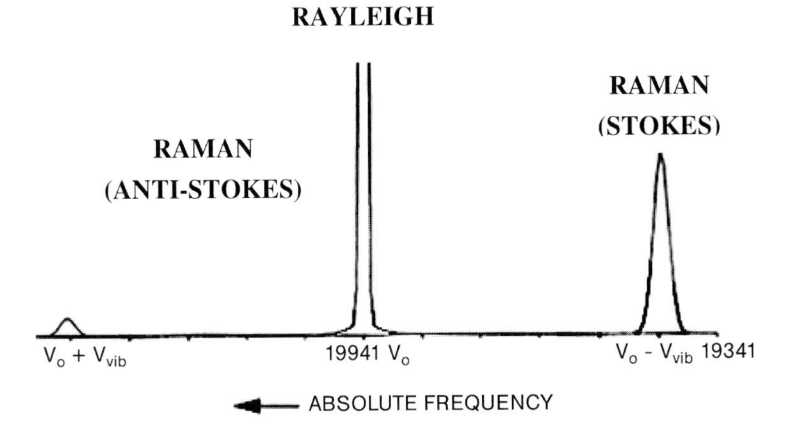

Fig. 6.2 Raman and Rayleigh scattering of excitation at a frequency ν_0. A molecular vibration in the sample is of frequency ν_{vib}.

Quantum mechanics requires that only certain atomic displacements are allowed for a given molecule. These are known as the *normal modes of vibration* of the molecule. There are several types of motion that contribute to the normal modes, some of which are

- stretching motion between two bonded atoms;
- bending motion between three atoms connected by two bonds; and
- out-of-plane deformation modes that change an otherwise planar structure into a nonplanar one.

Molecules or crystals can be classified according to symmetry elements or operations that leave at least one common point unchanged. This classification gives rise to the point group representation for the molecule, which is uniquely defined by a set of symmetry operations: rotations C_n, reflections (σ_h, σ_v, and σ_d), inversion i, and improper rotations $S_n = C_n\sigma_h$, that transform the molecule into itself [1]. Complete information of all symmetry transformations in a point group is given in the so-called character tables. Character tables not only give the number and degeneracy of normal modes, but they also tell us which of the normal modes will be infrared (IR)-active, Raman-active or both. It comes out that a fundamental transition will be Raman-active if the normal mode involved belongs to the same symmetry representation as any one or more of the Cartesian components of the polarizability tensor of the molecule.

6.2.2
Instrumentation

Raman microspectrometers are generally composed of several main parts: the excitation source (laser), the collection device (the incident light is focused on the sample through a microscope objective and the scattered light is collected by the same objective), the spectrograph (to separate the Raman scattered photons by wavelength), and the detector (which records the intensity of the Raman signal at each wavelength). A detailed description of these different parts can be found in [2, 3].

Significant advances in Raman spectroscopy have been afforded by the development of lasers, which constitute a coherent monochromatic light with very high power, and more recently, owing to the high detectivity of charge coupled device (CCD) detectors that are extremely sensitive to light and are able to take the whole spectrum at once in less than a second.

Because of different mechanisms, Raman spectroscopy is complementary to IR spectroscopy, and thus offers many advantages: little or no sample preparation is required. In normal excitation conditions, this technique is nondestructive. Raman is relatively unaffected by strong IR absorbers like water, CO_2, and glass (silica). No special accessories are needed for aqueous solutions because water is a weak scatterer. The visible excitation source can penetrate transparent container materials, and thus Raman measurements can be acquired through glass vials, envelopes, plastic bags, etc. In opaque materials, the value of the axial resolution, determined by the optical skin depth (δ) of the laser beam, is approximatively of 30–300 nm. Raman microspectrometry is particularly well suited for the study of heterogeneous materials. Spatial resolutions in the order of the micrometers, greater than those obtained using IR microscopy, can be achieved. Finally, as confocal Raman microspectrometry is able to analyze very small volumes in the order of the micrometer cubed, it is possible to perform Raman imaging from point-by-point analysis. Hence two-dimensional (2D) or three-dimensional (3D) chemical or structural mapping can be produced with a micrometric resolution.

6.3 Transition Metal-Oxide-Based Compounds

Raman spectroscopy is a well-adapted technique for the characterization of the local structure in crystalline solids such as transition metal oxides used as positive (or negative) electrode materials in lithium and lithium-ion batteries. From the analytical point of view, Raman spectroscopy technique can solve the problem of phase identification when various environments are present. Indeed, the wavenumbers and relative intensities of the Raman bands are very sensitive to crystal symmetry, coordination geometry, and oxidation states. Thus, this spectroscopic method is very effective in differentiating various kinds of metal oxides whose atomic arrangements are closely related to one another (e.g., MnO_2, $LiMn_2O_4$, and $LiCoO_2$-based compounds). Since it does not need a long-range structural order, Raman spectroscopy also constitutes an alternative structural tool as it allows the study of "amorphous" compounds, thin films, or cycled cathode materials that exhibit poor X-ray diffraction (XRD) information due either to their low crystallinity, their preferential orientation, or structural disorder. From a more fundamental point of view, Raman spectroscopy constitutes a local probe of great interest, complementary to long-range structural techniques such as X-ray or neutron diffraction to study the cathodic material under operation. The determination of frequencies of normal vibrations provides various useful data on the local structure variations induced by the lithium insertion/deinsertion process in the lattice host: changes in metal–oxygen bond lengths, lithium environment, lattice distortions, disorder, lithium–lithium and Li-host lattice interactions, cation ordering, etc., information that is of great interest to understand and then to improve the performance of a given compound. However, in spite of such advantages, Raman applications in the field of electrode materials for lithium batteries are rather few and concern mainly, as is discussed later, the analytical aspect. We present here a review of the most prominent data reported for various metal oxide-based materials, which have been selected to highlight the contribution of Raman spectroscopy in the area. Particular attention is paid to the $Li_xV_2O_5$ and the Li_xTiO_2 systems for which a thorough analysis has been provided using Raman microspectrometry, thanks to a careful and rigorous experimental approach combined with a theoretical analysis based upon lattice dynamics simulations.

6.3.1 $LiCoO_2$

This cathodic material deintercalates lithium at a very high voltage, about 4 V versus Li/Li^+. Only three lithiated systems based on transition metal oxides with high operating voltage are presently known. These are $LiCoO_2$ and $LiNiO_2$ with the pseudolayered α-$NaFeO_2$ structure, and the 3D spinel $LiMn_2O_4$. To improve their electrochemical properties, various substitutions on the transition metal site have been examined in these structures as the mixed compounds $LiM_{1-y}Co_yO_2$ (M = Ni, Al, etc.).

In 1980, Mizushima *et al.* [4] proposed using layered $LiCoO_2$ with α-$NaFeO_2$ structure as an intercalation cathode. It took around 10 years to put $LiCoO_2$ to commercial use. This oxide is now mainly used as cathode material in present lithium-ion batteries [5]. The layered structure of $LiCoO_2$ is shown in Figure 6.3. $LiNiO_2$, $LiCrO_2$, and $LiVO_2$ also adopt this structure. These $LiMO_2$ compounds, prepared at temperature ranges of 700–900 °C, are rock-salt-structured materials based on a close-packed network of oxygen atoms with Li^+ and M^{3+} ions in octahedral interstices in this packing in alternating (111) planes. This (111) ordering introduces a slight distortion of the lattice to hexagonal symmetry.

$LiCoO_2$ crystallizes in the hexagonal ($R\overline{3}m$) space group (D_5^{3d}) with a unit cell consisting of one formula unit ($Z = 1$) and unit cell parameters $a_{hex} = 2.82$ Å and $c_{hex} = 14.08$ Å. The atoms of the $LiCoO_2$ units are at sites with symmetries and coordinates given below:

$$\begin{array}{lll} \mathrm{Co}: & (3a, D_{3d}) & 0,0,0 \\ \mathrm{Li}: & (3b, D_{3d}) & 0,0,1/2 \\ \mathrm{O}: & (6c, C_{3v}) & 0,0,u, \quad 0,0,\overline{u} \end{array}$$

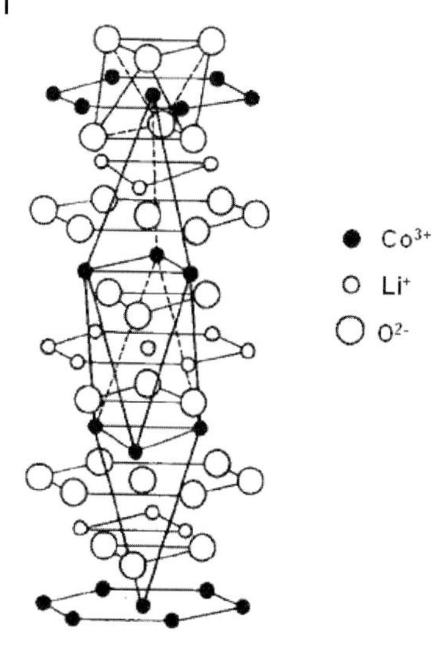

Fig. 6.3 Crystal structure of α-$NaFeO_2$-type compounds.

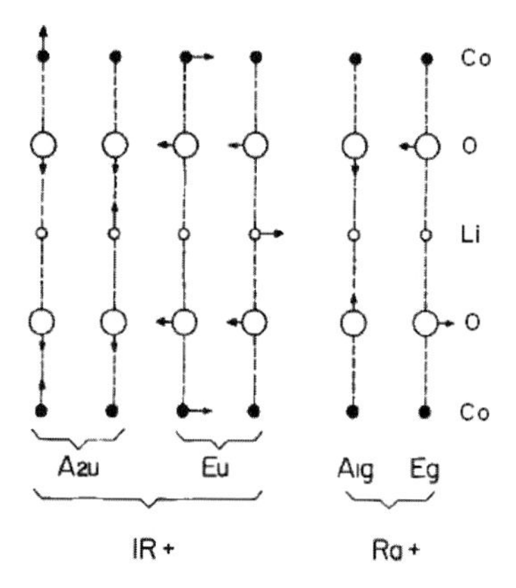

Fig. 6.4 Atomic displacements of the IR and Raman-active modes of hexagonal ($R\bar{3}m$) $LiCoO_2$.

By factor group analysis, the total irreducible representation for the vibrational modes of $LiMO_2$ is obtained as $A_{1g} + 2A_{2u} + E_g + 2E_u$.

The *gerade* modes are Raman active and the *ungerade* modes are IR active. The two Raman-active modes are especially simple. In the A_{1g} mode, two adjacent oxygen layers move rigidly against each other and parallel to the *c*-axis whereas the atomic displacements in the E_g mode are perpendicular to the *c*-axis (Figure 6.4).

The Raman spectrum of $LiCoO_2$ was first reported by Inaba *et al.*, [6], who studied the effects of replacing Co in $LiCoO_2$ by Ni. The Raman-active lattice modes were later assigned by the same team from polarized Raman measurements on a *c*-axis oriented $LiCoO_2$ thin film [7]. Two Raman-active modes have been found, which

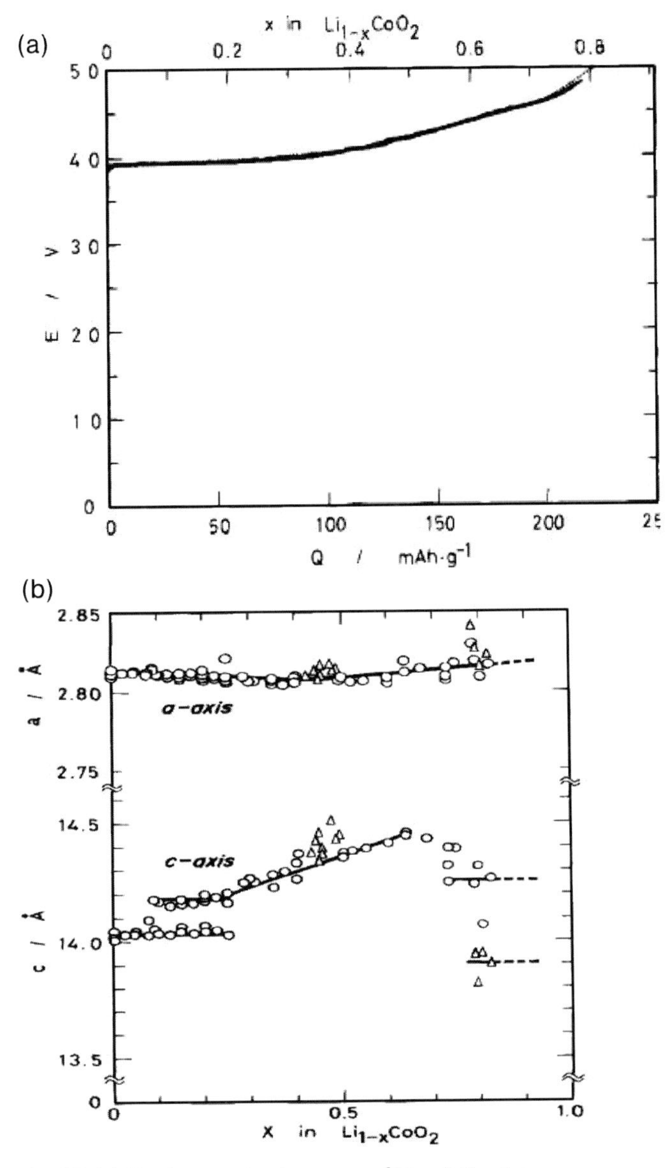

Fig. 6.5 (a) Continuous charging curve of $Li_{1-x}CoO_2$ at a rate of 0.17 mA cm^{-2} at 30 °C. (b) Lattice parameters of hexagonal unit cell of $Li_{1-x}CoO_2$ (from [5]).

correspond to oxygen vibrations involving mainly Co–O stretching, $\nu_1(A_{1g})$ at 595 cm^{-1}, and O–Co–O bending, $\nu_2(E_g)$ at 485 cm^{-1}.

The mechanism of lithium deintercalation and intercalation in $LiCoO_2$ has been investigated using XRD [5, 8, 9]. The discharge curve and the lattice-parameter changes in bulk $Li_{1-x}CoO_2$ are shown in Figure 6.5. The *a*-axis remains practically

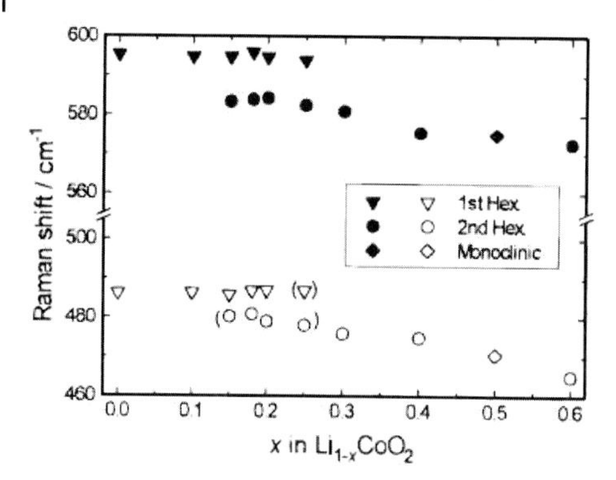

Fig. 6.6 Variations of the peak wavenumbers of the Raman bands with x in $Li_{1-x}CoO_2$ (from [7]).

constant, whereas the c-axis increases from approximately 14.08 to 14.45 Å for $0 < x < 0.5$, due to the production of a second hexagonal phase with an expanded c parameter [5]. The expansion in the c-axis has been ascribed to an increase in the electrostatic repulsion between adjacent CoO_2 layers because negatively charged oxygen–oxygen interactions increase with the removal of lithium ions [9]. Two monoclinic phases were also reported, one at about $x = 0.45$, and another for $0.75 < x < 1$ [5]. One study reports the Raman spectra of $Li_{1-x}CoO_2$ powder prepared by electrochemical lithium deintercalation [7]. The spectral changes were well correlated with the structural changes determined by XRD, namely as a series of phase transitions. In particular, a set of two new bands, located at lower wavenumbers, are observed for the second hexagonal phase (Figure 6.6). These downward shifts of both bands have been found in good agreement with the increase of the c-axis length as lithium ions are deinserted. However, the expected Raman peak splitting due to the distortion in the monoclinic phase was not observed. *In situ* Raman spectroscopy has been also conducted on thin film electrodes of pure $LiCoO_2$ [10, 11]. The authors report that applying potentials more positive than 4.7 V leads to a sudden increase of the Raman background signal, which has been ascribed to the formation of a film on the $LiCoO_2$ electrode surface in organic solution ($LiClO_4$/ propylen carbonate (PC) or ethylen carbonate (EC)) [11]. An interesting feature concerns the reported invariance of the peak frequency with the electrode potential, which is not discussed in [10] but can be correlated to the specific structural response of $LiCoO_2$ thin film reported later from XRD measurements, different from that usually known for the $LiCoO_2$ powder [12]. Indeed, when the cutoff voltage is limited to the conventional value of 4.2 V corresponding to the $Li_{0.5}CoO_2$ material, the XRD patterns of charged films at 4.2 V show negligible change in the c lattice parameter. Five or ten cycles are needed to provoke the appearance of the conventional $Li_{0.5}CoO_2$ expanded phase observed from the first charge for the bulk material.

Micro-Raman spectrometry constitutes a convenient and powerful technique for the qualitative microstructural analysis of $LiCoO_2$ cathodes. Because Raman

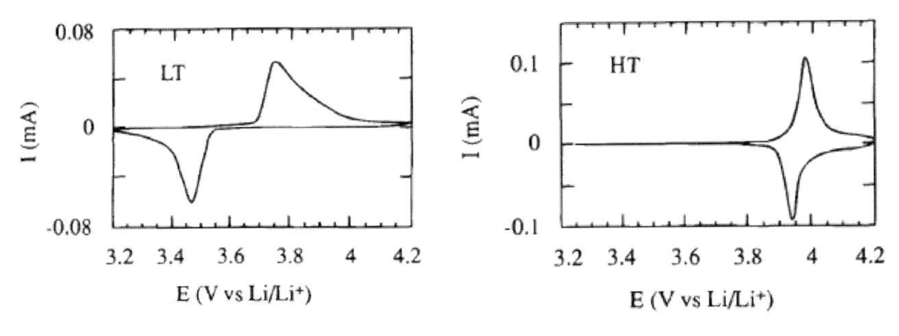

Fig. 6.7 Voltamperometric curves for HT-$LiCoO_2$ and LT-$LiCoO_2$ (from [28]).

spectroscopy is capable of detecting unambiguously the nature of the cobalt oxide phases when present lithiated or not, it has been extensively used for the characterization of powders [13–15] and thin film electrodes [16–25]. Considering the fabrication of the $LiCoO_2$ compound, especially as thin film material owing to the use of low-temperature synthesis conditions, Raman microspectrometry is thoroughly used as quality control tool to check the nature and the cristallinity of the $LiCoO_2$ active phase as a function of the synthesis conditions and to detect the presence of residual secondary phases, such as Co_3O_4 or Li_2CO_3 [16–25]. An important point concerns the formation of the layered $LiCoO_2$ structure, which is known to be crucial for obtaining a good rechargeability of the cell. Indeed, a low-temperature spinel $LiCoO_2$ phase (prepared at 400 °C and denoted by LT-$LiCoO_2$) has been reported, with different electrochemical features compared to the high-temperature layered $LiCoO_2$ phase (prepared at 850 °C and denoted by HT-$LiCoO_2$) [13, 26–28]. Comparison of cyclic voltammetric curves for LT- and HT-$LiCoO_2$ emphasizes a large difference in the electrochemical behavior (Figure 6.7). The LT $LiCoO_2$ is characterized by unusual broad anodic and cathodic peaks with a difference between the peak potentials greater than 200 mV, the latter being located at 3.75 and 3.45 V. A faradaic yield around 0.4 F mol^{-1} is involved in both cases.

Standard powder neutron diffraction and XRD cannot unambiguously distinguish between layered and spinel $LiCoO_2$ [26–29]. Conversely, from the spectroscopic viewpoint, the ideal spinel-type $LiCoO_2$ belongs to the space group $Fd3m$, the Bravais cell contains four molecules ($Z = 4$) and four Raman-active modes A_{1g}, E_g, and $2F_{2g}$ are predicted [29]. As shown in Figure 6.8, each $LiCoO_2$ crystal structure gives rise to a specific Raman fingerprint, i.e., four Raman bands are observed at ca 605, 590, 484, and 449 cm^{-1} for LT-$LiCoO_2$, whereas only two Raman bands at 597 and 487 cm^{-1} are observed for the layered HT-$LiCoO_2$.

This is consistent with the theoretical prediction given for a spinel $Fd3m$ and hexagonal ($R\bar{3}m$) crystal respectively. This peculiarity explains the systematic use of Raman spectroscopy as conclusive evidence for the structural determination of lithiated cobalt oxide phases.

Over the last few years, attention has been focused toward application of Raman spectroscopy as an *in situ* vibrational probe of electrode materials in an operating

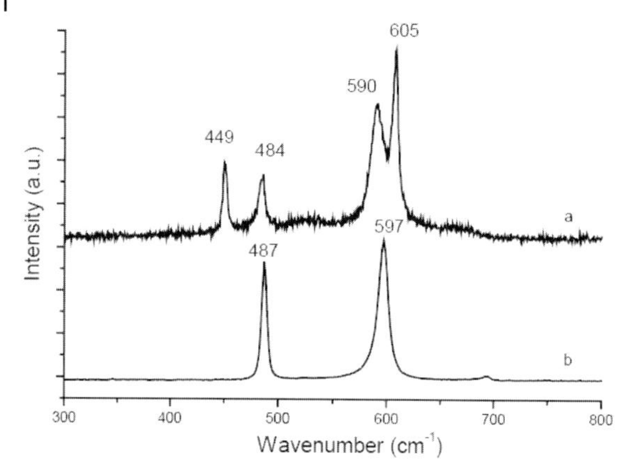

Fig. 6.8 Raman scattering spectra for (a) spinel LT-$LiCoO_2$ and (b) layered HT-$LiCoO_2$ (from 30).

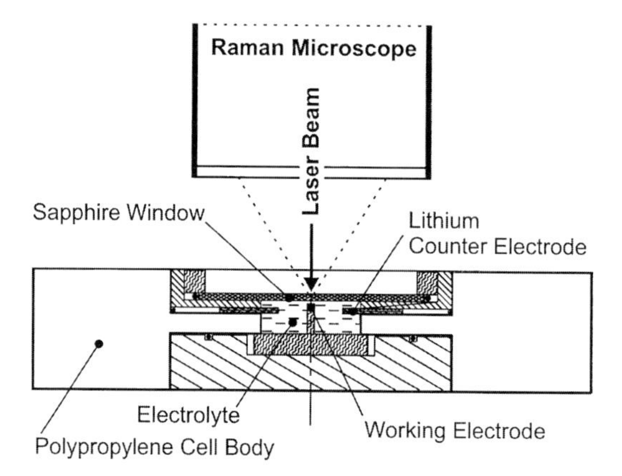

Fig. 6.9 An example of an electrochemical cell developed for *in situ* Raman microscopy (from [31]).

Li-ion battery during discharge/charge cycles. A specially designed *in situ* cell was developed by Novak *et al.* [31] with $LiCoO_2$ as cathode material and graphite-based compounds as anode (Figure 6.9). This approach allowed to obtain the Raman signature of $LiCoO_2$ particles randomly selected on the surface of the commercial electrode. It was shown that upon lithium insertion into the host material, the background intensity rises significantly, whereas upon lithium deinsertion, the background intensity is nearly constant. These results are found in accordance with those reported earlier for a three-electrode configuration [10]. More recently, *in situ* Raman measurements in an operating Li-ion battery during discharge have been carried out with the aim to construct time-resolved, 2D maps of the state of charge within the electrodes [32]. However, if clear evidence was obtained for changes in the amount of Li^+ within particles of graphite during battery discharge, this effect is observed in a lesser extent for the $LiCoO_2$ cathode.

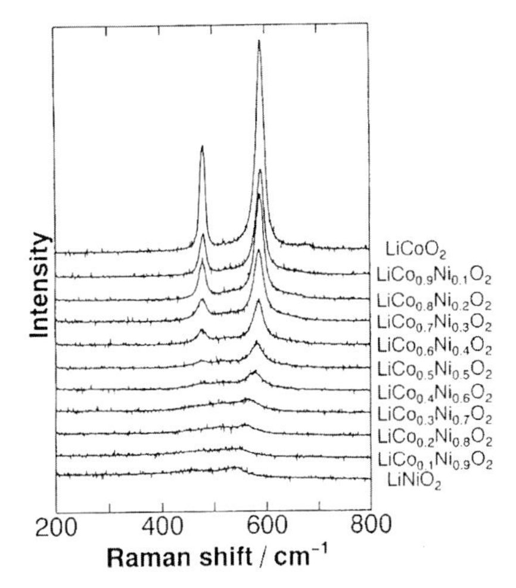

Fig. 6.10 Raman-scattering spectra of $Li_{1-y}Co_yO_2$ powder samples (from [6]).

6.3.2
$LiNiO_2$ and Its Derivative Compounds $LiNi_{1-y}Co_yO_2$ ($0 < y < 1$)

The limited capacity of $LiCoO_2$, its high cost, and its toxicity are considered as drawbacks. Another layered compound of interest, $LiNiO_2$, has been explored by several authors but suffers from a tendency to nonstoichiometry in relation to the presence of an excess of nickel [33] and from poor thermal stability in its highly oxidized state (Ni^{3+}/Ni^{4+}). Indeed the metastable layered structure $Li_{0.5}NiO_2$ transforms into the cubic spinel on heating to 300 °C. The substitution of cobalt by nickel has been reported to be an easy way to stabilize the 2D structure. The use of the solid solution $LiNi_{1-y}Co_yO_2$ ($0 < y < 1$), has therefore been explored by several workers and improved electrochemical properties have been reported [34]. The Raman spectra of lithium–cobalt–nickel oxides have been reported by many authors [6, 29, 35, 36]. It comes out that replacing Co by Ni does not change the space group, but both bands assigned to the E_g and A_{1g} Raman-active modes in ($R\bar{3}m$) symmetry are found to decrease drastically in intensity with increasing Ni content (Figure 6.10).

Hence the Raman scattering efficiency of $LiNiO_2$ appears to be very weak in comparison to that of other rock-salt compounds. The origin of these features has been ascribed to a reduction of the rhombohedral distortion by increasing Ni content or/and an increase in the electrical conductivity in $LiNiO_2$ [6]. Conversely, the Raman spectrum of a pulser laser deposited $LiNi_{0.8}Co_{0.2}O_2$ film recently reported [37] exhibits intense features at 478 and 587 cm^{-1} (which are not discussed) and they are contradictory toward previous data related with the same composition [6]. Another effect of cobalt substitution in $LiNi_{1-y}Co_yO_2$ powdered samples concerns the observed shift of both Raman bands toward higher frequencies, from 546 cm^{-1} in $LiNiO_2$ to 595 cm^{-1} in $LiCoO_2$ for the A_{1g} mode and from 470 cm^{-1} in $LiNiO_2$ to 485 cm^{-1} in $LiCoO_2$ for the E_g mode [35]. These frequency shifts are quite consistent with the observed decrease in the hexagonal unit-cell parameters as Co content increases. The a_{hex} parameter corresponds to the intralayer metal–metal distance in the MO_6 layer, and the c_{hex} parameter is equal to three times the interlayer distance. It is reported that a_{hex} varies from 2.86 to 2.82 Å and c_{hex} varies from 14.15 to 14.08 Å when y varies from 0.3 to 1 [36]. Hence the increase of the bond covalency inside the layers as cobalt content increases, suggested by the decrease of the metal–metal intralayer distance, could explain the frequency shifts toward higher frequency.

6.3.3 Manganese Oxide-Based Compounds

Manganese oxides (MO) with 3D and 2D crystal networks constitute a large family of porous materials that can accept foreign species in their tunnel or interlayer space. There is a wide variety of natural or synthetic MO, including the various allotropic forms of MnO_2 and ternary lithiated compounds Li_xMnO_y. Having good electrochemical performance, they are attractive as positive electrode materials for lithium cells because manganese has economical and environmental advantages over compounds based on cobalt or nickel [38, 39].

6.3.3.1 MnO_2-Type Compounds

MnO_2 was originally developed as the positive electrode for primary alkaline batteries [40]. Extensive research was carried out over the last decades to improve the reversibility of lithium insertion in manganese dioxide (MD) cathode for rechargeable Li–MnO_2 cells [41]. Since then, much effort has been devoted to the study of lithium insertion into various forms of manganese dioxides, especially synthetic products prepared by either electrolytic (EMD) or chemical (CMD) method that belong to the nsutite (γ-MnO_2) group, for their use as cathode materials in lithium batteries [42, 43]. Lithium accommodation in γ-MnO_2 occurs predominantly by insertion into the (2×1) tunnels of the ramsdellite (R-MnO_2) domains, while the β-MnO_2 domains only accommodate 0.2 Li in the (1×1) channels [44]. Besides the γ-MnO_2 form, which suffers from moderate capacity, low reversibility, and poor cycling stability, there are many MO materials under study such as $LiMn_2O_4$ spinel-like phases [45], layered $LiMnO_2$ [46], and new layered phases such as hexagonal $\alpha Li_{0.51}Mn_{0.93}O_2$ and orthorhombic $\beta Li_{0.52}MnO_2$ [47].

The common crystallographic unit building the lattice of MO and that of their lithiated products is the basal MnO_6 octahedron. Table 6.1 summarizes the crystallographic data of various MO compounds. Their structure can be described as a close-packed network of oxygen atoms consisting of edge- and corner-sharing MnO_6 octahedra forming tunnels of various sizes for the insertion of Li ions, leading to more or less compact structures in which Mn^{4+} and/or Mn^{3+} ions are distributed.

Figure 6.11 shows the Raman scattering spectra reported for various manganese dioxides compounds [48, 49]. The general peculiarity of the vibrational features of MOs is their low Raman activity. Three major regions can be distinguished: at

Table 6.1 Crystallographic data of some MO compounds (from [48]).

Compound	Mineral	Crystal symmetry	Lattice parameters (Å)	Features
MnO	Manganosite	Cubic ($Fm3m$)	$a = 4.44$	Rock salt
α-MnO_2	Hollandite	Tetragonal ($I4/m$)	$a = 9.96$; $c = 2.85$	(2 × 2) Tunnel
R-MnO_2	Ramsdellite	Orthorhombic ($Pbnm$)	$a = 4.53$; $b = 9.27$; $c = 2.87$	(1 × 2) Tunnel
β-MnO_2	Pyrolusite	Tetragonal ($P4_2/mnm$)	$a = 4.39$; $c = 2.87$	(1 × 1) Tunnel
γ-MnO_2	Nsutite	Complex tunnel (hex)	$a = 9.65$; $c = 4.43$	(1 × 1)/(1 × 2)
δ-MnO_2	Vernadite	Hexagonal	$a = 2.86$; $c = 4.7$	(1 × ∞) Layer
λ-MnO_2	Spinel	Cubic ($Fd3m$)	$a = 8.04$	(1 × 1) Tunnel
$MnO_x.H_2O$	Birnessite	Tetragonal	$a_{hex} = 2.84$; $c_{hex} = 14.64$	(1 × ∞) Layer
MnOOH	Groutite	Orthorhombic ($Pbnm$)	$a = 4.56$; $b = 10.70$; $c = 2.87$	
α-Mn_2O_3	Bixbyite	Cubic ($Ia3$)	$a = 9.41$	C-type
Mn_3O_4	Hausmannite	Tetragonal ($I4_1/amd$)	$a = 9.81$; $c = 2.85$	Spinel-like

200–450, 450–550 and 550–750 cm^{-1}. They correspond to spectral domains where skeletal vibrations, deformation modes of the metal-oxygen chain of Mn–O–Mn in the MnO_2 octahedral lattice and stretching modes of the Mn–O bonds in MnO_6 octahedra occur, respectively.

The birnessite-type MO constitute another class of materials with layered structure, water molecules, and/or metal cations occupying the interlayer region. It has been recently shown that attractive electrochemical performances, with stable capacities of 170 $mAh\,g^{-1}$ after 40 cycles at C/20 (Figure 6.12), could be reached for sol–gel prepared birnessite doped with Co (SGCo-Bir) with chemical formula $Co_{0.15}Mn_{0.85}O_{1.84}$, $0.6H_2O$ [50]. Raman features of several birnessite compounds have been reported for the first time in [51]. As shown in Figure 6.13, in spite of slight variation in band positions and relative band intensity, the general similarity of the spectra suggests that samples are characterized by the same basic structure. In fact, MnO_6 octahedral layers are separated by layers of lower-valent cations (Li^+, Na^+, Mn^{2+}, ...) and by layers of water. The highest Raman band is assigned to the symmetric vibration ν(Mn–O) of MnO_6 group, with A_{1g} symmetry in the O_h^7 spectroscopic space group. This mode is observed at 625 and 640 cm^{-1} for Li-Bir and Na-Bir, respectively, and at 646 and 638 cm^{-1} for SG-Bir and SGCo-Bir, respectively. A correlation between the wavenumber value of this stretching mode and the interlayer d-spacing has been proposed [51]. The band located at 575 cm^{-1}

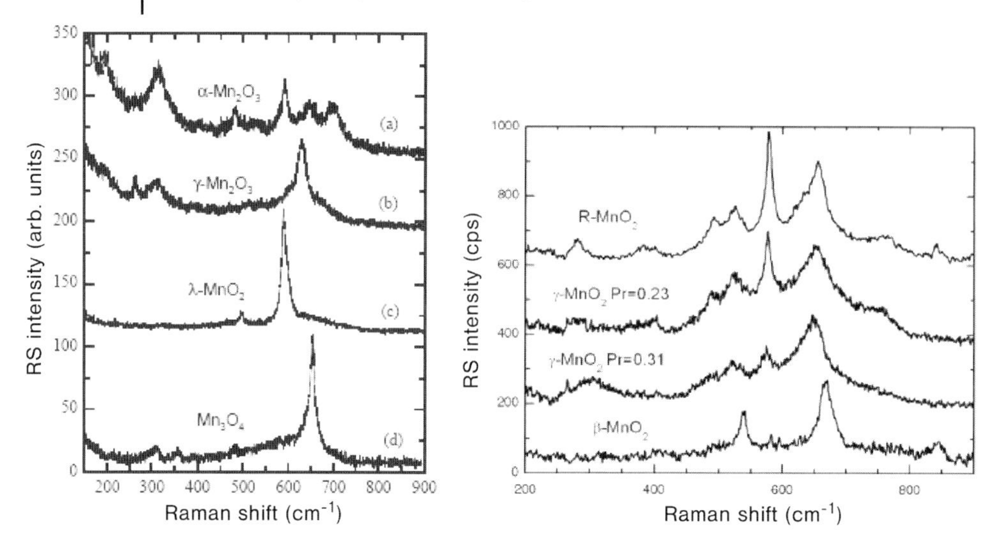

Fig. 6.11 Raman scattering spectra reported for various manganese dioxide frameworks. Pr is the intergrowth rate of the pyrolusite into the ramsdellite matrix (from [48, 49]).

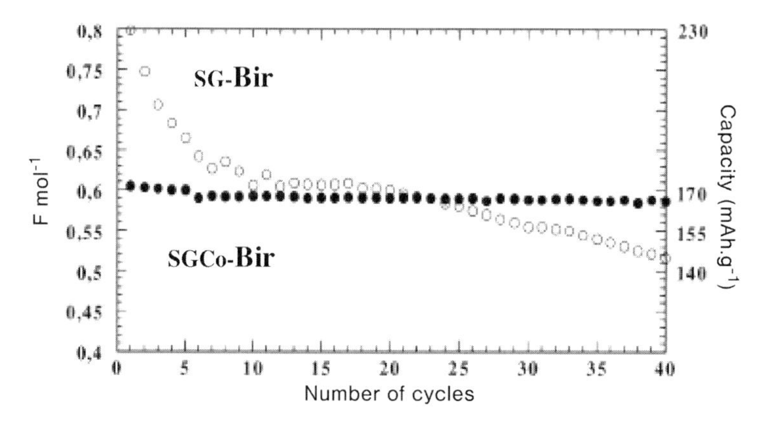

Fig. 6.12 Specific capacity as a function of the cycle number for SG-Bir and SGCo-Bir. C/20 rate (from [50]).

is attributed to the ν(Mn–O) stretching vibration with F_{2g} symmetry and is commonly related to the vibrational stretching frequency inherent to the presence of Mn^{4+} ions. Its intensity is particularly strong in birnessite compound compared with the literature data related to lithiated spinels due to the high rate of Mn^{4+} in the birnessite family [50–52]. The Raman spectrum of SGCo-Bir, where cobalt partly substitutes for manganese, displays similar features to SG-Bir, $MnO_{1.84}$, $0.6H_2O$. As a result of the substitution of Mn^{4+} by Co^{3+} ions in MnO_2 layers [50], a frequency shift of $10\,cm^{-1}$, from 575 to $585\,cm^{-1}$, was observed for the ν(Mn–O) stretching vibration with F_{2g} symmetry (Figure 6.13). This result has

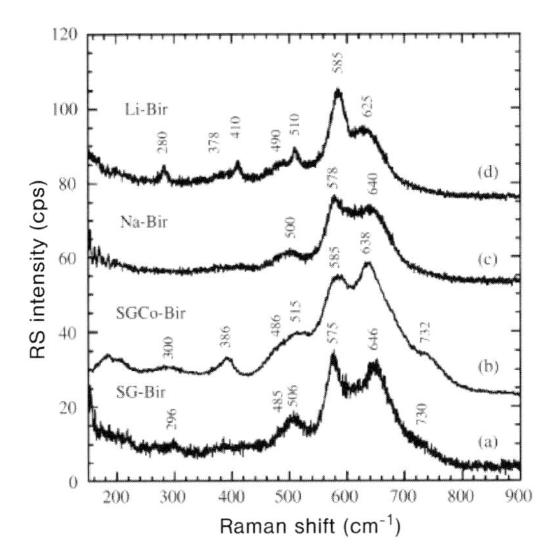

Fig. 6.13 Raman spectra of birnessite-type manganese oxides (a) $MnO_{1.84}$, $0.6H_2O$; (b) $Co_{0.15}Mn_{0.85}O_{1.84}$, $0.6H_2O$; (c) $Na_{0.32}MnO_2$, $0.6H_2O$; and (d) $Li_{0.32}MnO_2$, $0.6H_2O$ (from [51]).

been correlated to a strengthening of the Mn–O bond in the Co-doped SG-Bir. This finding is in good accordance with the better structural stability of the host lattice for the Co-doped material during cycling, as illustrated in Figure 6.12.

6.3.3.2 **Ternary Lithiated Li_xMnO_y Compounds**

Lithiated transition manganese oxides, used or involved as positive electrode materials in high-voltage lithium-ion batteries, exhibit different crystallographic structures. As a consequence, the number of active bands in the vibrational spectra of these compounds varies significantly, depending on their local symmetry (Table 6.2). Two classes of materials are currently studied: the spinel-type and the rock-salt-type compounds. The Li–Mn–O phase diagram in Figure 6.14 highlights the positions of spinel and rock-salt compositions within the λ-MnO_2, MnO, and Li_2MnO_3 tie-triangle. It emphasizes the wide range of spinel and rock-salt compositions that exist in the Li–Mn–O system.

Stoichiometric Spinel Phases in the Li–Mn–O System Stoichiometric spinels fall on the tie line between Mn_3O_4 and $Li_4Mn_5O_{12}$ (Figure 6.14). They are defined by a cation/anion ratio M/O of 3/4. The stoichiometric spinels of importance for lithium battery applications form a solid solution between $LiMn_2O_4$ and $Li_4Mn_5O_{12}$. They are considered currently of technological interest as insertion electrodes for rechargeable 4-V lithium batteries because of their diffusion pathways for Li ions, their low cost, low toxicity, and high energy density (due to the combination of high capacity and high voltage) [39, 54–56].

Table 6.2 Structure and Raman activity for various lithiated transition metal-oxide materials.

Compound	Type of structure	Space group	Raman activity
$LiCoO_2$	Layered hexagonal rock salt	$D_5^{3d} - R\bar{3}m$	$A_{1g} + E_g$
$mLiMnO_2$	Layered monoclinic rock salt	$C_{2h}^3 - C2/m$	$2A_g + B_g$
$LiMn_2O_4$	Normal cubic spinel	$O_h^7 - Fd3m$	$A_{1g} + E_g + 3F_{2g}$
λMnO_2	Modified cubic spinel	$O_h^7 - Fd3m$	$A_{1g} + E_g + 2T_{2g}$
$Li_2Mn_2O_4$	Normal tetragonal spinel	$D_{4h}^{19} - I4_1/amd$	$2A_{1g} + 2B_{1g} + 6E_g + 4B_{2g}$
$Li_{0.5}Mn_2O_4$	Ordered cubic spinel	$T_d^2 - F\bar{4}3m$	$3A_1 + 3E + 6F_2$

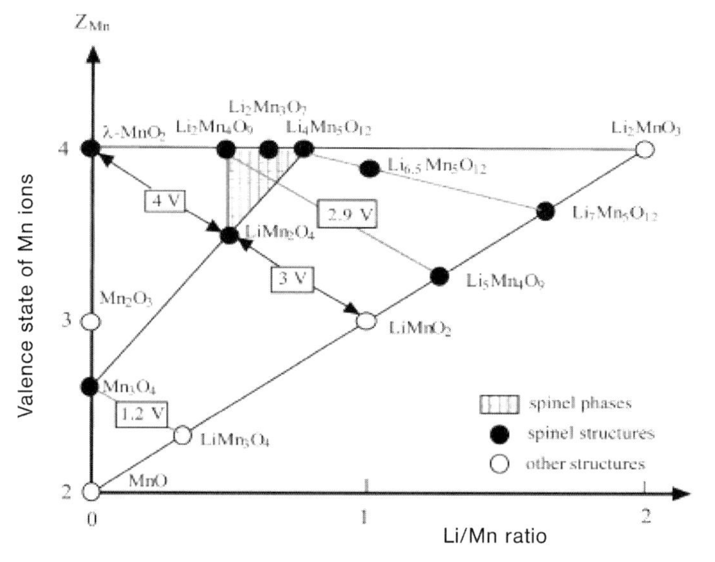

Fig. 6.14 The Li–Mn–O phase diagram showing the compositions of spinel, defect spinel, and rock-salt structures (from [53]).

The structural, chemical, and electrochemical properties of the $LiMn_2O_4$ system have been extensively reported [39, 57]. $LiMn_2O_4$ has a normal spinel structure, which is cubic with the space group *Fd3m* ($O^7{}_h$) containing eight AB_2O_4 units per unit cell. It can be represented by the formula $\{Li\}_{8a}[Mn_2]_{16d}O_4$, in which the subscript 8a indicates occupancy of tetrahedral sites by Li^+ ions, Mn^{3+} and Mn^{4+} ions (in 1 : 1 mixture) being randomly distributed over the octahedral 16 d sites and oxygen anions in 32e Wyckoff positions (Figure 6.15a). The approximatively cubic close-packed array of oxide ions incorporates MnO_6 octahedra, connected to one another in three dimensions by edge sharing, LiO_4 tetrahedra sharing each of their four corners with a different MnO_6 unit, and a 3D network of octahedral 16c holes and tetrahedral 8a sites in which lithium ions can move through the (1×1) channels of the spinel lattice (Figure 6.15b).

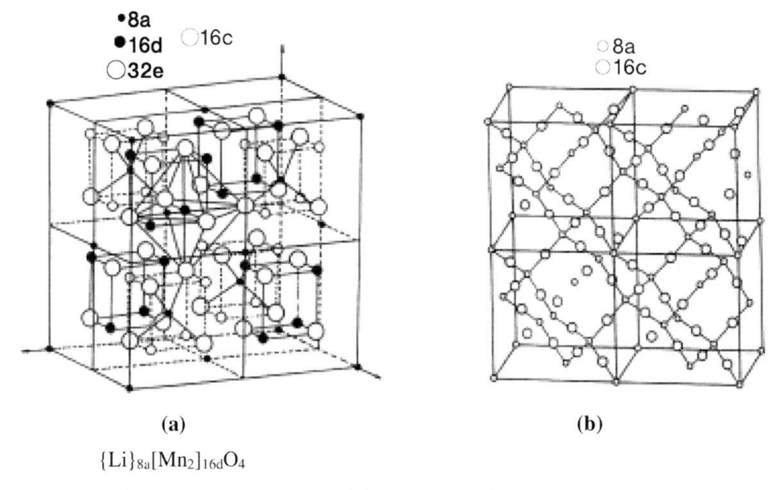

Fig. 6.15 Schematic representation of the structure of AB_2O_4 spinel lattices showing (a) the smallest (primitive) cubic unit cell of normal spinel (*Fd*3*m* space group) and (b) diffusion path of lithium.

Vibrational features of spinel oxide structures have been widely studied. Tarte and Preudhomme have published numerous IR data on spinels and solid solutions [58–60]. White and De Angelis have treated the vibrational spectra of spinel structures by a factor group analysis [61]. The various types of ordering were considered and the predicted changes were compared with literature data.

Early spectroscopic data on electrochemically prepared spinel-type $Li_xMn_2O_4$ composite electrodes were obtained from Fourier Transformed Infrared Spectroscopy (FTIR) measurements [62, 63]. Observation of $LiMn_2O_4$ Raman features at 593 and 628 cm^{-1} was first reported from *in situ* measurements of lithium-ion electroinsertion in a spinel-type Pt/λ-MnO_2 electrode [64]. Ammundsen *et al.* [65] have calculated the lattice dynamics of spinel-structured lithium manganese oxides using atomistic modeling methods, which allowed to predict and to assign the Raman spectra of lithiated, fully delithiated, and partially delithiated $Li_xMn_2O_4$ thin films. In good agreement with the calculations, five Raman-active modes are observed for the $LiMn_2O_4$ phase, one of A_{1g} symmetry at 625 cm^{-1}, one of E_g symmetry at 426–432 cm^{-1}, and three of T_{2g} symmetry at 365–380, 480–485, and 580–590 cm^{-1} (Table 6.3 and Figure 6.16).

The slight underestimation of the calculated A_{1g} mode for $LiMn_2O_4$ (598 instead of 625 cm^{-1}) has been ascribed to a small deviation from the model used to describe the electronic polarizability of oxygen ions in $LiMn_2O_4$ (transferred without any change from the λ-MnO_2 model) [65].

In a localized vibration analysis, the highest frequency A_{1g} mode is viewed as the symmetric Mn–O stretching vibration of MnO_6 group. Its broadness has been related with the cation–anion bond lengths and polyhedral distortions occurring

Table 6.3 Calculated and observed wavenumbers (in cm^{-1}) for Raman-active modes of $LiMn_2O_4$, λ-MnO_2, and $Li_{0.5}Mn_2O_4$ thin films, assuming a $F\bar{4}3m$ space group for the latest (from [65]).

$LiMn_2O_4$			λ-MnO_2			
Calculated	Observed		Calculated	Observed		Symmetry species
354	**365**	382[a]				$T_{2g}(1)$
434	**432**	426[a]	479	**463**	462[a]	E_g
455	**480**	483[a]	511	**498**	501[a]	$T_{2g}(2)$
597	**590**	580[a]	630	**647**	644[a]	$T_{2g}(3)$
598	**625**	625[a]	592	**592**	596[a]	A_{1g}

$Li_{0.5}Mn_2O_4$			
Calculated	Observed		Symmetry species
313	296	**296**[a]	T_2
317			E
385		**380**[a]	T_2
457		**425**[a]	E
481	493	**483**[a]	T_2
537		**522**[a]	T_2
559	560	**563**[a]	A_1
586			E
593	597	**596**[a]	A_1
615	612	**611**[a]	T_2
646	630	**648**[a]	T_2
665	**657**		A_1

[a] From [53].

in $LiMn_2O_4$ [53]. As the manganese cation of the spinel structure exhibits two valence states in $LiMn^{3+}Mn^{4+}O_4$, there are isotropic $Mn^{4+}O_6$ octahedra and locally distorted $Mn^{3+}O_6$ octahedra due to the Jahn–Teller effect. Hence, the broadness of the A_{1g} mode would originate from the stretching vibrations of both entities [53]. For the same author, the shoulder peak around 580 cm^{-1} of T_{2g} symmetry originates mainly from the vibration of the Mn^{4+}-O bond. Its intensity depends on the Mn^{4+} concentration in the spinel and reflects the Mn average oxidation state for this reason. The lowest energy $T_{2g}(1)$ phonon derives predominantly from a vibration of the Li sublattice and can be viewed as a Li–O stretching motion.

There have been already many reports on the charge/discharge characteristics of the $Li_xMn_2O_4$ cathode [55–57]. The theoretical value of the specific capacity of $LiMn_2O_4$ is 148 mA h g^{-1}. The open-circuit voltage (OCV) curve of $Li_xMn_2O_4$ ($0 < x < 2$) is shown in Figure 6.17 [55].

The structural changes of the $Li_xMn_2O_4$ powdered electrode during charge and discharge have been mainly characterized by XRD [55, 66–68] and neutron powder diffractometry [66]. However, due to the high symmetry of the spinel system, the low X-ray scattering power of lithium, and the occurrence of partial occupation of the various cation sites, the XRD patterns are not always easily interpreted. It is often difficult not only to determine the precise distribution of cations in a given pure phase but also to quantitatively analyze multiphase mixtures or to distinguish between different phases with similar lattice constants. This problem is especially acute for $Li_xMn_2O_4$ with $0.2 \leq x \leq 1$, as the cubic cell parameter changes by only 3% over the entire composition range. Diffraction data obtained from electrodes is commonly of poor quality due to degradation of crystallinity

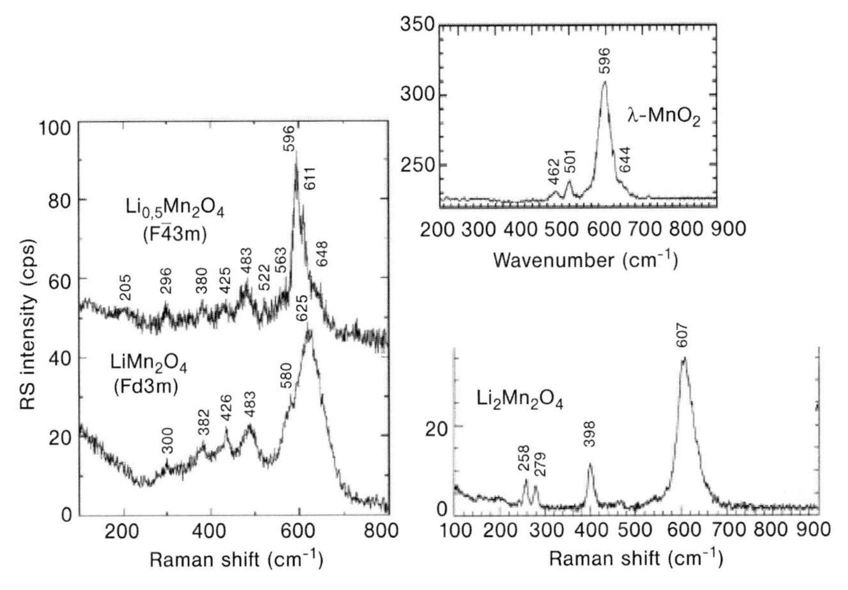

Fig. 6.16 Raman spectra of spinel $LiMn_2O_4$, $Li_{0.5}Mn_2O_4$, λ-MnO_2, and $Li_2Mn_2O_4$ (from [53]).

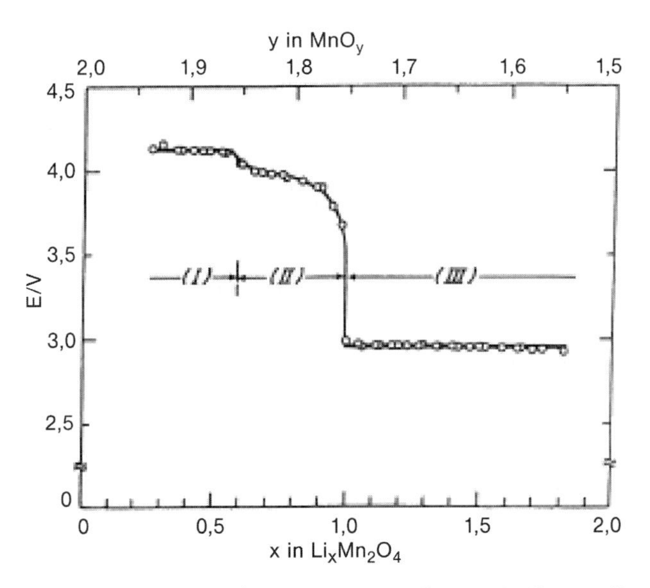

Fig. 6.17 Open-circuit voltage (OCV) curve of $Li_xMn_2O_4$ ($0 < x < 2$) at 30 °C (from [55]).

during cycling, the presence of other phases (carbon, electrolyte), and compositional inhomogeneity.

While vibrational spectroscopy cannot give the detailed structural information available from high-quality diffraction data, it is sensitive to the local environments of lithium and transition metal cations in the oxide lattice. The number, frequency, and relative intensities of the vibrational bands depend upon both coordination geometries and bond strengths (which may, in return, depend on occupancy factors and oxidation states). Poorly crystalline or amorphous components that make only broad background contributions to XRD patterns are, however, represented quantitatively in FTIR and Raman spectra.

Lithium may be electrochemically extracted from $LiMn_2O_4$, at a potential of about 4 V versus metallic lithium corresponding to the (I) + (II) region in Figure 6.17, according to the oxidation reaction:

$$LiMn_2O_4 \longrightarrow Li_xMn_2O_4 + (1-x)e^- + (1-x)Li^+ \quad 0.2 \leq x \leq 1$$

In this composition range, $Li_xMn_2O_4$ retains the cubic symmetry and appears to consist of a single phase for $0.5 < x < 1$. As lithium is removed from the tetrahedral 8a site, there is a monotonic decrease in the cubic-cell parameter. The OCV curve is nearly flat, showing two distinct plateaus above and below $x = 0.5$. The flat part of the discharge curve in the range $x < 0.5$ is considered to be a two-phase region in which a new cubic phase is produced and exists up to full electrochemical delithiation near a composition x almost equal to ≈ 0.015. The $Li_{0.015}Mn_2O_4$ phase (termed λ-MnO_2) has a slightly modified cubic symmetry, the removal of lithium ions resulting in the loss of the $T_{2g}(1)$ phonon at low wavenumber 365–382 cm^{-1} (see Tables 6.2 and 6.3). Phonon lattice calculations predict the T_{2g} and E_g modes to increase in wavenumber whereas the A_{1g} is predicted to decrease (Table 6.3) [65]. However, this prediction is not supported by any structural consideration. The experimental Raman spectrum of λ-MnO_2, reported in Figure 6.16, is in good accordance with these calculations, both in terms of number of bands and frequencies.

For intermediate compositions between $LiMn_2O_4$ and λ-MnO_2, the local symmetry no longer belongs to the $Fd3m$ space group. A lower symmetry has been proposed with $F\bar{4}3m$ (T_d^2) space group (Table 6.2). It comes out that a richer Raman fingerprint is expected for the partially delithiated compounds, with 12 expected Raman-active modes instead of 5 for $LiMn_2O_4$. This analysis is supported by reported experimental Raman data for the $Li_{0.5}Mn_2O_4$ phase, summarized in Table 6.3 and Figure 6.16. In this structure, every second lithium tetrahedral site is vacant, producing an ordered Li configuration, which has been experimentally observed for the $Li_{0.5}Mn_2O_4$ composition [69]. This subtle ordering transition could be responsible for the flat potential curve observed for $0.5 < x < 1$ (Figure 6.17).

Continuous charge/discharge cycling of $Li_xMn_2O_4$ electrodes in the high-voltage region (I + II) results in significant capacity fading, particularly when they are charged at potentials greater than 4 V. A FTIR spectroscopy study has demonstrated that overcharging causes a gradual conversion to a lithium poor defect spinel material via dissolution of manganese ions in the electrolyte [63].

Electrochemical insertion of lithium into $LiMn_2O_4$ proceeds in a very different manner for $1 \leq x \leq 2$, in the flat region (III), according to a reduction reaction involving a phase transition from cubic spinel to an ordered tetragonal, $Li_2Mn_2O_4$

phase:

$$LiMn_2O_4 + e^- + Li^+ \longrightarrow Li_2Mn_2O_4$$

For $1 < x < 2$, $Li_xMn_2O_4$ is a mixture of two distinct phases: the original cubic phase and a tetragonal spinel-type compound. The reason for the phase transition is an increase in concentration of Mn^{3+} ($3d^4$) ions that form during intercalation of lithium in $LiMn_2O_4$. This induces a Jahn–Teller distortion, which reduces the crystal symmetry from cubic to tetragonal and results in a volume expansion of about 6.4%. Mechanical degradation due to the large volume change during the cubic-to-tetragonal phase transition is a primary cause of capacity fading in the 3-V region ($1 < x < 2$).

The structure of $Li_2Mn_2O_4$ is a distorted spinel, which belongs to the $I4_1/amd$ space group (see Table 6.2). According to a neutron diffraction study, the added Li ions go into the previously vacant 16c octahedral sites, and about half of the lithium ions in the tetrahedral 8a sites move also in the 16c position [70]. However the cubic-to-tetragonal transition is detected only by the splitting of certain peaks in the XRD patterns [71]. Conversely, the Raman-active modes expected for the tetragonal $Li_2Mn_2O_4$ phase with spectroscopic D_{4h}^{19} symmetry are $2A_{1g}+ 2B_{1g}+ 6E_g+ 4B_{2g}$ (Table 6.2). Practically, lithium incorporation into $LiMn_2O_4$ leads to important changes in the vibrational features, as shown in Figure 6.16. The Raman spectrum of tetragonal $Li_2Mn_2O_4$ is dominated by four bands at 607, 398, 279, and 258 cm^{-1}. However, a greater number of bands (14) was expected from the spectroscopic analysis. A straightforward assignment of this Raman spectrum is not yet provided, even if some tentative attributions have been speculated from a localized vibration analysis [71].

Raman microspectrometry constitutes a very efficient probe for the identification of electrochemically produced spinel-like lithiated manganese oxide materials. *In situ* Raman measurements during the charge/discharge of a $LiMn_2O_4$ composite cathode have been performed [72]. However, only poor structural information can be drawn due to the limited frequency range, low-quality Raman spectra, and local inhomogeneity. *In situ* Raman investigation of pure $Li_{1-x}Mn_2O_4$ ($0 \le x \le 1$) thin films produced by electrostatic spray deposition did not allow a straightforward analysis since the Raman spectra are hampered by the lines of the sapphire optical window and the electrolyte [73].

In situ Raman experiments performed on a $LiMn_2O_4$ single crystal microelectrode polarized at 4.4, 4.04, and 3.58 V versus Li/Li^+ have allowed to clearly evidence the spectroscopic fingerprints of λ-MnO_2, $Li_{0.5}Mn_2O_4$, and $LiMn_2O_4$, respectively [74]. From the analysis of the Raman spectra, using classical least squares (CLS) curve resolution, the authors provide the fraction of the different $Li_xMn_2O_4$ components as a function of the potential (Figure 6.18).

One of the main routes considered to reduce the capacity fading of $LiMn_2O_4$ is partial substitution of manganese by transition metals M [44, 76, 77]. These $LiM_xMn_{2-x}O_4$ (M = Cr, Co, Ni, Al, Li, etc.) materials are regarded as attractive as cathodes for lithium batteries, allowing the cell voltage to be increased to

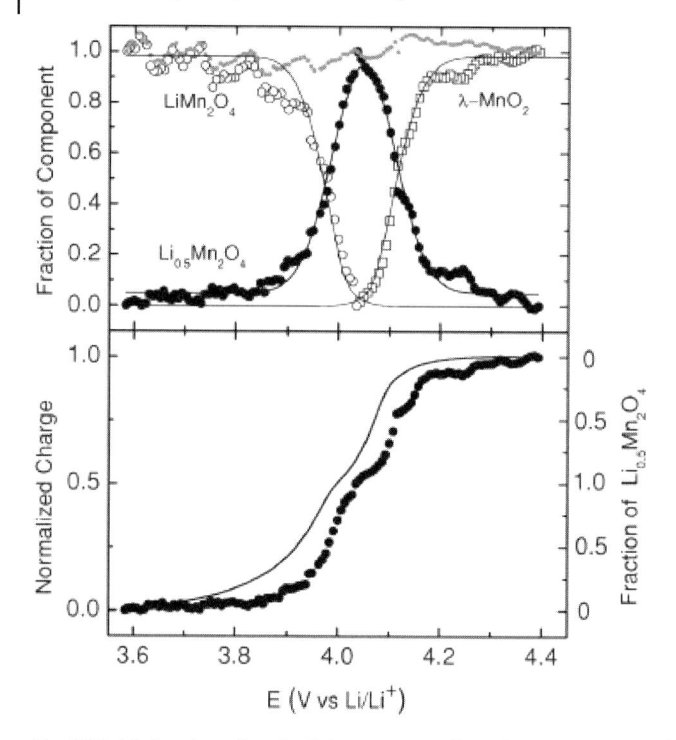

Fig. 6.18 (a) Fraction of each of the components of $Li_xMn_2O_4$ as a function of the potential extracted from Raman data using classical least squares (CLS). (b) Normalized charge determined by coulometric analysis of the voltammetric curve (left ordinate, solid curve) and fraction of $Li_{0.5}Mn_2O_4$ as a function of the potential (scattered symbols) (from [75]).

5 V with improved cycle life. The stabilization of the octahedral 16 d site is achieved by reducing the amount of Mn^{3+} causing the Jahn–Teller distortion. The enthalpy of the cubic-to-tetragonal phase transition in partially substituted $LiM_xMn_{2-x}O_4$ spinels gradually decreases with increasing amount of substituent, and is completely suppressed with 10–20 mol% of substitution in octahedral Mn site [76]. The variation of the specific capacity with cycle number for various $LiM_{1/6}Mn_{11/6}O_4$ spinels (M = Cr, Co, Ni, Al) is reported in Figure 6.19 [44]. It can be seen that the cells with substituted $LiMn_2O_4$ show better cycle characteristics than the one with undoped $LiMn_2O_4$.

Several studies have shown that the improvement of the cyclability for the substituted $LiM_xMn_{2-x}O_4$ was related to the enhanced stability of the local structure of the materials in comparison to that of $LiMn_2O_4$ [78]. Furthermore, when the cations involved have undistinguishable scattering power, vibration spectroscopy has proved to be very powerful in determining symmetry changes undetectable by XRD. In spite of this, only limited Raman studies have been devoted to this topic, presumably because of the difficulty in interpreting the Raman spectra of such materials [79–84].

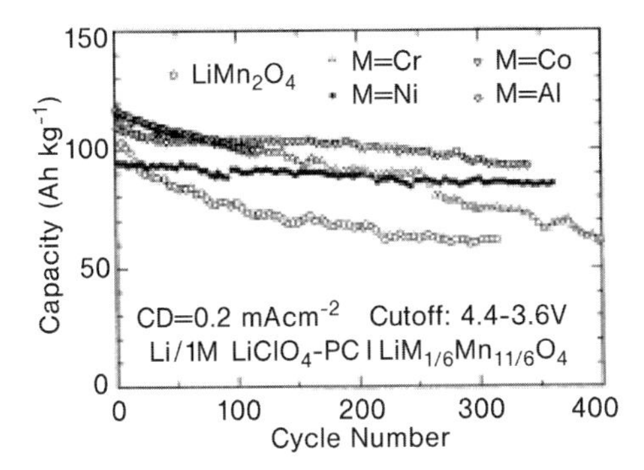

Fig. 6.19 Specific capacity as a function of the cycle number for $LiM_{1/6}Mn_{11/6}O_4$ (M = Cr, Co, Ni, Al) (from [44]).

Layered Rock Salt Phases in the Li–Mn–O System Stoichiometric rock-salt compounds are located on the tie line between MnO and Li_2MnO_3. (Figure 6.14). These layered lithiated manganese oxides are also under investigation for battery applications [46, 85]. Among them, α-$LiMnO_2$ exhibits the largest initial capacity for the 4-V region but it suffers from severe capacity fading after the first charge process [46]. In fact, most layered compounds have structural features in common with spinels and convert readily upon cycling.

The structure of α-$LiMnO_2$ is monoclinic ($C2/m$ space group, $C^3{}_{2h}$ spectroscopic symmetry, $Z = 2$) because the coordination polyhedron around the Mn^{3+} ions is distorted owing to the Jahn–Teller effect. This compound exhibits the cation ordering of the α-$NaFeO_2$ structure, in which Li^+ ions are located between the MnO_6 sheets, in the same octahedral 2 d interstices as Mn^{3+} ions, whereas oxygen anions are in 4i sites. According to the group theoretical calculations, the monoclinic $LiMnO_2$ oxide is predicted to show three Raman-active modes with $2A_g + B_g$ species. As shown in Figure 6.20(a), three main peaks are detected in the Raman spectrum of the monoclinic $LiMnO_2$ phase, at 422, 481, and 603 cm^{-1}, this latest peak being attributed to the Ag mode involving the symmetric stretching vibration of equatorial oxygen atoms while the shoulder at 575 cm^{-1} is attributed to the stretching mode of Mn ions bonded to axial oxygen atoms [71]. Indeed, in the case of layered $LiMnO_2$, a manganese ion possesses six neighboring oxygen ions with two different Mn–O distances, that is two $Mn-O_{equatorial} = 1.91$ Å and four $Mn-O_{axial} = 2.32$ Å [86].

Therefore, Raman spectroscopy is very effective in identifying closely related structures such as rock-salt (Figure 6.20a) and spinel lithium manganate (Figure 6.20b) phases. This has enabled examination of the local structural changes of lithium manganese oxide upon electrochemical cycling, leading to the conclusion that Mn migration into the interlayer lithium site from the first charge process is irreversible, resulting in the creation of the spinel-like cation ordering (Figure 6.20c–e) [87].

Work is underway in several groups to improve the metastability by admetal doping or by pillaring the layers. In fact less rigid manganese oxide structures, and

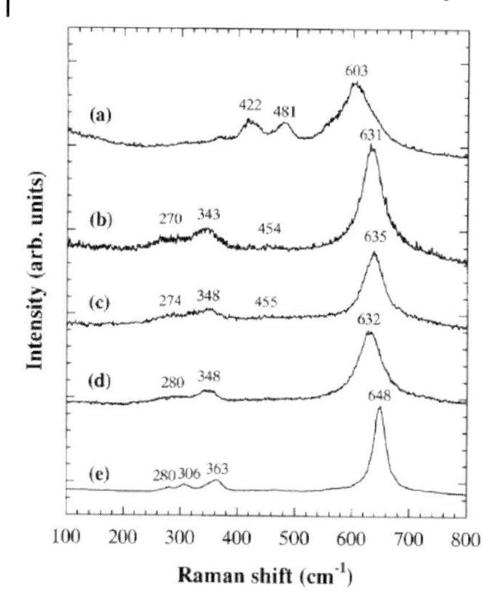

Fig. 6.20 Micro-Raman spectra for (a) monoclinic layered $LiMnO_2$, (b) cubic spinel $LiMn_2O_4$, (c) delithiated $LiMnO_2$, (d) relithiated $LiMnO_2$, and (e) electrochemically cycled $LiMnO_2$ (from [87]).

those not based on a cubic close-packed array of oxygen atoms, are much more likely to remain phase stable upon cycling in a lithium-cell configuration. As an example, Li_xMnO_2 derived from $Na_{0.44}MnO_2$ exhibits a remarkable stability attributed to the double tunnel structure, which cannot easily undergo rearrangement to spinel [88].

In conclusion, the wide variety of Mn–O and Li–Mn–O crystalline phases makes the Raman probe particularly relevant to identify the local signature of each family of compounds. Raman research in this field leads to two different sets of data. Considering the MO system, it is clear that apart from general considerations, there is, up to now, no clear and relevant trend in the data in the literature for a straightforward assignment of the complex and various vibrational features of MnO_2-based compounds. In particular, there is a lack of experimental approach devoted to the study of either chemically lithiated samples or cathode materials under operation. In contrast, for the more attractive spinel system $LiMn_2O_4$ working at 4 V, as well as for the layered $LiMnO_2$ material, detailed spectroscopic data are available with a relevant interpretation of the Raman spectra for the charged and discharged electrodes. Raman spectroscopy constitutes a particularly powerful technique in this field since it affords, in many cases, a clear identification of the phases XRD data analysis is hampered by their high structural similarity. It remains however that this complex system is far from being completely investigated. Other spinel compositions such as $Li_2Mn_3O_7$, $Li_2Mn_4O_9$, and $Li_4Mn_5O_{12}$ require additional efforts, both for obtaining reference Raman spectra and reliable assignments.

6.3.4 V_2O_5

Vanadium pentoxide is an attractive material for application as cathodic material in electrochromic thin film devices and lithium batteries owing to its capacity to accommodate up to three lithium ions per mole of oxide, hence providing a high specific capacity around 450 mA h g^{-1} in the voltage range 4–1.5 V [89–94]. The electrochemical performance being strongly related to the nature and the amplitude of the structural changes induced by the lithium insertion/deinsertion process, a great number of studies have focused on the structural features of V_2O_5 and its lithiated $Li_xV_2O_5$ phases. We present here the specific contribution of Raman spectroscopy to the knowledge of the Li–V_2O_5 system.

6.3.4.1 V_2O_5 Structure

V_2O_5 crystallizes in the *Pmmn* space group, with lattice parameters *a* and *c* of the orthorhombic cell equal to 11.50 and 4.40 Å, respectively. The point symmetry group of V_2O_5 is D_{2h}. The structure of the vanadium–oxygen layers in V_2O_5 is presented in Figure 6.21. It can be seen (Figure 6.21c) that the vanadium atom is located within the oxygen coordination polyhedron VO_5, and is shifted toward the plane formed by four oxygen atoms by a distance of 0.47 Å. Four types of V–O bonds, characterized by their particular bond length, can be distinguished:

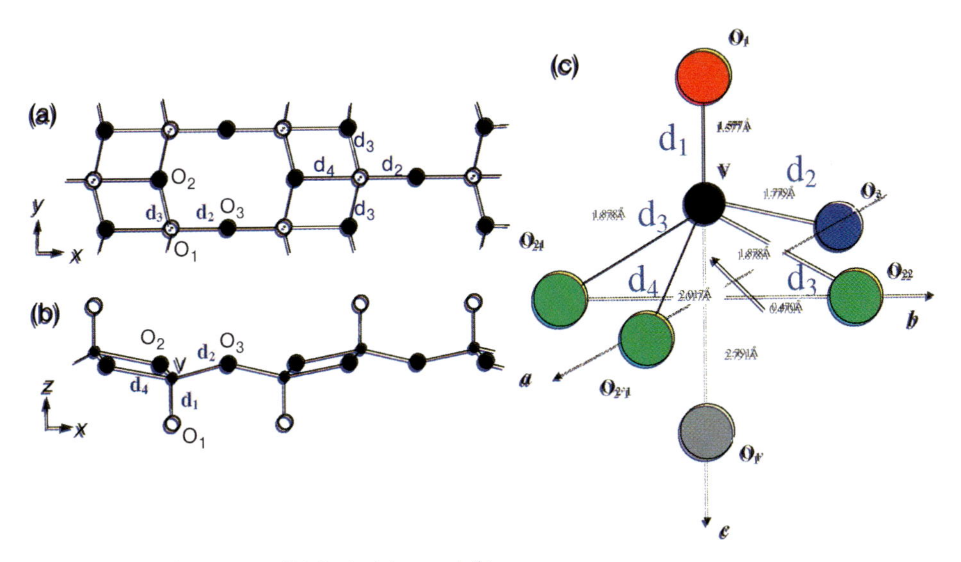

Fig. 6.21 Crystal structure of V_2O_5 in (a) *xy* and (b) *xz*-projections, and (c) local vanadium environments. The V atoms are shown by small black circles and the O_2 and O_3 atoms are shown by large black circles. The O_1 atoms are shown by large open circles.

- the short and strong apical $V{=}O_1$ bonds: $d_1 = 1.577$ Å;
- the bridge $V{-}O_3$ bonds: $d_2 = 1.779$ Å;
- the "ladder step" $V{-}O_2$ bonds (leading to two equivalent intraladder $V{-}O_{21}$ and $V{-}O_{22}$ bonds): $d_3 = 1.878$ Å;
- The interchain $V{-}O_2$ bonds (labeled as $V{-}O_{2'1}$ in Figure 6.21c): $d_4 = 2.017$ Å.

Polarized Raman [95] and IR reflection [96] spectra of the V_2O_5 crystal have been thoroughly studied and interpreted on the basis of the phonon state calculations. More recently, a qualitative characterization of the normal vibrations of the V_2O_5 lattice has also been done in terms of atomic displacement [97]. With all valence V–O bonds being oriented along coordinate axes (Figure 6.21a–b), any bond-stretching mode involves oscillations of particular oxygen atoms along particular Cartesians axes. Moreover, owing to the difference in bond lengths, the spectral lines characteristic for the bond-stretching oscillations of the different V–O bonds can be easily selected. There are four symmetry-equivalent atomic positions per unit cell for the V, O_1, and O_2 atoms and only two for the O_3 atoms, because they lie in the mirror plane perpendicular to the x-axis. All these positions correspond to atoms of the same layer, and all layers, which alternate with each other in z-direction, are equivalent. By using the standard table of characters of irreducible representations of D_{2h} group, 12 symmetric combinations can be built from Cartesian displacements of four equivalent atoms, six of which are IR-active and six others are Raman-active. The Raman-active combinations are shown in Figure 6.22, numbers 1, 2, 3, 4 referring to the four symmetry equivalent atomic positions for a given atom in the unit cell.

It is seen that the x and z-displacements give rise to the A_g, B_{2g} modes while the y-displacements give rise to B_{1g}, B_{3g} modes. Also taking also into account the IR B_{1u}, B_{3u} IR-active modes related to x, z-displacements, respectively, and the A_u, B_{2u} species related to the y–displacements, the vibrational species for V, O_1, and O_2 atomic motions can be represented as:

$$\Gamma(V) = \Gamma(O_1) = \Gamma(O_2) = 2A_g + 2B_{2g} + B_{1g} + B_{3g} + 2B_{1u} + 2B_{3u} + A_u + B_{2u}$$

Because of the special position of the O_3 atoms, the $x(O_3)$ displacements do not contribute to A_g and B_{1u} modes, the $z(O_3)$ displacements do not contribute to B_{2g} and B_{3u} modes, and the $y(O_3)$ displacements do not contribute to B_{1g} and A_u modes. Hence the vibrational species for the motion of this atom can be expressed as

$$\Gamma(O_3) = A_g + B_{2g} + B_{3g} + B_{1u} + B_{3u} + B_{2u}$$

In totality, the optically vibrational modes of V_2O_5 are obtained from the overall contributions of each atom after subtracting the acoustic modes

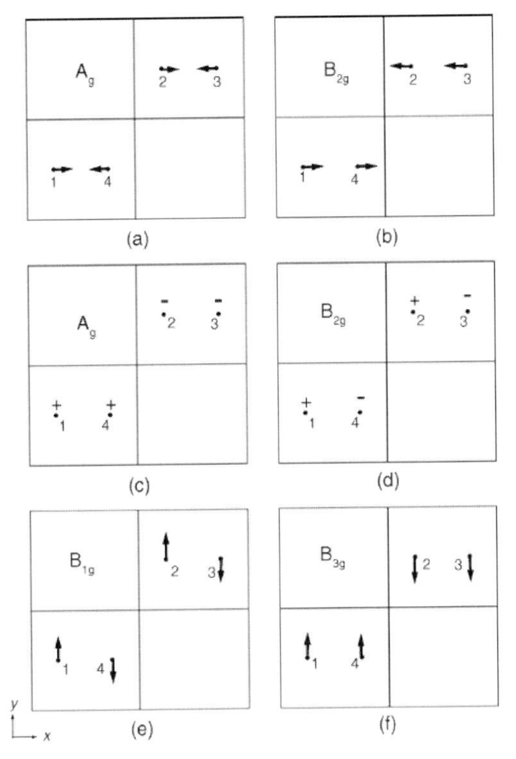

Fig. 6.22 Symmetric atomic displacement combinations for *Pmmn* space group (from [97]).

($\Gamma_{\text{acoustic}} = \text{B}_{1\text{u}} + \text{B}_{2\text{u}} + \text{B}_{3\text{u}}$):

$$\begin{aligned}
\Gamma(\text{V}_2\text{O}_5) &= \sum \Gamma(\text{i}) - \Gamma_{\text{acoustic}} \\
\Gamma(\text{V}_2\text{O}_5) &= [7\text{A}_\text{g} + 7\text{B}_{2\text{g}} + 3\text{B}_{1\text{g}} + 4\text{B}_{3\text{g}} + 7\text{B}_{1\text{u}} + 7\text{B}_{3\text{u}} + 3\text{A}_\text{u} + 4\text{B}_{2\text{u}}] \\
&\quad -[\text{B}_{1\text{u}} + \text{B}_{2\text{u}} + \text{B}_{3\text{u}}] \\
\Gamma(\text{V}_2\text{O}_5) &= 7\text{A}_\text{g} + 7\text{B}_{2\text{g}} + 3\text{B}_{1\text{g}} + 4\text{B}_{3\text{g}} + 6\text{B}_{1\text{u}} + 6\text{B}_{3\text{u}} + 3\text{A}_\text{u} + 3\text{B}_{2\text{u}}
\end{aligned}$$

It follows then that 21 Raman modes are expected for V_2O_5.

The Raman spectrum of V_2O_5 is shown in Figure 6.23. The frequency distribution of the normal vibrations of V_2O_5, with their assignments to particular atomic displacements, are presented in Table 6.4. The modes originated from the same atomic displacement are gathered in the rows of the Table 6.4. The modes of different symmetry are arranged in different columns of this table.

The bond-stretching modes cover the interval of 500–1000 cm^{-1}. First, z-displacements of O_1 atoms give rise to the highest frequency $\nu(d_1)$ mode at 994 cm^{-1}. It corresponds to the in-phase stretching vibration of all apical $V{-}O_1$ bonds.

The Raman-active $\nu(d_1)$ mode of B_{2g} symmetry expected around 976 cm^{-1} was not detected in our spectra. The low intensity of this mode is quite comprehensible. According to general theory of Raman intensities, the main contributions to

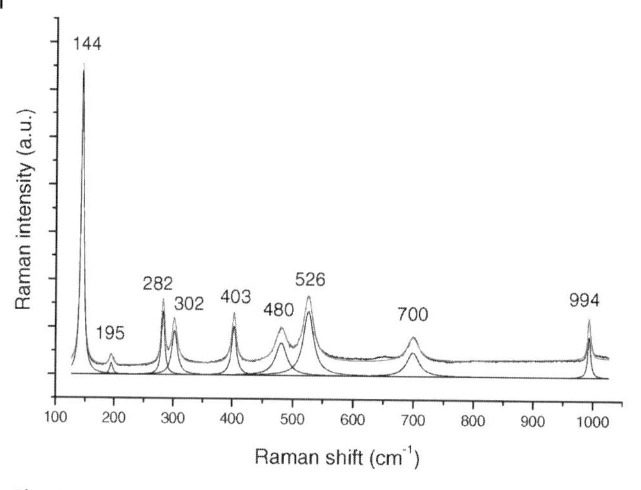

Fig. 6.23 Raman spectrum of V_2O_5, with the deconvolution shown below (from [97]).

Table 6.4 Symmetry and frequency distribution of normal vibrations of the V_2O_5 lattice with their assignment to particular atomic displacements[a] (from 96).

Atomic Displacement	Assignment	A_g	B_{2g}
Z(O1)	$\nu(d_1)$	994	976[a]
X(O3)	$\nu(d_2)$		848[a]
X(O2′)	$\nu(d_4)$	526	502[a]
Z(O3)	δ(V–O3–V)	480	
X(O1)	ρ(V=O1)	403	350[a]
Z(O2)		302	310[a]
X(V)	δ(O2–V–O2)	195	195
Z(V)	δ(O3–V–O2)	104	143[a]
Atomic Displacement	**Assignment**	B_{1g}	B_{3g}
Y(O2)	$\nu(d_3)$	700	700
Y(O1)	ρ(V=O1)	282	282
Y(O3)	δ(O2–V–O2)		220[a]
Y(V)	δ(O3–V–O2)	144	144

[a] Not observed experimentally.

Raman tensor for the bond-stretching modes are determined by derivatives of bond polarizability with respect to the bond lengths. Owing to symmetry of the B_{2g} representation, half of the V–O_1 bonds stretch whereas the other half shorten. So, this mode is permitted for Raman spectra by symmetry but is inactive, owing to its particular microscopic pattern. In [95], this mode was assigned to a very weak Raman peak observed at 976 cm^{-1}.

Next in the frequency scale is the $\nu(d_2)$ mode, which comes from $x(O_3)$ displacements and which corresponds to an antiphase stretching of the $V-O_3$ bonds forming the $V-O_3-V$ bridges. This mode of B_{2g} symmetry, calculated at 848 cm^{-1} [95], was also not detected experimentally. The low Raman intensity of this mode is caused by the fact that the $V-O_3-V$ bridge is pseudocentrosymmetric ($V-O_3-V$ angle is 148°).

Then follows the $\nu(d_3)$ mode (at 700 cm^{-1}) involving the $y(O_2)$ displacements. It corresponds to an antiphase stretching of the $V-O_2$ bonds. The $V-O_2$ bonds (1.88 Å) are longer than the $V-O_3$ bonds (1.78 Å). Correspondingly, frequencies of the O_2-modes are lower than that of the O_3-mode.

The Raman-active mode at around 526 cm^{-1} originates from the $\nu(d_4)$ stretching vibration involving x-displacements of $O_{2'}$ atoms. The A_g mode gives rise to the Raman line observed at 526 cm^{-1}. The B_{2g} line at 502 cm^{-1} was not detected because of its low intensity.

The angle-bending modes cover the interval of 200–500 cm^{-1}. It is more difficult to determine the frequency distribution for these modes because of considerable coupling. The Raman peak at 480 cm^{-1} can be characterized as bending vibrations of the $V-O_3-V$ bridge angle (Figure 6.21). Raman bands in the frequency region between 400–200 cm^{-1} correspond to the modes that involve the x- and y-displacements of O_1 atoms at 403 and 282 cm^{-1}, respectively, and z-displacements of O_{21} and O_{22} atoms at 302 cm^{-1}. These atomic displacements produce the $\delta(O_1-V-O_2)$ and the $\delta(O_1-V-O_3)$ bending deformations. Corresponding modes can be characterized as the $\rho(V{=}O_1)$ bond-rocking oscillations. The $\rho(V{=}O_1)$ in xz-plane, which involves the O_1 atoms oscillations along x-axis, gives rise to the A_g Raman peak at 403 cm^{-1}. The Raman features at around 300 cm^{-1} correspond to the modes involving y-oscillations of O_1 atoms accompanied with z-oscillations of O_2 atoms. Two peaks in the low-frequency region are associated with the modes involving displacements of the V atoms. The line at 195 cm^{-1} comes from A_g and B_{2g} modes with the atoms oscillating along the x-axis, $\delta(O_2-V-O_2)$. The most intense Raman line at 144 cm^{-1}, $\delta(O_3-V-O_2)$, corresponds to a mixture of the signals coming from B_{1g} and B_{3g}. The B_{1g} mode involves the shear motion of the ladders, whereas the B_{2g} consists of rotations of the ladders along their axes. The high intensity of this line reflects the long-range order in the plane of the vanadium–oxygen layers.

6.3.4.2 Structural Features of the $Li_xV_2O_5$ Phases

In spite of numerous studies devoted to the subject, the $Li-V_2O_5$ system is far from being completely elucidated, and this is mainly due to the complex nature of the lithiated phases involved during the lithium insertion/deinsertion process according to the following electrochemical reaction:

$$V_2O_5 + xe^- + xLi^+ \rightleftarrows Li_xV_2O_5 \quad 0 < x \leq 3$$

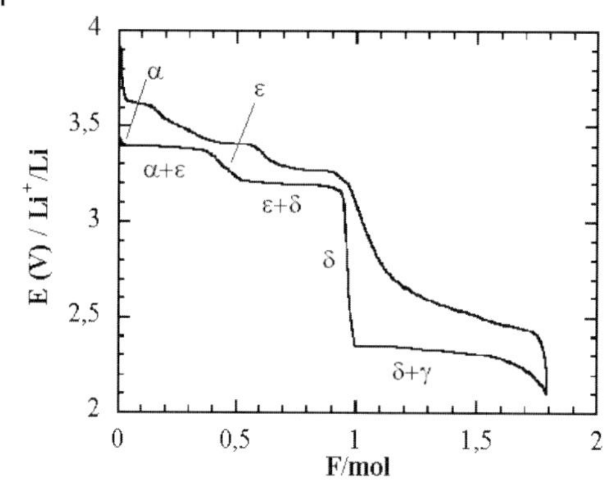

Fig. 6.24 First discharge/charge curve of a composite $Li_xV_2O_5$ electrode showing the different phases electrochemically produced in the $0 < x < 2$ composition range.

Table 6.5 Lattice parameters for α-V_2O_5, ε-$Li_{0.52}V_2O_5$, δ-LiV_2O_5 and γ-LiV_2O_5 phases.

	α-V_2O_5 (*Pmmn*)	ε-Phase (*Pmmn*)	δ-Phase (*Amma*)	γ-Phase (*Pnma*)
a (Å)	11.51	11.38	11.24	9.69
b (Å)	3.56	3.57	3.60	3.60
c (Å)	4.37	4.52	9.91	10.67

The typical discharge/charge curve of a composite V_2O_5 cathode exhibits several voltage plateaus corresponding to well-known phase transitions reported for the bulk $Li_xV_2O_5$ system (Figure 6.24). Depending on the amount of lithium (x) intercalated in V_2O_5, several structural modifications have been reported [89, 92, 9498–104]. The α-, ε-, and δ-$Li_xV_2O_5$ were identified in the $0 < x \leq 1$ composition range. Above 3 V, the α-$Li_xV_2O_5$ phase occurs with $x < 0.1$, whereas the pure ε-phase exists in the range $0.3 < x < 0.7$. Then the pure δ-phase appears with x between 0.9 and 1. The lithium content corresponding to the limit composition of the three solid solutions differs slightly from one report to another. The lattice parameters for the $Li_xV_2O_5$ phases are reported in Table 6.5.

The structure of the vanadium–oxygen layers is rather similar in pure V_2O_5 and in α, ε, and δ phases of $Li_xV_2O_5$ (Figure 6.25), with however an increased puckering of the layers revealed by the decrease in the a parameter in the ε and δ phases. Moreover, the increase in the number of inserted lithium ions between the layers is responsible for the increase in the c-parameter. All these phase transitions are fully reversible in this composition range ($x \leq 1$) and the structure of the pristine V_2O_5 phase is recovered upon deintercalation.

The situation becomes more problematic for lithium contents $x > 1$, the δ-phase being transformed into a γ-one on the third voltage plateau at 2.2 V via an irreversible reconstruction mechanism [92, 94]. The space group of the γ-LiV_2O_5

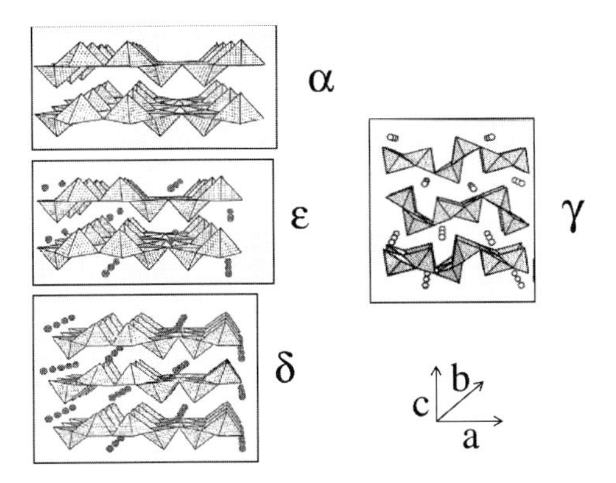

Fig. 6.25 Schematic representation of the structures of the different $Li_xV_2O_5$ phases electrochemically produced in the $0 < x < 2$ composition range (from [92]).

structure is *Pnma*. As in α and δ-phases, this structure still contains the V_2O_5 layers perpendicular to z-axis. However, the structure of the γ-phase differs from that of the δ-phase markedly (Figure 6.25). The Li atoms are shifted in the x-direction from their symmetric positions in the δ-phase. This is accompanied by important deformations of the V_2O_5 layers leading to irreversible symmetry loss and bond breaking. It follows that the γ-phase structure remains stable even after deintercalation of all Li atoms [94], leading to a new metastable γ' variety of V_2O_5 [102]. The intercalation of three lithium ions in V_2O_5 is possible through the formation of a weakly crystallized phase, namely ω-$Li_3V_2O_5$ with tetragonal [103] or cubic [104] symmetry. However some authors put in doubt the Li–V_2O_5 phase diagram for $x > 1$, reporting the appearance of a mixture of Li_xVO_2 and Li_3VO_4 compounds instead of the so-called "γ-" and "ω-phase" [105].

A number of structural investigations on the $Li_xV_2O_5$ system were carried out using different experimental techniques such as XRD [100, 105, 106], X-ray absorption [107], NMR [108], EPR [109], neutron, and electron diffraction [110–112]. Some researchers have also used Raman spectroscopy to characterize powdered [113–117] and thin film [118–122] $Li_xV_2O_5$ crystalline phases. The first vibrational *ex situ* study on chemically prepared lithium vanadium pentoxides has focused on the δ-, ε-, and γ-$Li_{0.95}V_2O_5$ compounds [113]. However, only the γ-Raman spectrum can be considered, the other spectra corresponding to degradation compounds locally produced under the laser beam. *In situ* experiments were performed by the same authors on a composite γ-$Li_{0.95}V_2O_5$ cathode in an operating rechargeable cell with the aim to study the transformation of γ-LiV_2O_5 into γ'-V_2O_5 and ξ-$Li_2V_2O_5$ [114]. Four $Li_xV_2O_5$ samples differing from their intercalation degree during the second discharge process were investigated but only poor spectral changes were detected, the Raman spectra recorded for the $x = 0.1$, 0.93, and 1.15 compositions being nearly identical, especially in the high-frequency range. *In situ* Raman microspectrometry has also been applied to the characterization of a lithium battery involving pure lithium metal at the anode and V_2O_5 powder at the cathode

[115, 116]. Rey *et al.* examined a functioning $Li/P(EO)_{20}Li(SO_2CF_3)_2/V_2O_5$ solid polymer electrolyte lithium cell with *in situ* confocal Raman microspectrometry but could observe only meager spectral changes in the V_2O_5 composite cathode [115]. Indeed *in situ* Raman spectra are seriously hampered in this configuration by fluorescence of the electrolyte. Furthermore, the existence of Li concentration gradient within the composite electrode prevents the obtaining of reference Raman spectra for the $Li_xV_2O_5$ phases. A recent spectroelectrochemical investigation has been carried out on a modified coin cell specially designed to facilitate routine *in situ* Raman measurements [116]; the reported Raman data were, however, limited to the 650–1000 cm^{-1} range, probably due to the high Raman activity of the organic solvent in the high-frequency region. This makes a straightforward assignment very difficult, especially in the absence of XRD data. Nevertheless, the authors tentatively assigned the Raman spectra observed during the first discharge at 2.5 V and 2.1 V to the ε- and γ- phases, respectively. Such approximate assignments are all the more surprising, for it seems that the Raman V_2O_5 fingerprint does not change up to 2.1 V, which is in disagreement with other works [97, 117, 118, 121].

In fact, the literature data related to the Raman features in the Li–V_2O_5 system are very disparate and controversial, essentially due to inappropriate experimental conditions and also because the efficiency of Raman spectroscopy depends on the availability of a relevant band assignment. The spectroscopic data on the lithiated phases have been recently enriched by Raman microspectrometry investigations which afforded reference Raman spectra for the ε-$Li_{0.5}V_2O_5$, δ-LiV_2O_5 and γ-LiV_2O_5 bulk phases chemically prepared [117] and allowed to follow the structural changes occurring in a $Li_xV_2O_5$ thin film under operation [97, 121].

We develop, in the following, the most prominent results extracted from these studies.

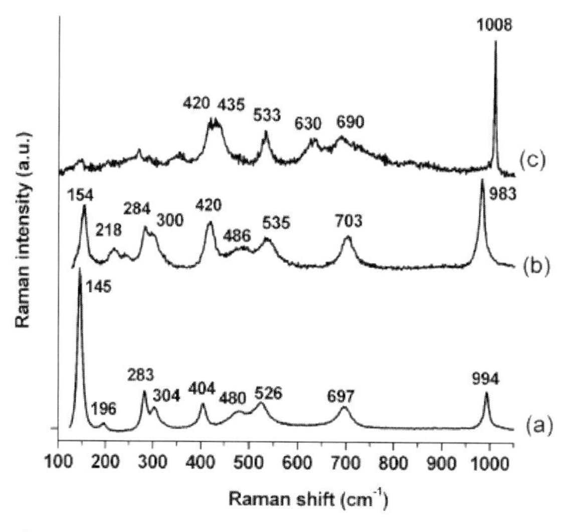

Fig. 6.26 Raman spectra of (a) V_2O_5, (b) ε-$Li_{0.52}V_2O_5$, and (c) δ-LiV_2O_5 (from [117]).

Reference Raman Spectra for the Different $Li_xV_2O_5$ Phases Figure 6.26 shows the Raman spectra obtained for the ε-$Li_{0.5}V_2O_5$ and δ-LiV_2O_5 phases prepared through soft chemistry reduction reactions according to the procedure described in [117], and compared to that of V_2O_5.

The Raman spectrum of ε-$Li_{0.52}V_2O_5$ exhibits several spectroscopic changes: the intensity of the translational mode is strongly quenched and its wavenumber is shifted from 145 to 154 cm^{-1}. Several modes in the 200–700 cm^{-1} range are also shifted toward higher wavenumber: 196 to 218 cm^{-1}, 404 to 420 cm^{-1}, 480 to 486 cm^{-1}, 526 to 535 cm^{-1}, and 697 to 703 cm^{-1}. Conversely, the V=O stretching mode along the c-axis decreases in frequency from 994 to 983 cm^{-1}. This shift in frequency has been shown to be consistent with the lengthening of the V=O bond from 1.58 Å for the α-phase to 1.6 Å for the ε-phase.

The Raman spectrum of δ-LiV_2O_5 can also be compared with that of V_2O_5 (Figure 6.26). Low intensity of the bands in the low-frequency part of Raman spectrum lines as well as the broadening of the bands over the entire spectrum of the δ-phase indicates that this phase is less ordered than the α- and ε-lattices. It is seen also that the same sole line, corresponding to the V=O stretching mode along the c-axis, dominates in the high-frequency part of the Raman spectra. In the spectrum of the δ-phase, this line is narrow and shifted up to 1008 cm^{-1} with respect to 994 cm^{-1}, typically observed for the α-phase. This line can serve as a spectral fingerprint of the δ-phase. Other marked distinctions of Raman spectrum of this phase are the presence of the peak at 630 cm^{-1} and the disappearance of the peak at 480 cm^{-1}. The Raman features of the δ-phase reflect the particularity of this structure. Indeed, the O_1–O_1 distances in the δ-phase (2.77 Å) are markedly shorter than in the α- and ε-phases (3 Å). The relatively high wavenumber of the V=O stretching mode could reflect the steric V=O repulsion, which becomes very strong in the δ-phase because of pronounced layer puckering.

Raman spectroscopy has allowed to investigate the $\delta \rightarrow \varepsilon \rightarrow \gamma$ phase transitions during appropriate heat treatment [117]. XRD measurements showed that the $\delta \rightarrow \varepsilon$ phase transition is initiated around 100 °C and that $\varepsilon \rightarrow \gamma$ transformation proceeds from 135 °C. The $\delta \rightarrow \varepsilon$ transformation ends at 170 °C, the ε- and γ-phases being simultaneously present. Finally at 250 °C, the complete transformation of ε- into γ-LiV_2O_5 is achieved. These results are consistent with those recently reported using a synchrotron X-ray powder analysis [123].

Hence, the thermal treatment of the δ-phase at 250 °C allows to obtain the pure γ-phase LiV_2O_5, which exhibits the Raman spectrum shown in Figure 6.27 and lattice parameters $a = 9.69$ Å, $b = 3.60$ Å, and $c = 10.67$ Å in good agreement with [100–102].

The space group of the γ-LiV_2O_5 crystal is $Pnma$ (D_{2h}^{16}). As in the α- and δ-phases, V_2O_5 layers are perpendicular to the z-axis, and built up of V–O_2 ladders connected by V–O–V bridges. Likewise, in δ-phase, the layers are alternatively shifted on the vector $\mathbf{b}/2$ and the unit cell of this structure is doubled in the z-direction. However the structure of the γ-phase differs from the structure of the δ-phase markedly. The Li atoms are shifted in the x-direction from their symmetric positions in the δ-phase. This is accompanied by important deformations of the V_2O_5 layers. Each

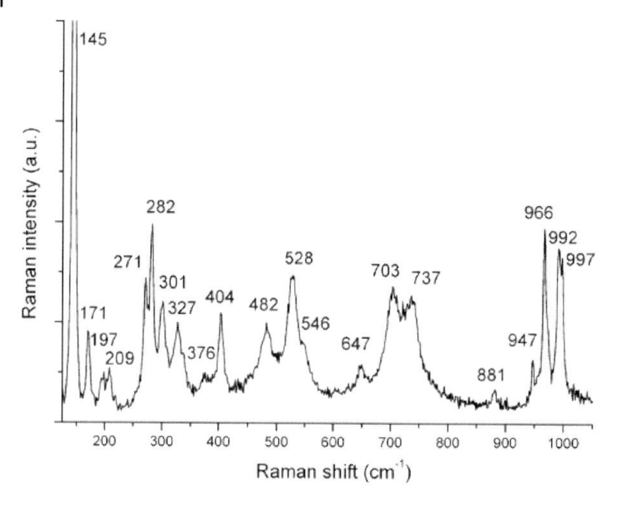

Fig. 6.27 Raman spectrum of γ-LiV_2O_5 (from [117]).

Table 6.6 Raman wavenumbers (in cm^{-1}) for Li_xTiO_2. s: strong; m: medium; w: weak; b: broad (from 125).

x = 0	x = 0.15	x = 0.3	x = 0.5
144s	144s	144s	144s
		166w	166 m
198	198	233vw	233vw
		320–335b	320–335b
		357	357
398	397	397s	397s
	450w	450b	450b
518	518	525	525
		559	559
639	637	637	634
	848	847	847
		897	894
934	930		

second ladder undergoes a strong transformation: it rotates along y-axis and the O atoms of neighboring vanadyl groups exchange their positions. As a result of such a deformation, two nonequivalent V positions appear and the $V1-O_3-V2$ bridges become strongly nonsymmetric.

The Raman wavenumbers observed for the γ-LiV_2O_5 polycrystalline powder are in good agreement with those previously measured for a LiV_2O_5 single crystal [124]. Furthermore, on comparing the Raman features of γ-LiV_2O_5 and V_2O_5 (Table 6.6), the following conclusions can be drawn: first, all the main spectral features observed in V_2O_5 can be clearly discriminated in the spectrum of γ-LiV_2O_5. Second, some bands, which have a singlet form in the Raman spectrum of V_2O_5, ε- and δ-V_2O_5,

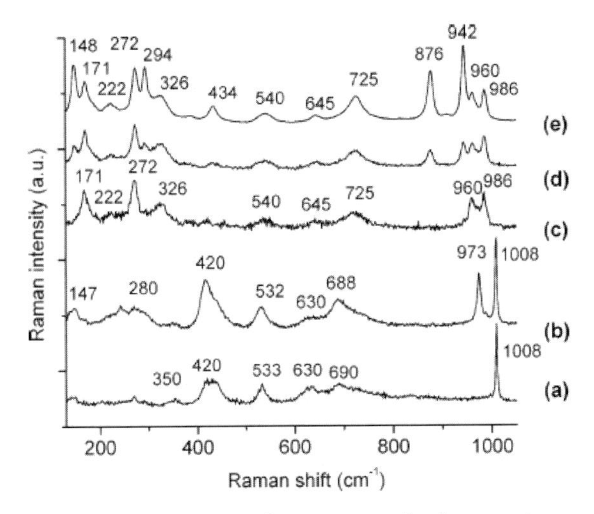

Fig. 6.28 Raman spectra of δ-LiV_2O_5 powder illuminated with (a) laser power of 0.2 mW; (b) laser power of 0.6 mW; (c) laser power of 0.8 mW, integration time: 20 seconds; (d) laser power of 0.8 mW, integration time: 90 seconds; (e) laser power of 0.8 mW, integration time: 900 seconds (from [117]).

are clearly twofold split in the spectrum of γ-LiV_2O_5. Third, some new spectral features not seen in the Raman spectrum of V_2O_5 can be detected in the spectrum of γ-LiV_2O_5 (bands at 209, 376, 546, 647, 881, and 966 cm^{-1}). At least two factors can account for the distinction between the spectra of these crystals. First, the nonequivalent character of the ladders in the lattice of γ-LiV_2O_{5s}, inducing two kinds of vanadium environments, must lead to the twofold splitting. Second, the Li-atom oscillations may couple with some modes of the V_2O_5 lattice. Our lattice dynamics calculations actually in progress confirm this suggestion, especially for the high-frequency modes.

A very interesting point that has to be outlined is the possibility for the Raman microprobe to provoke *in situ* phase transitions, solely by increasing the power of the incident laser beam illuminating the sample. This effect is well illustrated in Figure 6.28, where the spectral changes observed are related to the successive $\delta \rightarrow \varepsilon \rightarrow \gamma$ phase transitions. It is noteworthy that the γ-phase formation is achieved at a laser power less than 1 mW, as a result of the great instability of the δ-phase. This peculiarity is at the origin of numerous misinterpretations in the literature, the Raman response being strongly affected by this severe experimental artifact.

Application to the Raman Microspectrometry Study of Electrochemical Lithium Intercalation into Sputtered Crystalline V_2O_5 Thin Films Several studies carried out on polycrystalline films seem to show that the structural features of the lithiated $Li_xV_2O_5$ phases can be significantly different from that known for the bulk material

[106, 125–129]. *In situ* and *ex situ* XRD experiments performed on sol–gel [106, 129] and sputtered [125, 126] $Li_xV_2O_5$ films have shown the existence of an unexpected elongated *c*-axis for $x = 0.8$ as well as the nonappearance of the δ-phase usually described for $x = 1$ in bulk samples. Recent works reported attractive electrochemical properties for crystallized oriented V_2O_5 films prepared by magnetron reactive sputtering without any heat treatment [130, 131]. Once again, a different structural response of the thin film electrode was observed from XRD experiments when Li insertion proceeds in V_2O_5 ($0 < x < 1$ for x in $Li_xV_2O_5$) since the emergence of the δ-phase does not seem to take place as in the bulk material from $x = 0.6$ [94].

This specific behavior reported for polycrystalline V_2O_5 films requires deeper structural investigation. However, few Raman spectroscopy studies have been carried out on polycrystalline V_2O_5 films [118–122]. Fast amorphization of the sputtered oxide has been observed in. [118], the V_2O_5 lines vanishing nearly completely from the early lithium content, whereas in [119] practically no change seems to take place in the PLD film in the $0 < x < 1$ composition range. Moreover only two compositions are reported in [119], with their Raman spectra simply assigned on the basis of the phase diagram described for the bulk material.

Recent results obtained from a Raman spectroscopic analysis of electrochemically lithiated sputtered $Li_xV_2O_5$ thin films are described in [97].

The typical galvanostatic discharge/charge curves of the 0.6-μm sputtered thin film is reported in Figure 6.29 [97]. Two well-defined insertion steps located at about 3.4 and 3.2 V are observed as expected for the crystalline form of V_2O_5. The specific capacity obtained at C/15 rate is 22 μA h cm^{-2}, which corresponds to the accommodation of 0.94 Li^+ ion per mole of oxide. This insertion reaction is reversible as shown from the lack of polarization and the efficiency of 100% in the charge process. No influence of the current density on the discharge potential and capacity is seen from C/15 up to C/5 rate and the specific capacity decreases by only ≈18% when the C rate increases from C/15 to 8 C. Moreover, the polarization

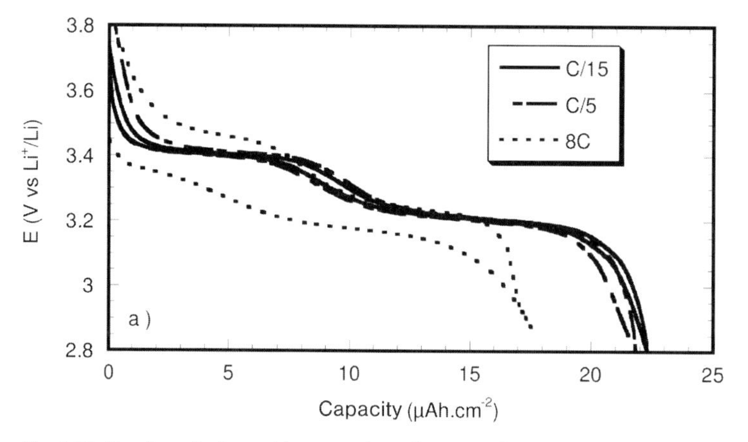

Fig. 6.29 The first discharge/charge cycles of a V_2O_5 thin film (thickness 0.6 μm) at different C rates (from [97]).

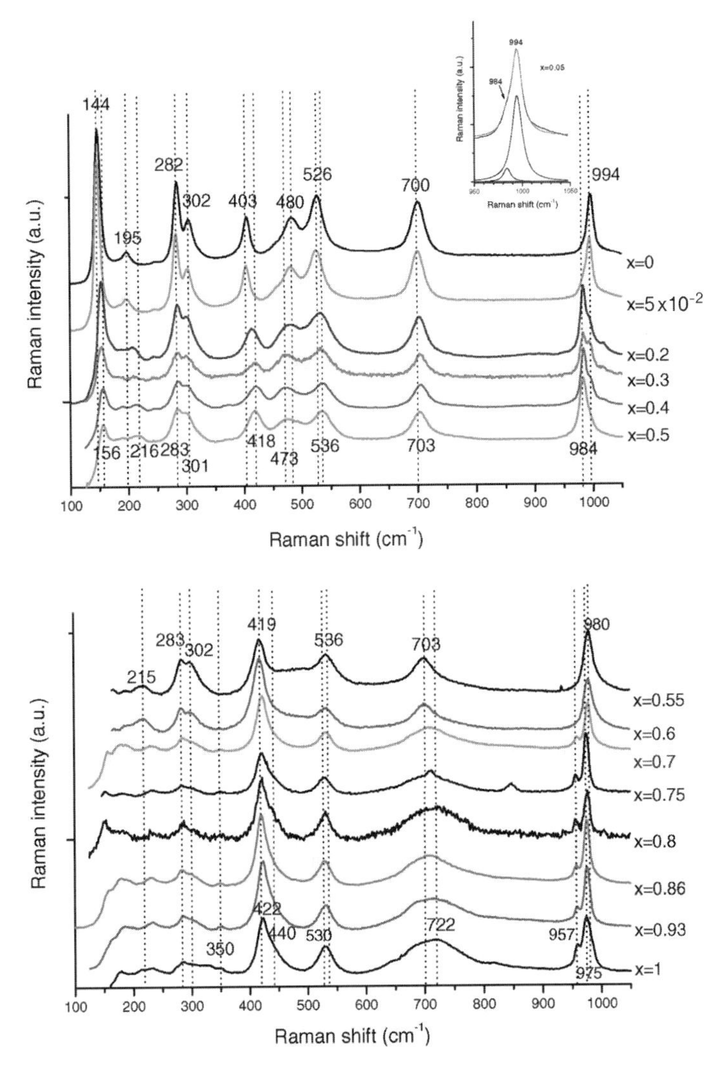

Fig. 6.30 Raman spectra of a galvanostatically lithiated $Li_xV_2O_5$ thin film, $0 \leq x \leq 1$ (from [97]).

observed at high rate never exceeds 150 mV, which suggests a high level of kinetics of lithium transport.

The evolution of the structural response of the V_2O_5 film as lithium accommodation proceeds has been investigated by *ex situ* XRD and Raman microspectrometry [97, 131]. Figure 6.30 pertains to Raman spectra with different x values in $Li_xV_2O_5$ thin films.

Examining the Raman spectra collected in the composition range $0 \leq x \leq 0.5$, several observations can be made. First, the intensity of the translational mode at low wavenumber is progressively quenched and its wavenumber is shifted from

144 to 156 cm^{-1} from $x = 0$ to $x = 0.5$. Several modes in the 200–700 cm^{-1} range are also shifted toward higher wavenumbers: 195 to 216 cm^{-1}, 403 to 418 cm^{-1}, 526 to 536 cm^{-1} and 700 to 703 cm^{-1}. Conversely, the 480 cm^{-1} band shifts progressively to 473 cm^{-1} for $x = 0.5$. In the vanadyl stretching frequency range, a new component at 984 cm^{-1} appears from the first lithium content value, and continuously increases in intensity at the expense of the 994 cm^{-1} band as x increases from 0.05 to 0.5.

For $0.5 < x \leq 1$, in frequency ranges lower than 400 cm^{-1}, a progressive loss of intensity is observed, indicating a significant increase of the local disorder. Several bands progressively disappear, in particular, the low-frequency modes at 156 and 216 cm^{-1}, whereas the band at 473 cm^{-1} completely vanishes from $x = 0.5$ to $x = 0.55$. The 419 cm^{-1} band progressively shifts to 422 cm^{-1} for $x = 1$ and becomes asymmetric owing to the emergence of a new component at 440 cm^{-1} from $x = 0.7$. The 536 cm^{-1} band progressively shifts to 530 cm^{-1}, whereas the 703 cm^{-1} line becomes broader and centered toward higher wavenumber of 722 cm^{-1}. In the high-wavenumber region, the vanadyl stretching mode shifts toward lower wavenumber, from 984 to 975 cm^{-1}, in the lithium content range $0.5 < x < 0.7$ and then remains constant up to $x = 1$. For $x \geq 0.7$, several new Raman features are detected at 350, 440, and 957 cm^{-1}.

It is clear that two main domains of variation are observed: For $x < 0.5$, the Raman spectra exhibit the same general features, in particular, the number of bands is conserved. The observed intensity and frequency changes are directly related to the existence of a solid solution of Li in V_2O_5 corresponding to the ε-phase. They vary to a limited extent because of the similarity of the α and ε-V_2O_5 related structures. This finding is supported by the corresponding XRD data, which clearly show a continuous increase of the c-parameter and a correlated decrease of the a-parameter [131]. The shift in frequency from 994 to 984 cm^{-1} for the apical $V-O_1$ stretching mode is consistent with the lengthening of the V=O bond from 1.58 Å for the α-phase to 1.60 Å for the ε-$Li_{0.5}V_2O_5$ phase. Conversely, continuous displacements toward higher wavenumbers are observed when x increases up to 0.5 for the 403 and 526 cm^{-1} modes, which shift respectively to 418 and 535 cm^{-1}. As these modes are all coming from oxygen displacements along the a-axis, $x(O_1)$, and $x(O_{2'1})$, respectively, it is plausible that the stiffening of the lattice observed here can be related to the x dependence of the a parameter in the $Li_xV_2O_5$ film, which has been found to decrease from 11.51 to 11.35 Å when the lithium content increases from 0 to 0.55 [131]. This result can be correlated to the existence of a compressive stress in V_2O_5-sputtered films evidenced by beam deflectometry [126] and varying in the same manner as the a-parameter, that is, increasing upon lithiation in the $0 < x < 0.5$ domain. For $x > 0.6$, the relative stability of these $x(O_1)$ and $x(O_{2'1})$-related modes can be correlated with that of the a-parameter [131].

For $0.5 < x \leq 1$, the spectral variations are less marked. The general features of the ε-phase are conserved. The continuous blue shift observed for the apical $V-O_1$ stretching mode from 985 to 975 cm^{-1} reflects a progressive weakening of the $V-O_1$ bond, which is probably related to the linear expansion of the interlayer spacing observed for the ε-phase, from 4.53 to 4.68 Å in the lithium composition range 0.5–0.95 [131].

These results are consistent with XRD data [125, 126, 129, 131], which report that the structural response of such thin films consists of a solid solution behavior instead of the successive phase transitions usually described for the bulk material [94] and corresponding to the α-, ε-, and δ-phases with the following compositions ranges $0 \leq x \leq 0.15$, $0.3 \leq x \leq 0.7$, and $0.9 \leq x \leq 1$, respectively, separated by two-phase regions. It is clear that the Raman fingerprint of the ε-phase appears here from the first lithium content value and is retained up to $x = 1$. At that instant, the characteristic Raman features of the δ-phase are never detected, in particular, the intense band at 1008 cm^{-1} [117].

The spectral variations reported here are related to the local distortions occurring in the ε-phase, which is able, in the thin film configuration, to accommodate up to 1 Li per mole of oxide. Although the lithium positions in the ε-phase remain to be determined accurately, a unique type of lithium site has been described for the whole ε-phase domain. It consists of elongated cubo-octahedra joining common faces, the symmetry of such sites remaining unchanged with x [132]. However, it has been reported that increasing the amount of lithium in the ε-phase leads to a lithium-rich phase (called ε') in which the lithium environment is modified [75, 100, 112, 132, 133]: for lithium amounts <0.5 the number of unoccupied cubo-octahedra exceeds the number of occupied octahedral, whereas above $x = 0.5$, neighbouring cubo-octahedra have to be occupied. This induces a change in the interatomic Li–Li distances: the shortest Li–Li distance decreases from 7.2 Å for $x < 0.5$ to 3.6 Å for $x > 0.5$ [100].

In contrast to the lithiated bulk material for which the $\varepsilon \rightarrow \delta$ phase transition occurs from $x = 0.6$ [94], strong repulsive Li^+–Li^+ forces probably prevail in the film for the ε-rich phase, but these are allowed without giving rise to significant structural modification of the framework. This probably explains the Raman band splitting at 975 and 957 cm^{-1} as observed for the highest lithium uptakes ($x \geq 0.7$). The existence of these two stretching vanadyl modes may indicate the presence of two kinds of vanadium atoms. This can be explained by a change in electron localization for a higher depth of discharge.

The Raman spectrum of the reoxidized compound clearly exhibits the characteristic Raman features of the pure V_2O_5 lattice. The recovery of the structure is in good agreement with the excellent electrochemical reversibility evidenced in the potential range 3.8–2.8 V.

As shown in Figure 6.31, *in situ* preliminary Raman microspectrometry experiments performed on sputtered thin films [134] confirm these whole Raman features.

For $x = 1.1$, in atomic layer deposited thin films [121], Raman spectra reveal a mixture of δ-phase (band at 1004 cm^{-1}) and an ε-rich phase. This difference in the structural response compared to the sputtered films probably stems from a lower crystallinity and nanosize effect in the case of sputtered films.

In conclusion, the Raman investigation of the electrochemical $Li_xV_2O_5$ system has been allowed in the first step through a rigorous and careful establishing

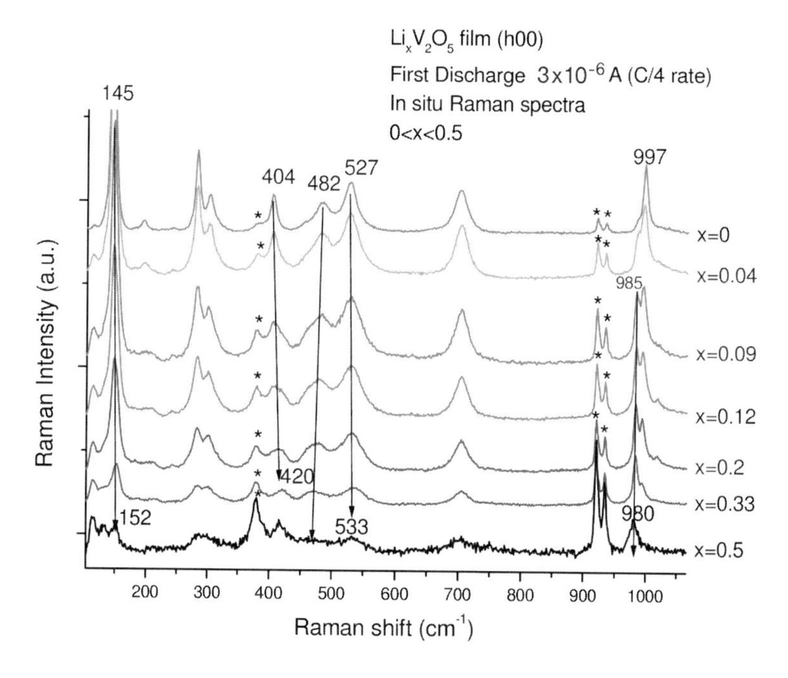

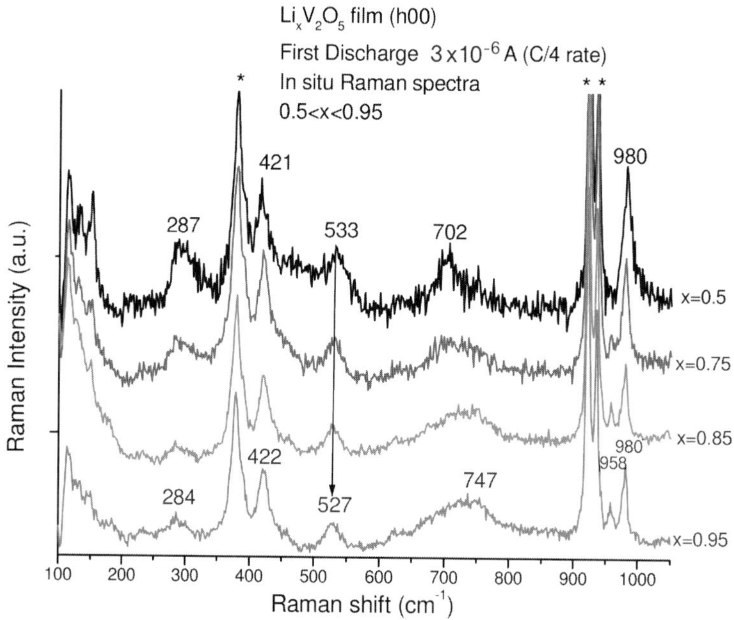

Fig. 6.31 *In situ* Raman spectra of a galvanostatically lithiated $Li_xV_2O_5$ thin film, $0 \leq x \leq 1$; asterisk, PC/$LiClO_4$ (from [134]).

of reference Raman data on the various lithiated phases involved in the lithium insertion mechanism governing the electrochemical behavior of the V_2O_5 positive electrode in secondary lithium batteries.

In the second step, Raman microspectrometry has been applied for the first time to investigate the local structure of sputtered V_2O_5 thin films, 600-nm thick, after electrochemical lithiation in the potential range 3.8–2.8 V. A qualitative characterization of the normal vibrations of the V_2O_5 lattice with their assignments to particular atomic displacements has allowed to discuss the changes occurring in the Raman spectra of the $Li_xV_2O_5$ films for $0 < x \leq 1$. In particular, the vanadyl stretching mode located in the high-wavenumber region (994 cm^{-1}) is of utmost importance since it constitutes a Raman band highly sensitive to Li interaction, which allows a meaningful and relevant discussion of the experimental data. Three main results are obtained from the Raman features of the $Li_xV_2O_5$ films:

1. For $x < 0.5$, all the Raman bands are consistent with the existence of the ε-phase as confirmed by the Raman spectra of the electrochemical and chemical phases that completely superimpose with a vanadyl mode at 984 cm^{-1}.
2. For $0.5 < x \leq 1$, a continuous shift of the apical stretching mode from 984 to 975 cm^{-1} and the emergence of a new vanadyl band at 957 cm^{-1} from $x = 0.7$ has been ascribed to Li insertion into the ε-phase with a continuous, expanded interlayer distance. This finding is consistent with the occurrence of two kinds of Li sites reported in literature for the Li rich ε-phase called ε'.
3. The presence of the δ-phase is never detected.

These Raman findings, in good accord with our XRD data on the sputtered V_2O_5 thin film oxide, indicate the structural response of such films consists in a single-phase behavior that strongly contrasts with the more pronounced structural changes described for bulk samples with the emergence of one- and two-phase domains as well as the formation of the δ phase from $x = 0.6$. It is thought that this difference mainly stems from a nanosize effect, the stacking of platelets occurring over only 40 nm $\times$ 200 nm. Hence a more homogeneous Li insertion reaction can take place in the present films, avoiding local overlithiation phenomenon and allowing a better relaxation of the host lattice and a lower structural stress. Actually, the degree of crystallinity and the morphology of the thin films strongly contribute to the structural changes as Li accommodation proceeds.

6.3.5 Titanium Dioxide

We focus now on TiO_2, which is an interesting candidate for electrode applications in photoelectrochemical solar cells [135] and rechargeable lithium batteries [136]. The tetragonal anatase polymorph of TiO_2 is of special interest, owing to its ability to store a significant amount of Li as well as its convenient insertion potential around 1.5 V. First *ex situ* XRD investigation of the anatase Li_xTiO_2 system reported the existence of a tetragonal-to-cubic phase transition [137], which has been proved to be irrelevant by further neutron diffraction [138, 139] and *in situ* powder XRD [140] investigations that rather invoked the existence of a phase transformation toward a Li-rich phase, here referred to as *lithium titanate* (*LT*) with orthorhombic symmetry.

As far as is known from the literature, spectroscopic data for the anatase bulk have been limited because of the poor availability of single crystals [141–143]. Nevertheless, the two Raman spectroscopic studies [141, 142] and one IR study [143] made it possible to definitely determine all the zone-center vibrational frequencies. Considering the anatase Li_xTiO_2 system, Raman data are limited to a few studies [144–148]. The main contributions have been recently reported in [146–148] on composite powdered electrodes.

We report here recent Raman data [146, 148] on Li_xTiO_2 composite electrodes ($0 \leq x \leq 0.6$) with the pure anatase powder being synthesized via the sol–gel process according to the procedure reported in. [148]. Structural changes and reversibility are discussed in relation to structural data drawn from XRD experiments. These results illustrate the usefulness interest of lattice dynamics simulation to perform a quantitative and thorough analysis of all the Raman features.

TiO_2 anatase has a tetragonal symmetry (space group D_{4h}^{19}, *I41/amd*, number 141) [149]. The unit-cell parameters are $a = b = 3.8$ Å; $c = 9.61$ Å. It is known that zigzag chains are formed in the close-packed planes owing to anion stacking and the half occupancy of the octahedral sites by titanium ions. Each $[TiO_6]$ octahedron shares two adjacent edges with other $[TiO_6]$ units in the (b, c) plane to form infinite, planar double chains parallel to a and b. These double chains share corners with identical chains above and below and are shifted along the c-axis with respect to each other. Finally, all $[TiO_6]$ octahedral share four edges and all oxygen ions are bonded to three titanium ions.

The typical galvanostatic curve of TiO_2 anatase is illustrated in Figure 6.32. At C/20 rate, one main discharge process located at 1.75 V is evidenced and corresponds to the insertion of 0.56 Li ion per mole of oxide according to the

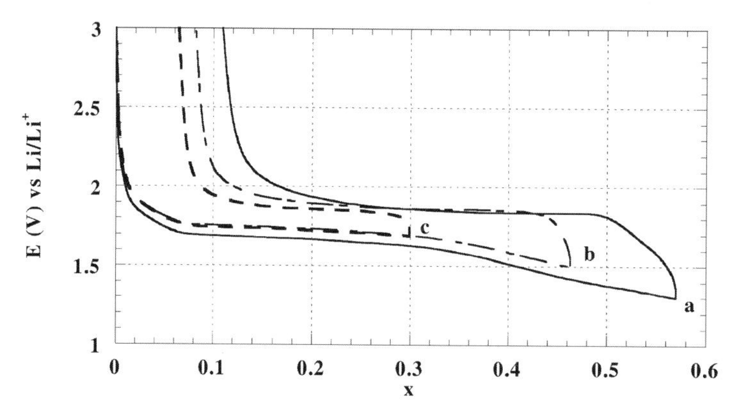

Fig. 6.32 First discharge/charge curve of TiO_2 anatase as a function of the depth of discharge in 1 M $LiClO_4$/PC (C/20). (a) $x = 0.56$; (b) $x = 0.45$; and (c) $x = 0.3$ (from [148]).

reaction:

$$TiO_2 + xe^- + xLi^+ + \Longleftrightarrow Li_xTiO_2$$

The charge process occurs at a higher voltage around 1.9 V. The efficiency (Q_{ox}/Q_{red}) of the intercalation process in the anatase is around 80%. These electrochemical data show that the extraction of lithium is only partially quantitative whatever the depth of discharge is.

Several authors have shown that chemical [138, 139] or electrochemical [137, 140] lithium insertion in the anatase structure induces from the first lithium contents a phase transformation leading to lithiated titanate $Li_{0.5}TiO_2$. Ohzuku *et al* reported from XRD powder experiments a cubic symmetry for the $Li_{0.5}TiO_2$ compound [137]. However, recent works based on XRD [140] and either neutron-diffraction data [138, 139] reported for LT $Li_{0.5}TiO_2$ an orthorhombic structure indexed with *Imma* space group.

In fact, the initial anatase structure with tetragonal symmetry and the new orthorhombic lithiated titanate structure have very close diffraction patterns. The orthorhombic lithiated titanate consists of the edge-sharing TiO_6 octahedra giving rise to an open structural framework. The lithium ions are located between the TiO_6 octahedra within the octahedral voids. The overall orthorhombic distortion of the atomic positions in the change from anatase to lithium titanate is small. XRD studies [138, 139, 148] showed the following changes in the unit cell dimensions of the anatase structure upon lithiation: from $a = b = 3.80$ Å; $c = 9.61$ Å to $a = 3.81$ Å, $b = 4.07$ Å, $c = 9.04$ Å. In lithium titanate, the a- and b-axes become different in length by about 7%, mainly increasing in the b-direction and decreasing in the c-direction by 9%. Examination of the interatomic distances reveals that the anatase structure is deformed upon lithiation [139]. This entails a reduction of symmetry by lost of C_4-axis on going from the pristine TiO_2 tetragonal $I41/amd$ to the $Li_{0.5}TiO_2$ *Imma* structures.

According to Cava *et al.* [138], the Li ions in the orthorhombic phase are randomly located in about half of the available interstitial octahedral 4e sites. This suggestion was also deduced from theoretical calculations [150]. Hence, only a maximum lithium uptake of 0.5–0.6 Li^+ ions per mole of TiO_2 can be inserted in the lithiated titanate structure at room temperature. This agrees with recent quantitative NMR results [151] and with the $Li_{0.56}TiO_2$ composition of the electrochemically formed orthorhombic phase found in [148].

The Raman spectra observed for different Li_xTiO_2 ($0 \leq x \leq 0.6$) compositions are shown in Figure 6.33 [148]. Corresponding wavenumbers of the Raman bands are reported in Table 6.6.

At low Li concentration (up to $x = 0.15$), the Raman spectrum is very close to that of the pure titanium oxide. Only a complex band structure in the high-frequency region seems to sign the presence of a new system. Along with increasing x up to $x = 0.3$ and $x = 0.5$, several spectroscopic changes are observed:

- The main bands of pure anatase (144, 398, 518, 639 cm^{-1}) are well detected. However, the band maximum position of the band at 518 cm^{-1} shifts to 525 cm^{-1} and that of the band at 639 cm^{-1} slightly shifts to 634 cm^{-1}.

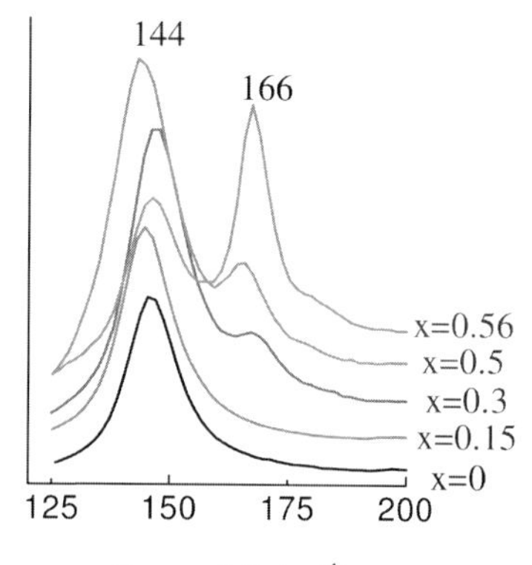

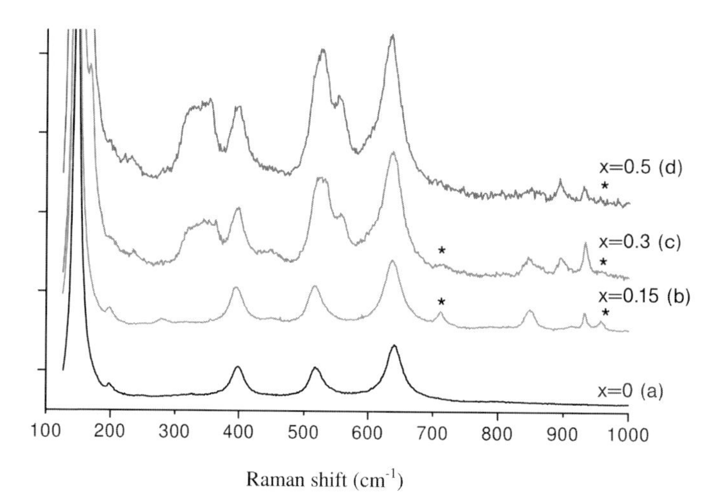

Fig. 6.33 Raman spectra of electrochemically lithiated Li_xTiO_2 ($x = 0$; 0.15; 0.3; 0.5; 0.56) (from [148]).

- Several new bands appear – a complex band structure in the high-wavenumber region, a band at 559 cm^{-1} whose intensity increases with x, another broad band structure around 320–330 cm^{-1} with a peak at 357 cm^{-1} and two bands in the low-wavenumber region at 166 and 233 cm^{-1}.

According to the factor group analysis, the 15 optical modes of TiO_2 anatase have the irreducible representation $1A_{1g} + 1A_{2u} + 2B_{1g} + 1B_{2u} + 3E_g + 2E_u$ [141]. The A_{1g}, B_{1g}, and E_g modes are Raman active and thus, six fundamental transitions are expected in the Raman spectrum of anatase:

$$\Gamma_{Raman} = 1A_{1g} + 3B_{1g} + 3E_g$$

The typical Raman fingerprint of anatase, shown in Figure 6.33 curve (a), exhibits five Raman bands at 144, 198, 398, 518, and 639 cm^{-1}. The band at 518 cm^{-1} is resolved into two components at 507 cm^{-1} and 519 cm^{-1} only at 73 K [141].

Upon increasing Li concentration in the TiO_2 host lattice, the evolution of the Raman spectra clearly indicates the emergence of a new phase. For low Li content ($x = 0.15$), the local symmetry of pure anatase is kept, as shown by curve (b) in Figure 6.33. Upon increasing Li concentration, the modification of the original anatase Raman spectrum indicates a break of the local symmetry, probably owing to the appearance of the lithiated phase $Li_{0.5}TiO_2$ with an orthorhombic symmetry (*Imma*). This new phase exhibits several typical Raman bands with wavenumbers cited in Table 6.6. A good agreement exists with the experimental data recently reported from an *in situ* Raman study on nanosized TiO_2 anatase powders [147]. A Raman study on TiO_2 anatase thin films reported several broad Raman bands, located at 176, 224, 316, 405, 531, and 634 cm^{-1} [144]. However, the vibrations of the propylene carbonate–$LiClO_4$ electrolyte and the ITO substrate hinder most of the Raman spectrum and the authors did not assign the observed new features to the $Li_{0.5}TiO_2$ phase.

The tetragonal-to-orthorhombic structural transition implies a reduction of symmetry of TiO_2 lattice from D_{4h} to D_{2h}. As a consequence, the A_{1g} and B_{1g} modes of tetragonal anatase lattice transform into A_g modes and E_g modes split into $B_{2g}(x)$ and $B_{3g}(y)$ modes of the orthorhombic LT lattice. It comes out that nine lattice modes are predicted for orthorhombic $Li_{0.5}TiO_2$:

$$\Gamma_{\text{Raman}}(\text{LT-lattice}) = 3A_g + 3B_{2g} + 3B_{3g}$$

Owing to small extent of the structural distortion, a fairly moderate perturbation of the phonon states of the TiO_2 lattice can be expected. However, the coupling between TiO_2 lattice modes and vibrations of the Li atoms will cause considerable changes in the Raman spectrum, giving rise to the emergence of new modes assigned to lithium vibrations and band splitting due to perturbated lattice modes [146].

Interpretation of the complex Raman spectrum of lithiated anatase requires the knowledge of Li positions. The NMR [151] and quasi-elastic neutron scattering data [139] suggest that the Li atoms have several positions within octahedral interstices and dynamic hopping between them occurs. The data of. [139] were interpreted with triply split Li positions, namely Li1, Li2, and Li3, one of which (Li2) was found considerably displaced from the center of the interstice with the shortest Li–O_{ax} distance of 1.67 Å (Figure 6.34).

Lattice dynamics simulations have been performed to determine the frequency region of vibrations of Li atoms within TiO_2 host lattice [146]. Within these simulations, the potential model of the TiO_2 lattice, borrowed from [141], was supplemented by the Li–O, Li–Ti and Li–Li potentials. Taking into account the variety of possible Li atom positions, the Li–O force constant K was represented as a function of Li–O distance R. The dependence $K(R)$ was retrieved from the available references on the Li–O force constant values in different molecules and crystals [152] and justified by the quantum mechanical simulations [146]. Spectra calculations have been carried out for each Li position, with R(Li–O) values shown in Figure 6.7–6.34 [139]. Taking into account the Li–O stretching modes along

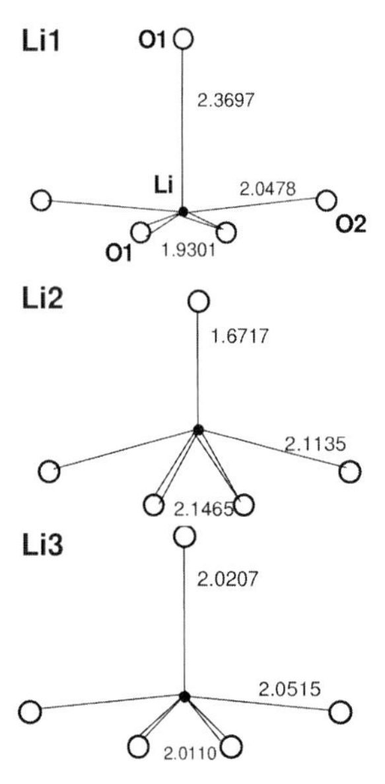

Fig. 6.34 The multiple Li environments in LT-$Li_{0.5}TiO_2$, with the different Li–O bond lengths indicated (in angstroms) (from [146]).

each of the three orientations, 12 Raman active modes are expected for each of the three Li positions in LT-$Li_{0.5}TiO_2$:

$$\Gamma_{Raman}(LT) = (3A_g + 3B_{2g} + 3B_{3g})_{lattice} + (A_g + B_{2g} + B_{3g})_{Li}$$

This leads to a complex pattern of simulated Raman spectra, which has made it possible to assign all the experimental observed features (Figure 6.35).

The Raman spectrum of the reoxidized material (Figure 6.36) is typical of the anatase fingerprint, which indicates the recovery of the tetragonal framework upon oxidation. However, some extra bands are still observed in the high-wavenumber region. This result suggests the presence of strong Li–O bonds that are not broken in the course of the reoxidation process and that are probably responsible for the partial loss of rechargeability of the material evidenced in Figure 6.32.

In conclusion, Raman experimental data reported in [148] supported by lattice dynamics simulation described in. [146] allow to evidence that

1. Li atoms in LT phase have multiple positions in the octahedral interstices of the orthorhombic LT phase. The multiplicity of possible Li positions manifests itself in the Raman scattering spectra as splitting of the bands originated from the Li atom oscillations along different orientations. This leads to the occurrence of multiple Raman bands originated from the Li atom vibrations, covering a wide wavenumber range from 450 up to 950 cm^{-1}. Simulated spectra show six Li-bands lying above 450 cm^{-1} and six lattice modes below 650 cm^{-1}.
2. Particular Li positions with rather short Li–O distances have been found to give rise to high-wavenumber features above 800 cm^{-1}.

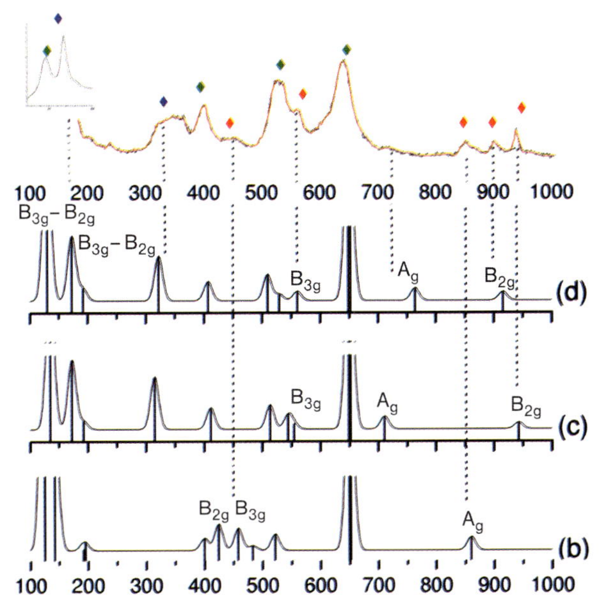

Fig. 6.35 (Top) Experimental Raman spectrum of $Li_{0.5}TiO_2$ showing the active lattice modes originating from anatase (in green), the new lattice modes perturbated by Li–Li interactions (in blue), and the new Li–O stretching modes (in red); (Bottom) Simulated Raman spectra of lithium titanate with Li atoms in Li2 position (b), in Li3 position (c), and in Li1 position (d). The x-coordinates are Raman shifts, values expressed in cm^{-1} (from [146]).

3. Partial electrochemical rechargability could be explained by the existence of these strong Li–O interactions observed in the high-frequency range.
4. Some Li interactions take place with the TiO_2 lattice, this perturbation leading to the splitting of the low-frequency lattice modes.

6.4 Phospho-Olivine $LiMPO_4$ Compounds

Over the past few years, phosphate-based materials $LiMPO_4$ (M = Fe, Mn, Co or Ni) isostructural with olivine have generated considerable interest as a new class of cathodes for lithium-ion rechargeable batteries [153–156]. Indeed, in contrast to Li_xCoO_2 materials which decompose at elevated temperature, leading to oxygen evolution, $LiMPO_4$ compounds have a highly stable 3D framework due to strong P–O covalent bonds in $(PO_4)^{3-}$ polyanion, which prohibits the liberation of oxygen [153]. These characteristics provide an excellent safety and a stable operation of battery even under unusual conditions [156].

Much attention has been focused on the low-cost and low-toxicity $LiFePO_4$ compound because of its attractive theoretical capacity ($170\,mA\,h\,g^{-1}$), based on the

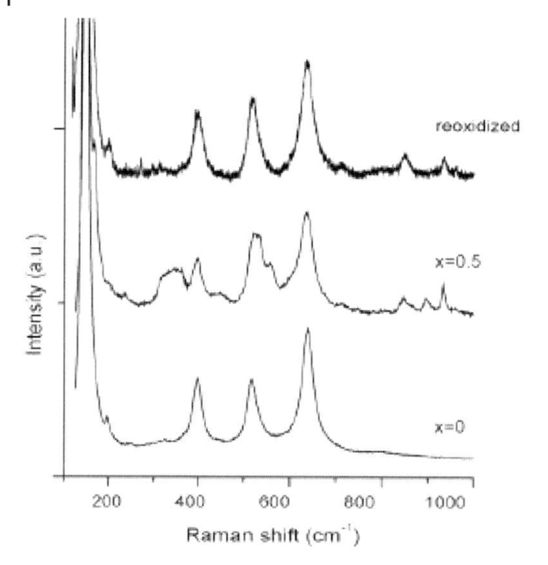

Fig. 6.36 Raman spectra of a Li_xTiO_2 electrode, $x = 0$ (TiO_2), $x = 0.5$ (LT-phase), and reoxidized electrode (from [147]).

two-phase reaction occurring at 3.4 V versus Li^+/Li working potential [153–156]:

$$LiFePO_4 \rightleftarrows FePO_4 + e^- + Li^+$$

Its excellent stability during normal cycling and storage conditions make this material particulary attractive for large scale applications such as hybrid and electric vehicles. However, the major drawbacks are the low capacities achieved even at moderate discharge and the poor rate capability, associated to limited electronic and/ or ionic conductivity. To address these issues, researchers have optimized synthesis techniques to minimize the particle size, which proved successful in achieving long-term cyclability of nanosized $LiFePO_4$ [157], or have incorporated additives to increase conductivity. The latter approach includes coating the olivine particles with carbon by incorporating an organic or polymeric component with the precursors before firing [158, 159], adding metal particles to the mix [160], or solid-solution doping by metals supervalent to Li^+ [161]. All have met with success in improving performance. An example of optimization is illustrated in Figure 6.37, where the quality of the carbon coating, achieved by additive incorporation, accounts mainly for the performance variations.

There have been several investigations into the structure of $LiMPO_4$ compounds. The majority have used Mössbauer spectroscopy and X-ray absorption spectroscopy to focus on the local structure around transition metal ions [156, 162, 163]. Tucker *et al.* used ^7Li and ^{31}P magic angle spinning nuclear magnetic resonance (MAS-NMR) to investigate the environment of the Li and P atoms in some $LiMPO_4$ materials [164, 165].

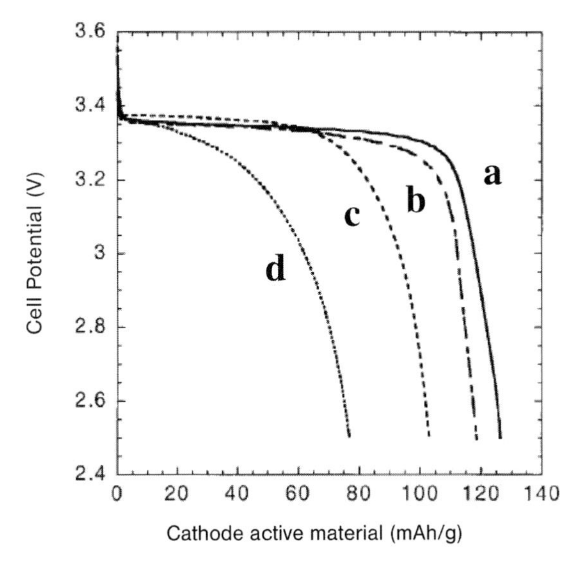

Fig. 6.37 Discharges of Li/1 M $LiPF_6$, EC-DMC/ $LiFePO_4$ cells at 0.055 mA cm^{-2}, showing the carbon-coating improvement achieved by incorporating small amounts of a polyaromatic additive during the sol–gel synthesis. (a) Sol–gel synthesis with additive, 1.15% total carbon; (b) solid-state synthesis from iron acetate, 1.5% carbon; (c) sol–gel synthesis with no additive, 0.69% total carbon; and (d) sol–gel synthesis with no additive, 0.4% total carbon (from [159]).

The vibrational features of olivine-type compounds are well documented. Early work by Paques-Ledent and Tarte focused on identifying the atomic contributions of vibrational modes in a series of $LiMPO_4$ olivine compounds [166, 167]. Using a computational simulation employing a normal coordinate analysis based on the Wilson's FG matrix method [168] Paraguassu *et al.* provided for the first time a complete set of Raman wavenumbers for $LiMPO_4$ (M = Fe, Co, Ni) with their assignments to vibrational motions and symmetry species of the crystal group [169]. Their predicted values are in good agreement with the experimental Raman features reported for single crystals [170, 171] and polycrystalline powder.

$LiFePO_4$ is characterized by a true olivine structure (space group ${D_{2h}}^{16}$ – *Pnma*). The primitive unit cell is centrosymmetric, with four formula units in the cell. There are two types of octahedral sites; the divalent cations Fe^{2+} are randomly located on the largest 4c site with C_s symmetry, the monovalent cation Li^+ being located on the smaller 4a site with C_i symmetry. The pentavalent P^{5+} are located in "isolated" tetrahedra PO_4 in 4c site with C_s symmetry. They bridge the FeO_6 chains to form an interconnected 3D structure (Figure 6.38). The arrangement of the different coordinated groups was given in [172].

Group theory [170, 171] showed that the normal vibrational modes of $LiMPO_4$ are distributed on the irreducible representation of the D_{2h} point group as

$$\Gamma_{\text{vibr}} = 11\text{A}_\text{g} + 7\text{B}_{1\text{g}} + 11\text{B}_{2\text{g}} + 7\text{B}_{3\text{g}} + 10\text{A}_\text{u} + 14\text{B}_{1\text{u}} + 10\text{B}_{2\text{u}} + 14\text{B}_{3\text{u}}$$

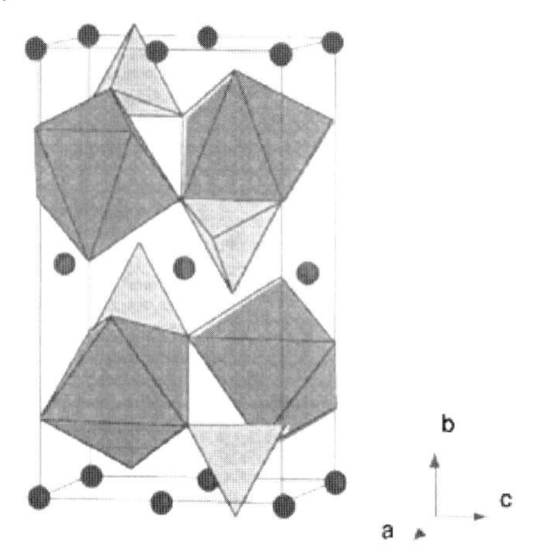

Fig. 6.38 Schematic representation of the polyhedral structure of $LiFePO_4$, showing the arrangement of tetrahedral PO_4 and octahedral FeO_6 entities (from [153]).

The number of predicted active fundamentals is rather large: 36 for the Raman spectrum, but the number of observed bands is smaller than the number of predicted active fundamentals. It is usual, although not strictly correct, to classify these vibrations into internal and external modes [167]. Internal modes refer to vibrations occurring in the $PO_4{}^{3-}$ tetrahedra and external to pseudorotations and translations of the units. The translations include motions of the center-of-mass $PO_4{}^{3-}$ and M^{2+}. It should be mentioned that this separation is a guide to discussion only, because the vibrations may be coupled. The assignment of the $LiMPO_4$ internal modes is made using the simplified Herzberg's notation for the vibrational modes of a free PO_4 tetrahedron [173]. According to this, the free PO_4 group representations decompose into the irreducible representations of the T_d symmetry group as $A_1 + E + F_1 + 3F_2$. Among these, the internal vibrations in terms of symmetry species are $A_1(\nu_1)$ symmetric P–O stretching, $E(\nu_2)$ symmetric O–P–O bond bending, $F_2(\nu_3)$ antisymmetric P–O stretching, and $F_2(\nu_4)$ antisymmetric O–P–O bond bending. The Raman wavenumbers and assignments observed for $LiFePO_4$ are summarized in Table 6.7. The Raman spectrum of $LiFePO_4$ is shown in Figure 6.39.

Raman spectroscopy was carried out to follow the chemical delithiation process in Li_xFePO_4 ($0 \leq x \leq 1$) [174]. Important changes in the spectral features were observed in the first delithiation step: the presence of new lines with increased intensities upon Li_xFePO_4 delithiation, several band splitting, and frequency shifts (Figure 6.40). In particular, new Raman bands were observed in the ν_1 and ν_3 stretching region at 911, 962, 1064, 1080, and 1124 cm^{-1}. In spite of a lack of thorough vibrational analysis and symmetry assignments, the authors assign these new features to the $FePO_4$ compound. The Raman positions were observed to be independent of the lithium content, and the authors found their results to be in accordance with the generally adopted two-phase process [153].

Table 6.7 Raman wavenumbers and assignments for $LiFePO_4$.

Raman wavenumbers from [248]	Raman wavenumbers from [243]	Assignment	
107	106	Translations of Fe^{II}, PO_4 and rotations	External modes
151	147		
161	157		
199	195		
226	238		
245	249		
292			
410			
447	442	$\nu_2(PO_4)$	Internal modes essentially
499			
513			
572	571	$\nu_4(PO_4)$	
595	586		
612	628		
631			
657			
953	949	$\nu_1(PO_4)$	
999	995	$\nu_3(PO_4)$	
1071	1067		
	1079		

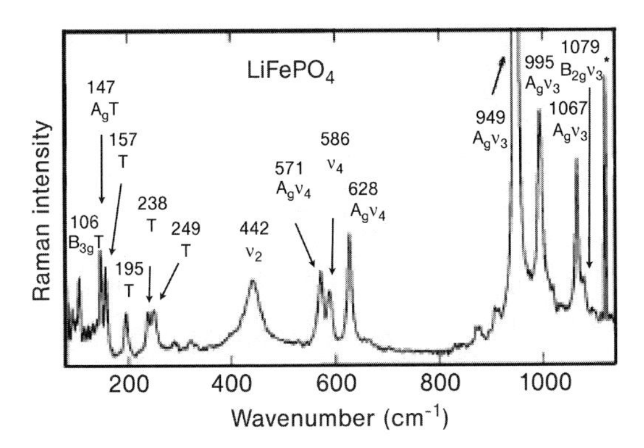

Fig. 6.39 Raman spectrum of $LiFePO_4$ microcrystals (from [169]).

Conversely, a recent vibrational study reported a contrasting finding failing to support the well-accepted two-crystalline-phase mixture model for lithium extraction [175]. In this work, the main Raman bands observed for delithiated $Li_{0.11}FePO_4$ are in good agreement with those reported in [174]. However, a deconvolution analysis of the vibrational spectra is provided, which allows to evidence an additional significant broadband contribution both in the Raman and FTIR spectra of $Li_{0.11}FePO_4$. The authors claim that the two-phase process for

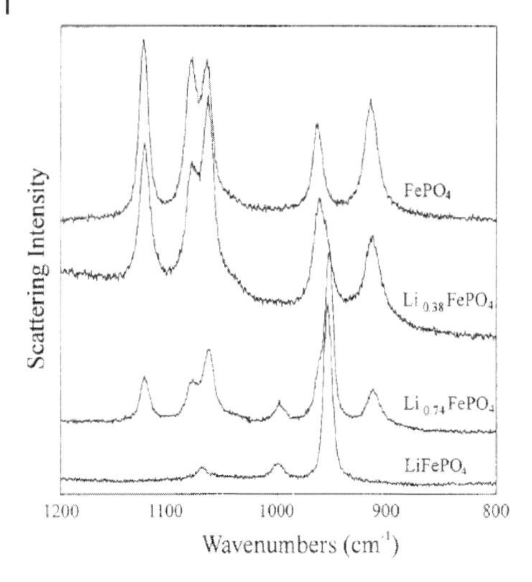

Fig. 6.40 Raman spectra in the 800–1200 cm^{-1} frequency range obtained for chemically delithiated Li_xFePO_4 samples (from [174]).

tranformation of $LiFePO_4$ into $FePO_4$ upon lithium extraction probably includes a pathway where a highly disordered phase is generated.

The carbon coating of the $LiFePO_4$ matrix required for the electrochemical application of this material hinders any bulk Raman data on the Li_xFePO_4 cathode material to be obtained. This is illustrated in Figure 6.41, where the typical Raman spectrum of a $LiFePO_4$ powder exhibits only a sharp band at 953 cm^{-1} and two weaker bands at 997 and 1098 cm^{-1} assigned to the internal modes of the $(PO_4)^{3-}$ anion [159]. These weak features are explained by the fact that Raman spectroscopy allows to examine phenomena on a carbon surface within a limited penetration depth (about 50–100 nm with a 514.5-nm green laser beam [176]. The two intense broad bands observed at ~1590 and 1350 cm^{-1} are assigned to carbon vibrations on the basis of the corresponding features in the spectrum of graphite. The structure at 1590 cm^{-1} mainly corresponds to the G-line associated with the optically allowed E_{2g} zone-center mode of crystalline graphite. The structure at 1350 cm^{-1} mainly corresponds to the D-line associated with disorder-allowed zone-edge mode of graphite.

In so far as the structural properties of carbons are known to be strongly related to the shape of the G and D bands [177–188], Raman spectroscopy, in the field of $LiFePO_4$-based cathode materials, is mainly considered as a diagnostic tool to evaluate the quality of the carbon coating. This property is of utmost importance since several observations indicated that the structure of carbon coating influences greatly the electrochemical performance of $LiFePO_4$ powders and films [159, 189–191]. Several studies have shown that the shape of the D and G carbon bands change substantially with the pyrolysis temperature and the nature of the precursor material. This effect is illustrated in Figure 6.41, where the Raman spectra of the residual carbon are in fact resolved into four components located at ~1190, 1350,

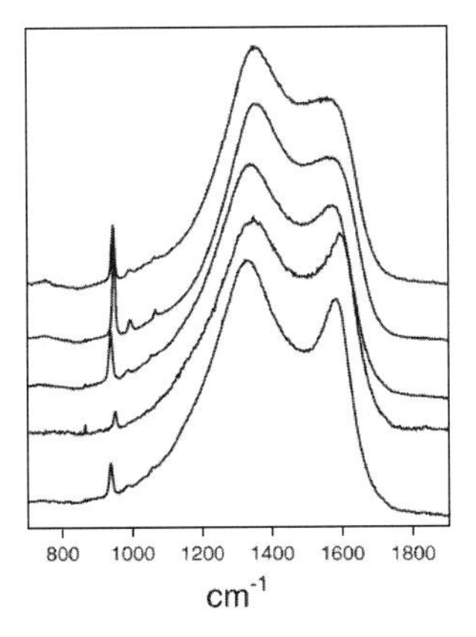

Fig. 6.41 Microraman spectra for different processed LiFePO4 powders. The corresponding electrochemical discharge capacities are (from top to bottom): 58, 77, 106, 120, 126 mA h g^{-1} (from [159]).

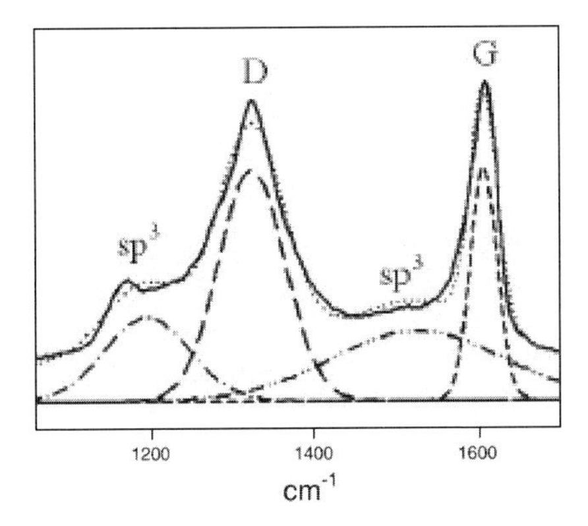

Fig. 6.42 Raman spectrum of the residual carbon of a $LiFePO_4$ powder, with the deconvolution shown below (from [189]).

1518, and 1590 cm^{-1} (Figure 6.42). The bands at 1190 and 1518 cm^{-1} have uncertain origin but have been assigned to short-range vibrations of sp^3-coordinated carbons, as already observed in disordered carbon blacks and diamond-like carbons [159, 189–191]. Some authors have found a relationship between the electrochemical capacity and the integrated Raman intensity ratios of D/G bands and sp^3/sp^2 coordinated carbons [189–191]. As shown in Figure 6.43, the lower both ratios, the higher the performance exhibited by the $LiFePO_4$ composite electrode. This trend was interpreted in terms of the increasing amount of larger graphene clusters in the very disordered carbon structure, and consequently, an improved electronic conductivity of the carbon deposit [159, 189].

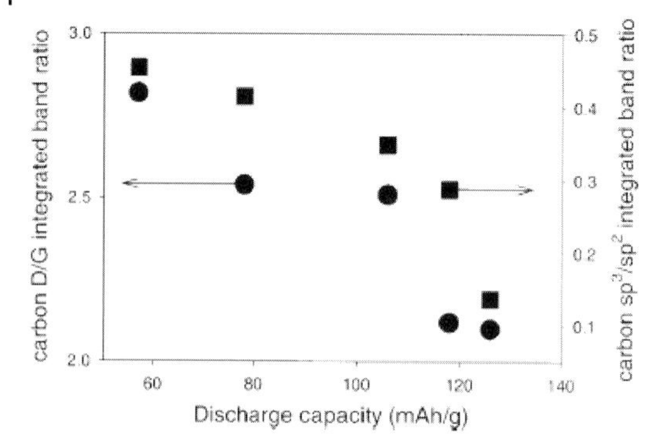

Fig. 6.43 Electrochemical discharge capacity of $LiFePO_4$ electrodes in lithium cells versus structure of residual carbon (from [159]).

6.5 General Conclusion

The results analyzed and discussed in this work clearly indicate that it is now possible to obtain a lot of helpful structural data from the Raman spectra of various Li-intercalated host lattices, from transition metal oxides to phosphates. This approach is proved to considerably enrich the understanding of the relationship between structure and electrochemistry of cathode materials for lithium batteries. Though Raman spectroscopy is a nondestructive analysis and does not require any specific conditioning of the sample, a great part of Raman data on cathode materials mainly deals with the strating materials and the usual approach remains very qualitative, owing to the lack of a clear and justified assignment of the Raman bands. In particular, very few significant data are available on electrochemically reduced, oxidized, or cycled $LiCoO_2$, $LiNiO_2$ compounds in spite of their large impact in the field of Li-ion batteries.

Most of the contributions devoted to the $LiCoO_2$, Mn_xO_y, Li–Mn–O, $Li_xV_2O_5$ and $Li_xTi_yO_z$ systems suffer from a lack of understanding the complex Raman features observed. We have demonstrated in the case of $Li_xV_2O_5$ and Li_xTiO_2 phases that an appropriate experimental approach that enables obtaining reliable Raman fingerprints on lithiated samples, combined with a theoretical approach based upon lattice dynamics simulation, could afford a complete picture of the local structural deformations upon lithium insertion.

Raman spectroscopy constitutes an excellent local probe to enrich the knowledge of the structure of Li intercalation compounds. Structural informations such as local disorder, changes in bond lengths, bond angles, coordination, Li dynamics, cation ordering etc. can be provided by this technique. A dynamic picture of the electrochemical insertion reaction (phase transitions, local structural changes vs. Li content, C rate, number of cycles, grain size, temperature etc.) can then be obtained. In other words, we have shown that specific electrochemical questions could be solved using the Raman microprobe. As an example, the loss of electrochemical reversibility for Li_xTiO_2 can be related to the existence of strong Li–O interactions. In the case of $Li_xV_2O_5$, we have shown that the structural response was strongly influenced by the nanosized electrode morphology, whereas for $LiFePO_4$, the electrochemical performance of the carbon-coated cathode is strongly related with the amount of graphene clusters evaluated from the Raman analysis of the carbon bands.

As of now, two fundamental aspects need to be developed: the first one consists in the extension of the use of Raman spectroscopy to various electrode materials, the second is to build up reliable and appropriate theoretical models allowing a quantitative assignment of the Raman features. This should lead to a comprehensive view of the local structure and Li-induced distortions and prompt the interest of Raman microspectrometry for many researchers involved in material science applied to the field of energy storage.

References

1 Cotton, F.A. (1963) *Chemical Applications of Group Theory*, Wiley Interscience, New York.

2 Turrell, G. and Corset, J. (1996) *Raman Microscopy Developments and Applications*, Academic Press Incs.

3 Weber, W.H. and Merlin, R. (2000) *Raman Scattering in Materials Science*, Springer Verlag.

4 Mizushima, K., Jones, P.C., Wiseman, P.J., and Goodenough, J.B. (1980) *Mater. Res. Bull.*, **15**, 783.

5 Ohzuku, T. and Ueda, A. (1994) *J. Electrochem. Soc.*, **141**, 2972.

6 Inaba, M., Todzuka, Y., Yoshida, H., Grincourt, Y., Tasaka, A., Tomida, Y., and Ogumi, Z. (1995) *Chem. Lett.*, **10**, 889.

7 Inaba, M., Iriyama, Y., Ogumi, Z., Todsuka, Y., and Tasaka, A. (1997) *J. Raman Spectrosc.*, **28**, 613.

8 Reimers, J.N. and Dahn, J.R. (1992) *J. Electrochem. Soc.*, **139**, 2091.

9 Amatucci, G.G., Tarascon, J.M., and Klein, L.C. (1996) *J. Electrochem. Soc.*, **143**, 1114.

10 Itoh, T., Sato, H., Nishina, T., Matue, T., and Uchida, I. (1997) *J. Power Sources*, **68**, 333.

11 Itoh, T., Anzue, N., Mohamedi, M., Hisamitsu, Y., Umeda, M., and Uchida, I. (2000) *Electrochem. Commun.*, **2**, 743.

12 Kim, Y.J., Lee, E.K., Kim, H., Cho, J., Cho, Y.W., Park, B., Oh, S.M., and Yoon, J.K. (2004) *J. Electrochem. Soc.*, **151**, 1063.

13 Kang, S. G., Kang, S.Y., Ryu, K.S., and Chang, S.H. (1999) *Solid State Ionics*, **120**, 155.

14 Santiago, E.I., Andrade, A.V.C., Paiva-Santos, C.O., and Bulhoes, L.O.S. (2003) *Solid State Ionics*, **158**, 91.

15 Ra, D.I. and Han, K.S. (2006) *J. Power Sources*, **163**, 284.

16 Julien, C., Camacho-Lopez, M.A., Escobar-Alarcon, L., and Haro-Poniatowski, E. (2001) *Mater. Chem. Phys.*, **68**, 210.

17 Tang, S.B., Lai, M.O., and Lu, L. (2006) *J. Alloys Compd.*, **424**, 342.

18 Park, H.Y., Lee, S.R., Lee, Y.J., Cho, B.W., and Cho, W.I. (2005) *Mater. Chem. Phys.*, **93**, 70.

19 Jeon, S.W., Lim, J.K., Lim, S.H., and Lee, S.M. (2005) *Electrochim. Acta*, **51**, 268.

20 Pracharova, J., Pridal, J., Bludska, J., Jakubec, I., Vorlicek, V., Malkova, Z., Dikonimos Makris, Th., Giorgi, R., and Jastrabik, L. (2004) *J. Power Sources*, **108**, 204.

21 Liao, C.L., Wu, M.T., Chen, J.H., Leu, I.C., and Fung, K.Z. (2006) *J. Alloys Compd.*, **414**, 302.

22 Lee, T., Cho, K., Oh, J., and Shin, D. (2007) *J. Power Sources*, **174**, 394.

23 Koike, S. and Tatsumi, K. (2007) *J. Power Sources*, **174**, 976.

24 Kushida, K. and Kuriyama, K. (2002) *J. Cryst. Growth*, **237–239**, 612.
25 Lee, J.H., Han, K.S., Lee, B.J., Seo, S.I., and Yoshimura, M. (2004) *Electrochim. Acta*, **50**, 467.
26 Rossen, E., Reimers, J.N., and Dahn, J.R. (1993) *Solid State Ionics*, **62**, 53.
27 Gummow, R.J., Liles, D.C., Tackeray, M.M., and David, W.I.F. (1993) *Mater. Res. Bull.*, **28**, 1177.
28 Garcia, B., Farcy, J., Pereira-Ramos, J.P., and Baffier, N. (1997) *J. Electrochem. Soc.*, **144**, 1179.
29 Huang, W. and Frech, R. (1996) *Solid State Ionics*, **86–88**, 395.
30 Mendoza, L., Baddour-Hadjean, R., Cassir, M., and Pereira-Ramos, J.P. (2004) *Appl. Surf. Sci.*, **225**, 356.
31 Novak, P., Panitz, J.C., Joho, F., Lanz, M., Imhof, R., and Coluccia, M. (2000) *J. Power Sources*, **90**, 52.
32 Luo, Y., Cai, W.B., Xing, X.K., and Scherson, D.A. (2004) *Electrochem. Solid-State Lett.*, **7**, 1.
33 Goodenough, J.B., Wickham, D.G., and Croft, W.J. (1958) *J. Phys. Chem. Solids*, **5**, 107.
34 Delmas, C. and Saadoune, I. (1992) *Solid State Ionics*, **53–56**, 370.
35 Julien, C. (2000) *Solid State Ionics*, **136–137**, 887.
36 Rougier, A., Nazri, G.A., and Julien, C. (1997) *Ionics*, **3**, 170.
37 Ramana, C.V., Zaghib, K., and Julien, C.M. (2006) *J. Power Sources*, **159**, 1310.
38 Le Goff, P., Baffier, N., Bach, S., and Pereira-Ramos, J.P. (1996) *Mater. Res. Bull.*, **31**, 63.
39 Thackeray, M.M. (1997) *Prog. Solid State Chem.*, **25**, 1.
40 McBreen, J. (1975) *Electrochim. Acta*, **20**, 221.
41 Nardi, J.C. (1985) *J. Electrochem. Soc.*, **132**, 1787.
42 Ohzuku, T., Kitagawa, M., and Hirai, T. (1989) *J. Electrochem. Soc.*, **136**, 3169.
43 Hill, L.I., Portal, R., La Salle, A.L., Verbaere, A., and Guyomard, D. (2001) *Electrochem. Solid-State Lett.*, **4**, 1.
44 Wakihara, M. (2001) *Mater. Sci. Eng.*, **R33**, 109.
45 Tackeray, M.M., De Picciotto, L.A., De Kock, A., Johnson, P. J., Nicholas, V. A., and Andendorf, K. T. (1987) *J. Power Sources*, **21**, 1.
46 Armstrong, A.R. and Bruce, P. G. (1996) *Nature*, **381**, 499.
47 Bordet-Le Guenne, L., Deniard, P., Biensant, P., Siret, C., and Brec, R. (2000) *J. Mater. Chem.*, **10**, 2201.
48 Julien, C.M., Massot, M., and Poinsignon, C. (2004) *Spectrochim. Acta, Part A*, **60**, 689.
49 Julien, C.M. (2006) *Solid State Ionics*, **177**, 11.
50 Franger, S., Bach, S., Farcy, J., Pereira-Ramos, J.P., and Baffier, N. (2002) *J. Power Sources*, **109**, 262.
51 Julien, C., Massot, M., Baddour-Hadjean, R., Franger, S., Bach, S., and Pereira-Ramos, J.P. (2003) *Solid State Ionics*, **159**, 345.
52 Franger, S., Bach, S., Pereira-Ramos, J.P., and Baffier, N. (2000) *J. Electrochem. Soc.*, **147**, 3226.
53 Julien, C.M. and Massot, M. (2003) *Mater. Sci. Eng. B*, **97**, 217.
54 Thackeray, M.M., Johnson, P., De Piciotto, L., Bruce, P.G., and Goodenough, J.B. (1984) *Mater. Res. Bull.*, **19**, 179.
55 Ohzuku, T., Kitagawa, M., and Hirai, T. (1990) *J. Electrochem. Soc.*, **137**, 769.
56 Guyomard, D. and Tarascon, J.M. (1992) *J. Electrochem. Soc.*, **139**, 937.
57 Yamada, A., Miura, K., Hinokuma, K., and Tanaka, M. (1995) *J. Electrochem. Soc.*, **142**, 2149.
58 Tarte, P. (1967) *J. Inorg. Nucl. Chem.*, **29**, 915.
59 Tarte, P. and Preudhomme, J. (1970) *Spectrochim. Acta*, **26A**, 747.
60 Preudhomme, J. and Tarte, P. (1971) *Spectrochim. Acta*, **27A**, 845.
61 White, W.B. and De Angelis, B.A. (1967) *Spectrochim. Acta*, **23A**, 985.
62 Richardson, J.T., Wen, S.J., Striebel, K.A., Ross, P.N., and Cairns, E.J. (1997) *Mater. Res. Bull.*, **32**, 609.
63 Wen, S.J., Richardson, J.T., Ma, L., Striebel, K.A., Ross, P.N., and Cairns, E.J. (1996) *J. Electrochem. Soc.*, **143**, L136.

64 Kanoh, H., Tang, W., and Ooi, K. (1998) *Electrochem. Solid-State Lett.*, **1**, 17.

65 Ammundsen, B., Burns, G.R., Islam, M.S., Kanoh, H., and Roziere, J. (1999) *J. Phys. Chem. B*, **103**, 5175.

66 Liu, W., Kowal, K., and Farrington, G.C. (1998) *J. Electrochem. Soc.*, **145**, 459.

67 Mukerjee, S., Thurston, T.R., Jisrawi, N.M., Yang, X.Q., McBreen, J., Daroux, M.L., and Xing, X.K. (1998) *J. Electrochem. Soc.*, **145**, 466.

68 Xia, Y. and Yoshio, M. (1996) *J. Electrochem. Soc.*, **143**, 825.

69 Gao, Y., Reimers, J.N., and Dahn, J.R. (1996) *Phys. Rev. B*, **54**, 3878.

70 David, W.I.F., Tackeray, M.M., De Picciotto, L.A., and Goodenough, J.B. (1987) *J. Solid State Chem.*, **67**, 316.

71 Julien, C.M. and Massot, D. (2003) *Mater. Sci. Eng. B*, **100**, 69.

72 Huang, W. and Frech, R. (1999) *J. Power Sources*, **82**, 616.

73 Anzue, N., Itoh, T., Mohamedi, M., Umeda, M., and Uchida, I. (2003) *Solid State Ionics*, **156**, 301.

74 Shi, Q., Takahashi, Y., Akimoto, J., Stefan, I.C., and Scherson, D.A. (2005) *Electrochem. Solid-State Lett.*, **8**, A521.

75 Katzke, H., Czank, M., Depmeier, M., and van Smaalen, S. (1997) *J. Phys. Condens. Matter.*, **9**, 6231.

76 Song, D., Ikuta, H., Uchida, T., and Wakihara, M. (1999) *Solid State Ionics*, **117**, 151.

77 Li, G.H., Ikuta, H., Uchida, T., and Wakihara, M. (1996) *J. Electrochem. Soc.*, **143**, 178.

78 Kim, K.W., Lee, S.W., Han, K.S., Chung, H.J., and Woo, S.I. (2003) *Electrochim. Acta*, **48**, 4223.

79 Strobel, P., Ibarra-Palos, A., Anne, M., Poinsignon, C., and Crisci, A. (2003) *Solid State Sci.*, **5**, 1009.

80 Wei, Y., Kim, K.-B., and Chen, G. (2006) *Electrochim. Acta*, **51**, 3365.

81 Dokko, K., Mohamedi, M., Anzue, N., Itoh, T., and Uchida, I. (2002) *J. Mater. Chem.*, **12**, 3688.

82 Hwang, S.-J., Park, D.-H., Choy, J.-H., and Campet, G. (2004) *J. Phys. Chem. B*, **108**, 12713.

83 Ramana, C.V., Massot, M., and Julien, C.M. (2005) *Surf. Interface Anal.*, **37**, 412.

84 Dokko, K., Anzue, N., Mohamedi, M., Itoh, T., and Uchida, I. (2004) *Electrochem. Commun.*, **6**, 384.

85 Zhang, F. and Whittingham, M.S. (2000) *Electrochem. Solid-State Lett.*, **3**, 7.

86 Park, H.S., Hwang, S.J., and Choy, J.H. (2001) *J. Phys. Chem. B*, **105**, 4860.

87 Hwang, S.J., Park, H.S., Choy, J.H., Campet, G., Portier, J., Kwon, C.W., and Etourneau, J. (2001) *Electrochem. Solid-State Lett.*, **4**, A213.

88 Doeff, M.M., Anapolsky, A., Edman, L., Richardson, T.J., and De Jonghe, L.C. (2001) *J. Electrochem. Soc.*, **148**, A230.

89 Murphy, D.W., Christian, P.A., Disalvo, F.J., and Waszczak, J.V. (1979) *Inorg. Chem.*, **18**, 2800.

90 Whittingham, M.S. (1976) *J. Electrochem. Soc.*, **126**, 315.

91 Wiesener, K., Schneider, W., Ilic, D., Steger, E., Hallmeir, K.H., and Brackmann, E. (1978) *J. Power Sources*, **20**, 157.

92 Delmas, C., Cognac-Auradou, H., Cocciantelli, J.M., Ménétrier, M., and Doumerc, J.P. (1994) *Solid State Ionics*, **69**, 257.

93 Bates, J.B., Gruzalski, G.R., Dudney, N.J., Luck, C.F., and Xiaohua, Y. (1994) *Solid State Ionics*, **70–71**, 619.

94 Cocciantelli, J.M., Doumerc, J.P., Pouchard, M., Broussely, M., and Labat, J. (1991) *J. Power Sources*, **34**, 103.

95 Abello, L., Husson, E., Repelin, E., and Lucazeau, G. (1983) *Spectrochim. Acta*, **39A**, 641.

96 Clauws, P. and Vennik, J. (1980) *Phys. Stat. Sol. B*, **59**, 469.

97 Baddour-Hadjean, R., Pereira-Ramos, J.P., Navone, C., and Smirnov, M. (2008) *Chem. Mater.*, **20**, 1916.

98 Dickens, P.G., French, S.J., Hight, A.T., and Pye, M.F. (1979) *Mater. Res. Bull.*, **14**, 1295.

99 Galy, J., Darriet, J., and Hagenmuller, P. (1971) *Rev. Chim. Min.*, **8**, 509.
100 Rozier, P., Savariault, J.M., Galy, J., Marichal, C., Horschinger, J., and Granger, P. (1996) *Eur. J. Solid State Inorg. Chem.*, **33**, 1.
101 Galy, J. (1992) *J. Solid State Chem.*, **100**, 229.
102 Cocciantelli, J.M., Gravereau, J.M., Doumerc, J.P., Pouchard, M., and Hagenmuller, P. (1991) *J. Solid State Chem.*, **93**, 497.
103 Leger, C., Bach, S., Soudan, P., and Pereira-Ramos, J.P. (2005) *J. Electrochem. Soc.*, **152**, A236.
104 Delmas, C., Brethes, S., and Ménétrier, M. (1991) *J. Power Sources*, **34**, 113.
105 Rozier, P., Savariaut, J.M., and Galy, J. (1997) *Solid State Ionics*, **98**, 133.
106 Meulenkamp, E.A., van Klinken, W., and Schlatmann, A.R. (1999) *Solid State Ionics*, **126**, 235.
107 Prouzet, E., Cartier, C., Vilain, F., and Tranchant, A. (1996) *J. Chem. Soc., Faraday Trans.*, **92**, 103.
108 Hirschinger, J., Mongrelet, T., Marichal, C., Granger, P., Savariault, J.M., Déramond, E., and Galy, J. (1993) *J. Phys. Chem.*, **97**, 10301.
109 Pecquenard, B., Gourier, D., and Baffier, N. (1995) *Solid State Ionics*, **78**, 287.
110 Cava, R.J., Santoro, A., Murphy, D.W., Zahurak, S.M., Fleming, R.M., Marsh, P., and Roth, R.S. (1986) *J. Solid State Chem.*, **65**, 63.
111 Savariault, J.M. and Rozier, P. (1997) *Physica B*, **234–236**, 97.
112 Katzke, H., Czank, M., Depmeier, W., and van Smaalen, S. (1997) *Phil. Mag. B*, **75**, 757.
113 Zhang, X. and Frech, R. (1997) *Electrochim. Acta*, **42**, 475.
114 Zhang, X. and Frech, R. (1998) *J. Electrochem. Soc.*, **145**, 847.
115 Rey, I., Lassègues, J.C., Baudry, P., and Majastre, H. (1998) *Electrochim. Acta*, **43**, 1539.
116 Burba, M. and Frech, R. (2006) *Appl. Spectrosc.*, **60**, 490.
117 Baddour-Hadjean, R., Rackelboom, E., and Pereira-Ramos, J.P. (2006) *Chem. Mater.*, **18**, 3548.
118 Cazzanelli, E., Mariotto, G., Passerini, S., and Decker, F. (1994) *Solid State Ionics*, **70–71**, 412.
119 Ramana, C. V., Smith, R.J., Hussain, O.M., Massot, M., and Julien, C.M. (2005) *Surf. Interface Anal.*, **37**, 406.
120 Julien, C., Ivanov, I., and Gorenstein, A. (1995) *Mater. Sci. Eng. B*, **33**, 168.
121 Baddour-Hadjean, R., Golabkan, V., Pereira-Ramos, J.P., Mantoux, A., and Lincot, D. (2002) *J. Raman Spectrosc.*, **33**, 631.
122 McGraw, J.M., Perkins, J.D., Zhang, J.G., Liu, P., Parilla, P.A., Turner, J., Schulz, D.L., Curtis, C.J., and Ginley, D.S. (1998) *Solid State Ionics*, **113–115**, 407.
123 Satto, C., Sciau, P., Dooryhee, E., Galy, J., and Millet, P. (1999) *J. Solid State Chem.*, **146**, 103.
124 Popovic, Z.V., Gajic, R., Konstantinovic, M.J., Provoost, R., Moshchalkov, V.V., Vasil'ev, A.N., Isobe, M., and Ueda, Y. (2000) *Phys. Rev.*, **B61**, 11454.
125 Talledo, A. and Granqvist, C.G. (1995) *J. Appl. Phys.*, **77**, 4655.
126 Scarminio, J., Talledo, A., Andersson, A., Passerini, S., and Decker, F. (1993) *Electrochim. Acta*, **38**, 1637.
127 Ptitsyn, M.V., Tikhonov, K.I., and Rotinyan, A.L. (1981) *Sov. Electrochem.*, **17**, 1297.
128 Andrukaitis, E., Jacobs, P.W.M., and Lorimer, J.W. (1988) *Solid State Ionics*, **27**, 19.
129 Vivier, V., Farçy, J., and Pereira-Ramos, J.P. (1998) *Electrochim. Acta*, **44**, 831.
130 Navone, C., Pereira-Ramos, J.P., Baddour-Hadjean, R., and Salot, R. (2006) *J. Electrochem. Soc.*, **153**, A2287.
131 Navone, C., Baddour-Hadjean, R., Pereira-Ramos, J.P., and Salot, R. (2005) *J. Electrochem. Soc.*, **152**, A1790.
132 Katzke, H. and Depmeier, W. (1996) *Phase Transitions*, **59**, 91.

133 Galy, J., Satto, C., Sciau, P., and Millet, P. (1999) *J. Solid State Chem.*, **146**, 129.

134 Baddour-Hadjean, R., Navone, C., and Pereira-Ramos, J.P. (2009) *Electrochim. Acta*, in press.

135 O'Regan, B. and Gratzel, M. (1991) *Nature*, **353**, 737.

136 Bonino, F., Busani, L., Manstretta, M., Rivolta, B., and Scrosati, B. (1981) *J. Power Sources*, **6**, 261.

137 Ohzuku, T., Takehara, Z., and Yoshizawa, S. (1979) *Electrochim. Acta*, **24**, 219.

138 Cava, R.J., Murphy, D.W., Zahurak, S., Santoro, A., and Roth, R.S. (1984) *J. Solid State Chem.*, **53**, 64.

139 Wagemaker, M., Kearley, G.J., van Well, A., Mutka, H., and Mulder, F.M. (2003) *J. Am. Chem. Soc.*, **125**, 840.

140 de Krol, R. V., Goossens, A., and Meulenkamp, E. (2003) *J. Electrochem. Soc.*, **146**, 3150.

141 Ohsaka, T., Izumi, F., and Fujiki, Y. (197) *J. Raman Spectrosc.*, **7**, 321.

142 Sekiya, T., Ohta, S., Kamei, S., Hanakawa, M., and Kurita, S. (2001) *J. Phys. Chem. Solids*, **62**, 717.

143 Gonzales, R.J., Zallen, R., and Berger, H. (1997) *Phys. Rev. B*, **55**, 7014.

144 Dinh, N.N., Oanh, N.Th.T., Long, P.D., Bernard, M.C., and Hugot-Le Goff, A. (2003) *Thin Solid Films*, **423**, 70.

145 Lindsay, M.J., Blackford, M.G., Attard, D.J., Luca, V., Skyllas-Kasacos, M., and Griffith, C.S. (2007) *Electrochim. Acta*, **52**, 6401.

146 Smirnov, M. and Baddour-Hadjean, R. (2004) *J. Chem. Phys.*, **121**, 2348.

147 Hardwick, L.J., Holzapfel, M., Novak, P., Dupont, L., and Baudrin, E. (2007) *Electrochim. Acta*, **52**, 5357.

148 Baddour-Hadjean, R., Bach, S., Smirnov, M., and Pereira-Ramos, J.P. (2004) *J. Raman Spectrosc.*, **35**, 577.

149 Murphy, D.W., Greenblatt, M., Zahurak, S.M., Cava, R.J., Waszczak, J.V., Hull, G.W., and Hutton, R.S. (1982) *Rev. Chim. Miner.*, **19**, 441.

150 Lunell, S., Stashans, A., Ojamae, L., Lindstrom, H., and Hagfeldt, A. (1997) *J. Am. Chem. Soc.*, **119**, 7374.

151 Wagemaker, M., van de Krol, R., Kentgens, A.P.M., van Well, A.A., and Mulder, F.M. (2001) *J. Am. Chem. Soc.*, **123**, 11454.

152 Lazarev, A.N., Mirgorodsky, A.P., and Ignatiev, I.S. (1975) *Vibrational Spectra of Complex Oxides*, Nauka, Leningrad.

153 Padhi, A.K., Nanjundaswamy, K.S., and Goodenough, J.B. (1997) *J. Electrochem. Soc.*, **144**, 1188.

154 Amine, K., Yasuda, K., and Yamachi, M. (2000) *Electrochem. Solid-State Lett.*, **3**, 178.

155 Okada, S., Sawa, S., Egashira, M., Yamaki, J., Tabuchi, M., Kageyama, H., Konishi, T., and Yoshino, A. (2001) *J. Power Sources*, **97–98**, 430.

156 Yamada, A., Chun, S.C., and Hinokuma, K. (2001) *J. Electrochem. Soc.*, **148**, A224.

157 Prosini, P.P., Karewska, M., Scaccia, S., Wisniewski, P., and Pasquali, M. (2003) *Electrochim. Acta*, **48**, 4205.

158 Huang, H., Yin, S.C., and Nazar, L. (2001) *Electrochem. Solid-State Lett.*, **4**, A170.

159 Doeff, M.M., Hu, Y., McLarnon, F., and Kostecki, R. (2003) *Electrochem. Solid-State Lett.*, **6**, A207.

160 Croce, F., D'Epifanio, A., Haussoun, J., Deptula, A., Olczac, A., and Scrosati, B. (2002) *Electrochem. Solid-State Lett.*, **5**, A47.

161 Chung, S.Y., Bloking, J.T., and Chiang, Y.M. (2002) *Nat. Mater.*, **2**, 123.

162 Anderson, A.S., Kalska, B., Häggström, L., and Thomas, J.O. (2000) *Solid State Ionics*, **130**, 41.

163 Yamada, A., Takei, Y., Koizumi, H., Sonoyama, N., and Kanno, R. (2006) *Chem. Mater.*, **18**, 804.

164 Tucker, M.C., Doeff, M.M., Richardson, T.J., Finones, R., Reimers, J.A., and Cairnes, E.J. (2002) *Electrochem. Solid-State Lett.*, **5**, A95.

165 Tucker, M.C., Doeff, M.M., Richardson, T.J., Finones, R.,

Cairnes, E.J., and Reimers, J.A. (2002) *J. Am. Chem. Soc.*, **124**, 3832.

166 Paques-Ledent, M.T. and Tarte, P. (1973) *Spectrochim. Acta*, **29A**, 1007.

167 Paques-Ledent, M.T. and Tarte, P. (1974) *Spectrochim. Acta*, **30A**, 673.

168 Wilson, E.B., Decius, J.C., and Cross, P.C. (1955) Molecular vibrations, in *The Theory of Infrared and Raman Vibrational Spectra*, Dover, New York.

169 Paraguassu, W., Freire, P.T.C., Lemos, V., Lala, S.M., Montoro, L.A., and Rosolen, J.M. (2005) *J. Raman Spectrosc.*, **36**, 213.

170 Fomin, V.I., Gnezdilov, V.P., Kumosov, V.S., Peschanskii, A.V., and Yeremenko, V.V. (1999) *Low Temp. Phys.*, **25**, 829.

171 Fomin, V.I., Gnezdilov, V.P., Kumosov, V.S., Peschanskii, A.V., and Yeremenko, V.V. (2002) *Low Temp. Phys.*, **28**, 203.

172 Geller, S. and Durand, J.L. (1960) *Acta Crystallogr.*, **13**, 325.

173 Herzberg, G. (1975) *Infrared and Raman Spectra of Polyatomic Molecules*, Van Nostrand, New York.

174 Burba, C.M. and Frech, R. (2004) *J. Electrochem. Soc.*, **151**, A1032.

175 Lemos, V., Guerini, S., Mendes Filho, J., Lala, S.M., Montoro, L.A., and Rosolen, J.M. (2006) *Solid State Ionics*, **177**, 1021.

176 Solin, S.A. (1990) *Graphite Intercalation Compounds* (eds I.H. Zabel and S.A. Solin), Springer-Verlag, Berlin, p. 165.

177 Lespade, P., Marchand, A., Couzi, M., and Cruege, F. (1984) *Carbon*, **22**, 375.

178 Knight, D.S. and White, W.B. (1989) *J. Mater. Res.*, **4**, 385.

179 Ghodbane, S., Deneuville, A., Tromson, D., Bergonzo, P., Bustarret, E., and Ballutaud, D. (2006) *Phys. Stat. Sol.*, **A203**, 2397.

180 Haouni, A., Mermoux, M., Marcus, B., Abello, L., and Lucazeau, G. (1999) *Diamond Relat. Mater.*, **8**, 657.

181 Ramamurti, R., Shanov, V., Singh, R.N., Mamedov, S., and Boolchand, P. (2006) *J. Vac. Sci. Technol. A*, **24**(2), 179.

182 Irish, D.E., Deng, Z., and Odziemkowski, M. (1995) *J. Power Sources*, **54**, 28.

183 Spahr, M.E., Paladino, T., Wilhelm, H., Würsig, A., Goers, D., Buqa, H., Holzapfel, M., and Novak, P. (2004) *J. Electrochem. Soc.*, **151**(9), 1383.

184 Tuinstra, F. and Koenig, J.L. (1976) *J. Chem. Phys.*, **53**, 1126.

185 Panitz, J.C. and Novak, P. (2001) *J. Power Sources*, **97–98**, 174.

186 Goers, D., Buqua, H., Hardwick, L., Würsig, A., and Novak, P. (2003) *Ionics*, **9**, 258.

187 Nakajima, T. (2007) *J. Fluorine Chem.*, **128**, 277.

188 Reynier, Y., Yazami, R., Fultz, B., and Barsukov, I. (2007) *J. Power Sources*, **165**, 552.

189 Wilcox, J.D., Doeff, M.M., Marcinek, M., and Kosteki, R. (2007) *J. Electrochem. Soc.*, **154**, A389.

190 Hong, J., Wang, C., Dudney, N.J., and Lance, M.J. (2007) *J. Electrochem. Soc.*, **154**, A805.

191 Julien, C.M., Zaghib, K., Mauger, A., Massot, M., Ait-Salah, A., Selmane, M., and Gendron, F. (2006) *J. Appl. Phys.*, **100**, 63511.

7
Development of Lithium-Ion Batteries: From the Viewpoint of Importance of the Electrolytes

Masaki Yoshio, Hiroyoshi Nakamura, and Nikolay Dimov

7.1
Introduction

Lithium has a low atomic weight and a high ionization potential that results in high energy density compared to lead and zinc electrode materials widely used in the traditional batteries. These facts have been often highlighted by many scholars since the early 1970s. However, the development of rechargeable lithium systems working at ambient temperature has been retarded owing to the safety issues related with the secondary lithium anodes.

Lithium-ion batteries (LIBs), employing carbon materials instead of lithium metal as a negative electrode, were developed by Asahi Kasei Co., Japan [1] and on the basis of their patent, commercialized by Sony Co. and A&T Battery Co. in 1991 and 1992, respectively. Later on, LIBs were widely accepted as a lightweight, compact electric power sources for mobile electric devices because of their high energy density, availability in a wide temperature range owing to organic solvent electrolytes, and the lack of memory effect. LIBs have been applied to cellular phones, laptop computers, camcorders, digital cameras, etc. Recently, the area of application of LIBs has been extended to power tools, battery-assisted bicycles, hybrid electric vehicles (HEVs), etc.

Figure 7.1 depicts the increase in capacity of the cylindrical 18650 cells (diameter 18 mm, length 650 mm) from 1992 to 2006. Initially developed carbonaceous materials exhibited capacity of roughly 200 mA h g^{-1} and $LiCoO_2$ usable capacity was nearly 130 mA h g^{-1} because the charging voltage was limited to 4.1 V. Because of these limitations, the overall capacity of these early batteries was smaller than 1 A h. The energy density of the LIBs has gradually increased by employing carbon materials having higher capacity. At that time it was widely believed that the highly crystallized graphite with larger capacity cannot be used as an anode for LIBs because of the severe catalytic electrolyte decomposition on its surface in the course of repetitive charge/discharge cycles. Notably such decomposition occurs in all alkyl carbonate solutions, that is, not only in propylene carbonate (PC)-based but also in ethylene carbonate (EC)-based electrolytes. However, the use of EC electrolyte

Lithium Ion Rechargeable Batteries. Edited by Kazunori Ozawa

ISBN: 978-3-527-31983-1

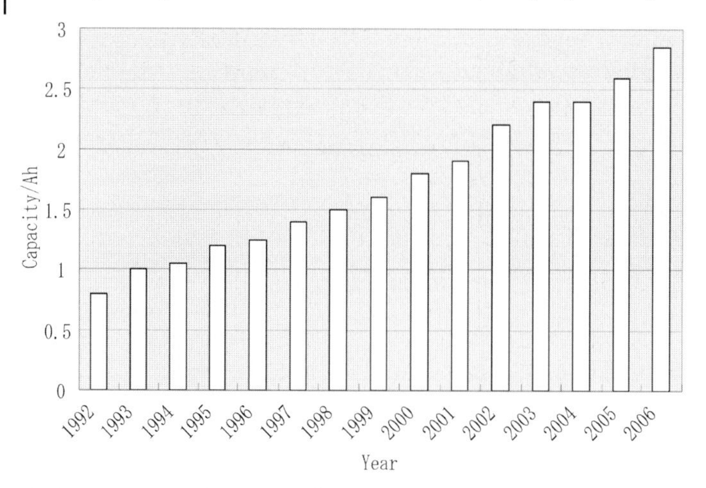

Fig. 7.1 Increase in the 18650 cell capacity year by year.

solvent to replace PC electrolyte solvent greatly suppresses solvent decomposition on the surface of the graphite anodes and was the key to developing higher capacity cells, which was achieved in the mid-1990s. Since most of the highly graphitized graphite blends increase solvent decomposition to unacceptably high levels, artificial graphite with lower degree of graphitization such as graphitized mesophase carbon (first generation MCMB 6–28) was chosen as an anode. Its practical capacity is around 280 mA h g^{-1}.

In 1997–1998, the energy density of the LIBs increased rapidly. Functional electrolytes were introduced by H. Yoshitake. They contain several additives and all electrolyte ingredients are highly purified to overcome the low battery performance of highly graphitized graphite anodes [2, 3]. Following the introduction of the functional electrolytes, the hard carbon anodes or graphite with lower degree of graphitization were quickly replaced by graphite anodes, which resulted in the quick increase of the LIB energy density shown in Figure 7.1. The electrochemical mechanism explaining the role of the additives is as follows: all compound additives decompose during the first charging prior to electrolyte decomposition. Decomposition products of the additive cover the active points of the graphite anode and suppress further electrolyte decomposition at the active sites. Therefore, the surface film formed by the additives on the anode plays an important role and partly replaces the solid electrolyte interface (SEI) film formed by the decomposition of the electrolyte solvents. This research result allows increase in the degree of graphitization of the anode carbon and now the practical capacity of the most anodes used in LIBs is close to the theoretical value (372 mA h g^{-1}). Functional electrolytes contain various additives such as 1,3-propanesulton (PS) [4], vinylene carbonate (VC) [5], and some other anode-film formers that improve not only the cyclability of LIBs but also their safety. The amount and the type of the additives vary and depend on the degree of crystallization of the graphite. Among the most suitable graphite blends that show superior battery performance in the presence of these electrolyte additives is the artificial graphite called "MAG," which is produced by Hitachi Chemical Co. Superior performance of this material is explained by its morphology, which allows the electrolyte to penetrate inside the MAG particles through a set of tiny channels [6]. The MAG market share in Japan is more than 60% in the field of LIBs for cellular phones. Soon after H.Yoshitake (Ube chemical Co.), M. Ue (Mitsubishi Chemical Co.) commercialized another electrolyte blend containing various additives under the name Role Assigned Electrolytes [7, 8].

Production of LIBs with graphite anodes is now widespread. At the same time, laminated film-packaged cells have been developed with the invention of additives.

In 2003, novel additives for improving the cathode capacity were developed by H. Yoshitake [16, 17] and the capacity of the 18650 cells reached 2.4 A h. It corresponds to an energy density of over $200\,\mathrm{W\,h\,kg^{-1}}$ or $500\,\mathrm{W\,h\,l^{-1}}$. These values were reached in part by increasing the cell operating voltage to 4.2–4.3 V, which, in turn, increases the practical capacity of $LiCoO_2$ in the LIB. However, charging voltage exceeding 4.2 V leads to electrolyte oxidation on the surface of $LiCoO_2$. To suppress the decomposition of the cell electrolyte at the active sites on the surface of $LiCoO_2$, new additives should be firstly decomposed at the active points of the cathode; these additives should suppress further electrolyte decomposition in a way similar to that at the anodes. Surface films formed at high voltage/high oxidation potential were called *conductive membranes* by Yoshitake. This new finding is particularly useful in the case of new cathode materials with higher energy density. For example, $LiCo_{1/3}Mn_{1/3}Ni_{1/3}O_2$ has comparatively low capacity when the charging voltage is limited to 4.2 V.

There are at least two ways to further increase the energy density of the LIB. The first is to work in a wider voltage window, which means increasing the charging cutoff voltage, or decreasing the discharging cutoff voltage, or changing both. The second is to adopt a new generation of anode materials, based on the alloying type of electrochemical reaction. Alloying of lithium with certain elements is very attractive because the Li : M mole ratio in the Li_xM alloy at the end of charge might be much higher than in the case of the intercalation hosts, which generally cannot accommodate and release large amount of Li^+ to maintain a stable crystal lattice over the cycles. The most promising elements that form binary alloys with Li in an appropriate potential window are Al, Si, Sn, and their alloys. Silicon is the most attractive because of its low atomic weight, high lithium uptake, and low alloying potential versus Li/Li^+. Replacing the graphite anodes widely used at present with silicon may result in a next generation of LIBs with energy densities approaching that of the primary lithium cells. Successful development of such LIBs is supposed to be closely related to the development of new binding systems, electrode fabrication techniques, and electrolyte additives specifically designed to protect the moving phase boundaries of the alloying type of anodes.

We describe the importance and the various roles of the additives, which are now indispensable materials in the LIB industry.

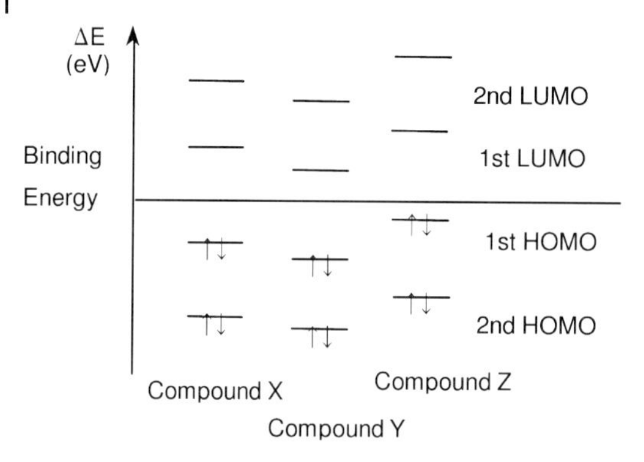

Fig. 7.2 LUMO and HOMO values of compounds, X, Y, and Z.

7.2
General Design to Find Additives for Improving the Performance of LIB

The fundamental process for finding appropriate electrolyte additives consists of the following steps: first, appropriate software package (in our case, Win-MOPAC V3.5) that implements molecular orbital (MO) theory is employed to calculate the energies of the lowest unoccupied molecular orbital (LUMO) and the highest occupied molecular orbital (HOMO) of the respective molecule under consideration.

LUMO is the orbital that acts as an electron acceptor, because it is the lowest energy orbital that has room to accept an electron. Having lower LUMO energy means that the respective molecule is a better electron acceptor and the corresponding compound will tend to decompose at higher voltage. The latter is a favorable property for additives at the anode side. If we consider three compounds X, Y, and Z with HOMO and LUMO energies as shown in Figure 7.2, they will tend to decompose on the anode site in the following order: Y > X > Z.

On the other hand, high HOMO energy relates to the tendency for oxidation. Therefore, the same compounds will tend to decompose on the cathode site of the cell in the following order: Z > X > Y.

It is interesting to classify the electrolyte solvents for 3- and 4-V nonaqueous batteries from the standpoint of MO theory. Figure 7.3 shows the calculated LUMO and HOMO for various solvents used to prepare 3- and 4-V electrolytes. These are the most popular electrolyte solvents for LIBs at present. They are PC, EC, dimethyl carbonate (DMC), diethyl carbonate (DEC), ethyl methyl carbonate (EMC), and $\gamma\gamma$–butyrolactone (γ-BL) for the case of 4-V electrolytes, and 1,2-dimethoxy ethane (DME), diethoxy ethane (DEE), tetrahydrofuran (THF), and 2-methyl tetrahydrofuran (2MeTHF) for the case of 3-V cells. It is clearly seen (Figure 7.3) that the solvents for 4-V electrolytes have lower HOMO energies than those for 3-V electrolytes, which means that it is rather difficult to oxidize them. High resistance toward oxidation of these 4-V electrolyte solvents is a natural requirement, keeping in mind that the LIB cathode materials in their charged state consist of metal oxides that are strong oxidizers. On the contrary, electrolyte solvents for 3-V cells possess

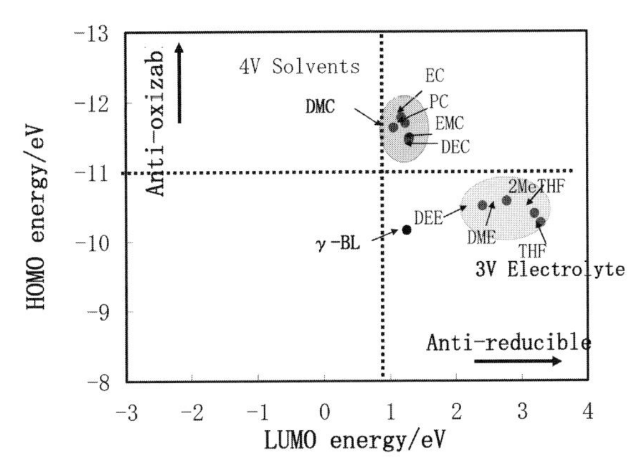

Fig. 7.3 Classification of electrolyte solvents for 3- and 4-V battery electrolyte using MO theory.

higher LUMO energies, making them reduction-resistant and compatible with the Li metal used as an anode.

It is noteworthy that appropriate range of HOMO/LUMO energies of the candidate electrolyte additive molecule is a necessary, but by no means sufficient, condition that determines whether a given compound will act as an appropriate electrolyte additive. Besides possessing HOMO/LUMO orbital lying within the appropriate energy range, the candidate compound should satisfy at least the following conditions:

1. The additive should be stable in the presence of the electrolyte.
2. The additive should be reduced prior to the solvent decomposition and form alternative SEI on the anode.
3. The additive should be electrochemically stable on the cathode site in the course of repetitive charge/discharge cycles.

Note that requirement (2) relates to the MO concept, but is not equivalent to it, because some additives may not decompose at a sufficiently high rate though their HOMO/LUMO values may lie within the appropriate energy range. Therefore, requirement (2) represents the need of sufficiently fast kinetics of the alternative SEI layer formation process.

Therefore MO theory gives us the necessary first step of the research, namely, estimating the reduction and oxidation potential of chemicals and screening the appropriate compounds as anodic or cathodic additives for LIB electrolyte according to their decomposition potentials. The relation between the LUMO energies and the reduction potentials of some chemical compounds and solvents measured using platinum electrode in $LiPF_6$ solutions is shown in Figure 7.4.

The following compounds have been studied in this manner: PS, VC, vinyl acetate (VA), PC, EC, allyl methyl carbonate (AMC), ethylene sulfite (ES), methyl acetate (MA), divinyl adipate (ADV), and catechol carbonate (CC) and its derivatives [10–12]. Figure 7.4 shows that the decomposition potential of these compounds

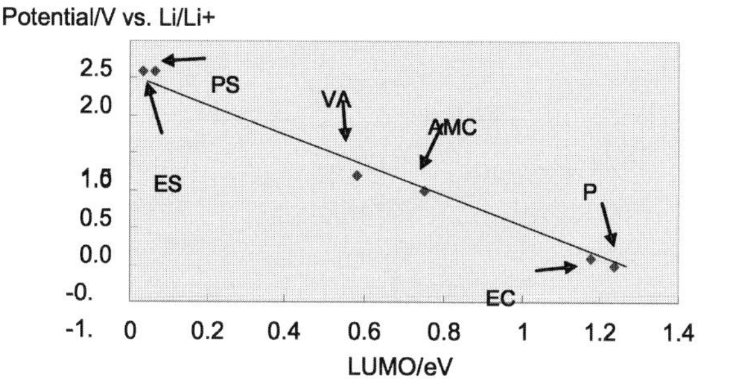

Fig. 7.4 Relation of LUMO values and reduction potential measured on a platinum electrode.

correlates very well with their LUMO energies. Such a straightforward relationship allows us to estimate the reduction potential of the candidate additives by means of the calculated LUMO values. It is noteworthy that PC or EC decompose at around −0.5 V (versus Li/Li^+) on platinum electrodes, but at around 0.5–0.8 V versus Li/Li^+ on graphite electrodes owing to their catalytic action. Therefore, appropriate additives should decompose at potentials higher than 0.5 V versus Li/Li^+, form stable films on the graphite anodes, and prevent electrolyte decomposition due to catalytic decomposition of the solvents. In accordance with Figure 7.4, appropriate LUMO values of the candidate anode additives should lie below 1 eV, i.e., higher reduction potential than the PC or EC decomposition potential at the graphite anode. We have determined temporarily that the appropriate LUMO energies of the candidate anodic additives should fall within −1 to 1 eV.

The relationship between the HOMO energies and the oxidation potential of the same compounds measured using platinum electrode in $LiPF_6$ solutions is shown in Figure 7.5. HOMO energies correlate in a linear fashion with the oxidation potential and we are able to estimate the oxidation potential of the anodic candidate additives by their HOMO energy values in a way similar to the case of LUMO energies and reduction potentials.

As mentioned above, the anode additives should not get decomposed by the cathode oxides and therefore their HOMO energies should be distributed between −9 and −11.5 eV. The value of −11.5 eV stands for the oxidation potential of PC/EC.

Summarizing the considerations above, we finally arrive at the following rule of thumb for screening anodic electrolyte additives: HOMO energies of the suitable candidates should fall between −11.5 and −9 eV, while their LUMO energies should fall between −1 and 1 eV. This rule takes into account the fact that the additive should decompose prior to the electrolyte solvent on the anode surface and at the same time remain resistant against oxidation at the cathode site.

On the contrary, suitable cathode additives should decompose on the surface of the cathode site, cover the active points of the respective metal oxides, and remain

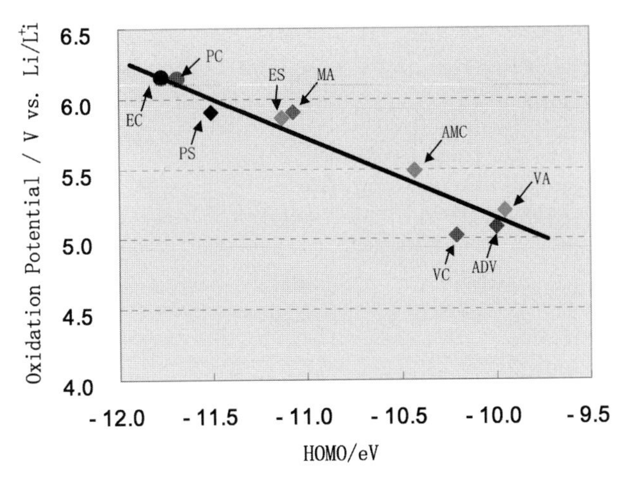

Fig. 7.5 The relation between HOMO energies and oxidation potential of several anodic additives and solvents.

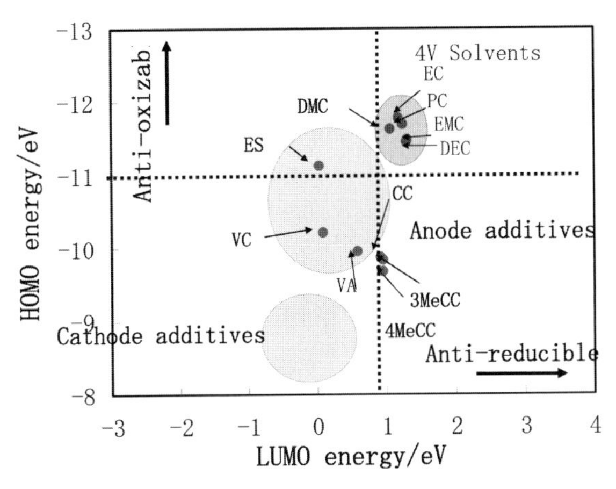

Fig. 7.6 The suitable HOMO and LUMO energy range for anode and cathode candidate electrolyte additives.

resistant toward reduction on the anode site. Therefore HOMO energies of these candidates should fall between −8 and −9.5 eV and their LUMO energies between −1 and 1 eV. These regions are shown in Figure 7.6.

7.3 A Series of Developing Processes to Find Novel Additives

After the invention of the functional electrolyte, the usage of highly crystalline graphite as an anode material for LIBs has become possible and this has led to

Table 7.1 Successive steps for developing the electrolyte additives.

- ♦ First screening: evaluation through MO calculation (calculation of HOMO and LUMO energies)
- ♦ Second screening: analysis of physical properties of the surface layer (oxidation/reduction potential, conductance, viscosity of electrolyte)
- ♦ Third screening: charge/discharge and battery test using various electrolytes
- ♦ Analysis of the used cell and characterization (electrolyte and SEI characterization by GC, HPLC, NMR, FT-IR)

an increase in the energy density of the cells. These types of electrolytes suppress gas evolution from the electrodes, especially during the first charging and greatly improve the cycle performance of the cells. Functional electrolytes have also contributed to the usage of PC with graphite anodes. Each functional electrolyte contains several kinds of additives and highly purified electrolyte ingredients. Owing to the high purity of all electrolyte ingredients, HF formation is greatly suppressed. Its concentration in the electrolyte is limited to less than 20 ppm for more than three months and the electrolyte remains stable and colorless. The functional electrolyte contains at least PS and VC plus typically two or three extra additives taken in small quantities to improve the performance of the highly graphitized anodes. These additives have been developed through several successive steps shown in Table 7.1.

After the first screening of the candidate additives, their effectiveness should be examined directly during the second screening step. Decomposition of PC-based solvents was employed as a criterion for estimating the effectiveness of the candidate additives. Since PC solvents are easily decomposed on the graphite anode, PC-based electrolytes were used as fingerprint solvent systems by changing the PC solvent ratio and the second solvent type [13]. Thus electrolytes based on the mixed solvent PC/DMC can passivate graphite anodes more effectively than PC/DEC mixed solvent because the reduction products of DMC-containing electrolytes have lower solubility than those of DEC-containing electrolytes [14]. Results of the cycle experiments of several additive candidates have been published [12].

Charge/discharge curves of a graphite anode in PC/DMC (1 : 2 by volume) mixed solvent systems are shown in Figure 7.7. PC decomposes at a 0.7–0.8 V (versus Li/Li^+), which is clearly seen as a long plateau in voltage/capacity coordinates. Such a plateau always appears in the absence of additives due to severe decomposition of PC at the graphite anode. In this case, PC molecules decompose by means of a one-electron reaction [10].

The intercalation/deintercalation of Li^+ into graphite seems to be very smooth in the presence of a PS additive as shown in Figure 7.7(b). PS seems to form a protective layer on the graphite during the first charging process, and Li^+ intercalation occurs even in PC-based electrolyte. This example clearly shows that PC-based electrolytes are a convenient and effective means of evaluating the additives and are widely used as a fingerprint electrolyte. PS, VC, VA, CC, and succinimides

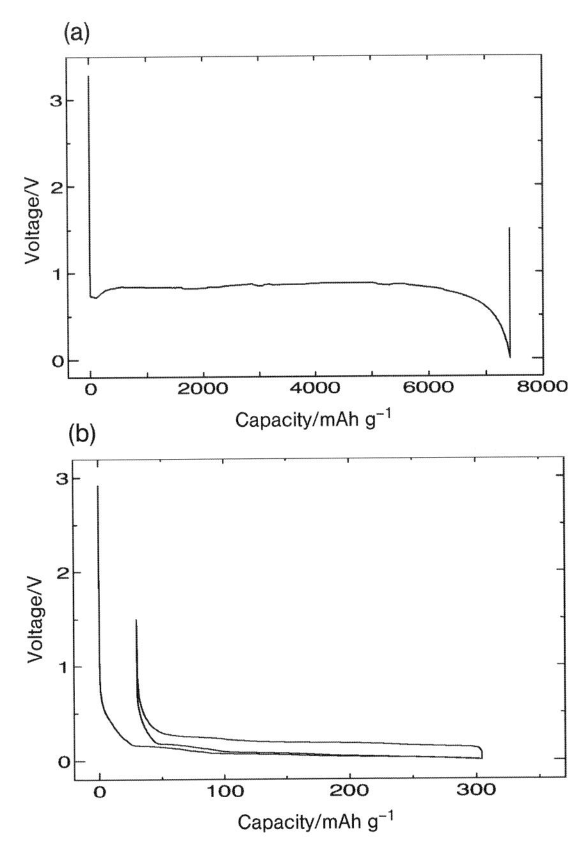

Fig. 7.7 Charge/discharge curves of natural graphite in 1 M $LiPF_6$-PC/DMC (1 : 2): (a) in the absence of additive and (b) in the presence of 1% PS.

studied in such a way proved to be effective additives forming protective layers with appropriate properties on the surface of the graphite anodes.

Results obtained using PC-based electrolytes served as a basis for the third screening shown in Table 7.1. It consists of a practical life-cycle test performed with an EC-based electrolyte in the presence of the additive that has passed the second screening. Figure 7.8 shows the cyclability of $LiCoO_2$/graphite cell with VC and PS additives. The electrolyte containing PS shows better cyclability than that containing VC.

However, the selection of additives is not so simple, because their effectiveness depends also on the nature of the graphite anode. AMC seemed to form a protect layer film as other additives do, but the mobility of Li^+ within the SEI film derived from AMC reduction is too low to maintain adequate lithium insertion into graphite at a sufficient rate. Therefore the last two steps shown in Table 7.1 are absolutely necessary to approve the suitability of candidate additives.

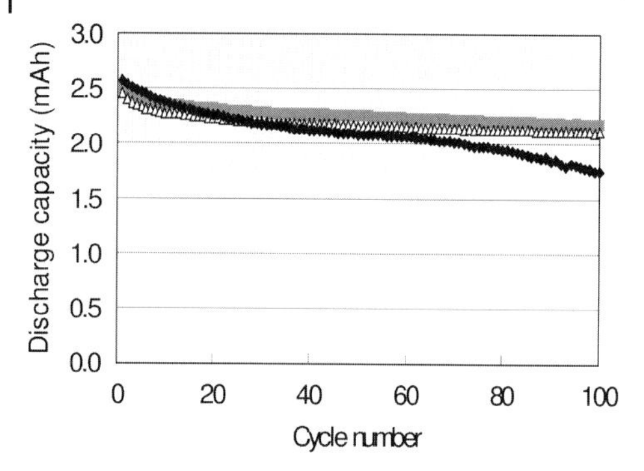

Fig. 7.8 Cyclability of 18650 $LiCoO_2$/graphite cells in the EC-based electrolyte: (a) without additive (gray dots, the best), (b) 1% PS addition (empty triangles, middle), and (c) 1% VC addition (black dots, the worst).

7.4 Cathodic and the Other Additives for LIBs

The energy density of an LIB that consists of $LiCoO_2$ as a cathode and graphite as an anode had almost approached its theoretical maximum in 2004. Trials to use new cathode materials such as layered Ni-based materials, for example, $LiNi_xCo_yMn_zO_2$, where $x + y + z = 1$, were initiated to increase the energy density of the LIB. Although the capacity of $LiNi_xCo_yMn_zO_2$ is higher than that of the $LiCoO_2$, it is necessary to charge it to higher than 4.2 V or discharge it to less than 3.0 V to achieve higher energy density. However, Ni-based materials tend to deteriorate when the charging voltage becomes higher than 4.2 V. Since the potential distribution on the surface of the cathode is not homogeneous, the electrolyte would decompose at the active points of the cathode when charging potential exceeds 4.2 V. Because the principles behind designing cathode additives resemble those of anode additives, new types of cathode additives were introduced in 2004. One of the proposed additives is biphenyl (BP) and its derivatives. BP was originally developed as an overcharge protector because it starts releasing substantial amount of gas at around 4.5 V versus Li/Li^+. The gas released at such a high voltage opens the safety cap at a certain pressure and then the battery operation stops [15]. To develop a gas pressure that is sufficient to open the safety vents, at least several percent addition of BP is needed. This large BP excess leads to the formation of a protective film that is too thick. Such a thick film covers the cathode so tightly that it makes the cathode inert and no longer usable.

To use BP as a cathode additive, one should add it in much smaller quantities. Thus addition of 0.1–0.2% BP yields a very thin film on the cathode surface formed during the initial charging process. It covers the cathode-active sites and improves the cyclability of the cathode [16, 17].

The effect of BP content on the cyclability of $LiCoO_2$/graphite cells was studied under the following conditions: charging to 4.3 V at 45°C, which is a hard cycle test condition for the cells. The cyclability of the cell is poor without additives, but

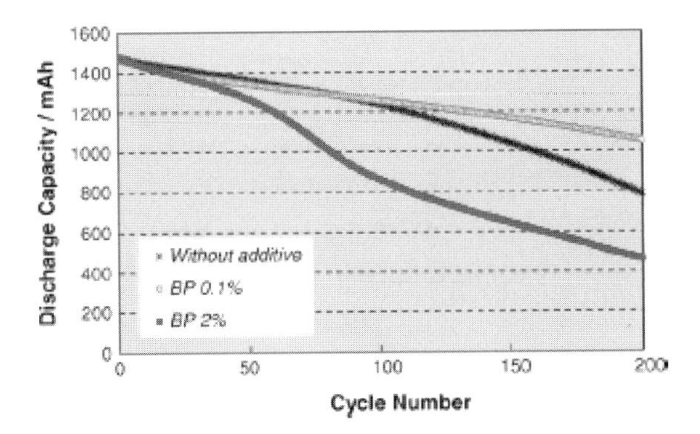

Fig. 7.9 Cycle performance of $LiCoO_2$/graphite cylindrical cells with 1 M $LiPF_6$-EC/MEC 3/7 electrolyte for three cases: without additive, with BP 0.1 wt %, and with BP 2 wt %. Voltage window 2.7–4.3 V, test performed at 45°C [17].

the electrolyte containing 0.1% of BP shows the best results as seen in Figure 7.9. Thick film with low conductivity was formed at the first charge/discharge in the case of high BP content (2%), which suppressed the smooth reaction of Li^+ with the cathode. On the contrary, polymerized thin conductive film was formed on the surface of $LiCoO_2$ in the case of BP 0.1% addition. The conductive polymerized thin film is called *electroconducting membrane* (*ECM*) by Abe *et al.* Novel type batteries using Ni–Mn–Co materials as a cathode are now being developed by means of cathode additives.

The cycle performance of $LiCoO_2$/graphite cells was tested using several cathode additives in the presence of the popular anode additive, VA. The results are shown in Figure 7.10. In the presence of VA alone, capacity degradation develops roughly 100 cycles after the beginning of the test because of the electrolyte decomposition taking place at the cathode site owing to the higher charging voltage. However, electrolytes that contain heterocyclic compounds in addition to the anodic additive (VA) show satisfactory cycle performance, similar to the case of BP addition.

As discussed above, the recent high-quality LIBs contain several different types of additives. One of the important additives is the overcharge protector. This additive releases hydrogen gas at higher voltages and activates the current interrupt device and safety vent to prevent fire and smoke. Cyclohexylbenzene (CHB) is a commonly used overcharge protection additive along with BP.

By virtue of the important role of electrolytes in LIBs, there are different ways of modification from different viewpoints. For instance, there are electrolytes designed for better SEI performance of carbon anode material as mentioned above and additives such as trioctyl phosphate, alkyl and aryl esters R-COOR′, for improving the wettability of the separator. Fluorinated cyclotriphosphazene has been developed as a novel flame-retarding additive in the electrolyte for LIBs. Phosphazenes have a good effect on the thermal stability of $LiCoO_2$ cathode without affecting the performance of the LIB. The possible mechanism for the influence of the additive on the thermal stability of $LiCoO_2$ has been presented in Ref. [18].

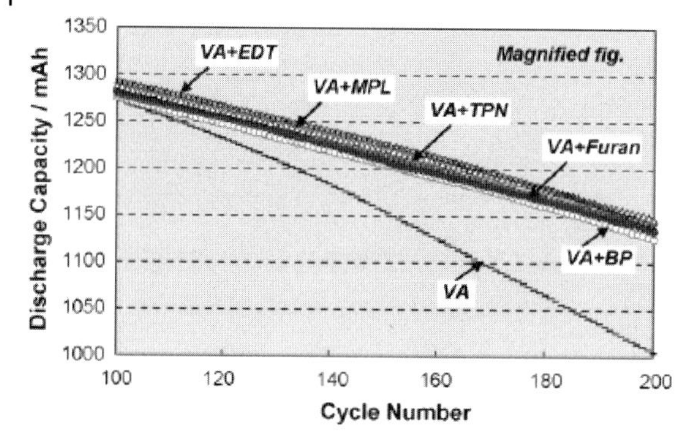

Fig. 7.10 Cycle performance of $LiCoO_2$/graphite cylindrical cells in 1 M $LiPF_6$-EC/MEC 3/7 in the presence of VA (1 wt %) as a base electrolyte. Various cathode additives have been tested (BP, *N*-methylpyrrole (MPL), Furan, thiophene (TPN), and 3,4-ethylenedioxythiophene (EDT)). Amount of the additives maintained at 0.1 wt %. Test conditions: 2.7–4.3 V at 45°C [17].

7.5 Conditioning

Formation of SEI caused by electrochemical reaction on the carbonaceous electrode/electrolyte interface has been debated for a long time. SEI film strongly affects the gas evolution during charge/discharge cycles. Such a film always forms on the surface of the graphite electrodes in the presence of various electrolytes. This is why controlling the SEI film formation is very important for the LIB industry. The process of SEI film formation and growth is called *conditioning* or *aging* in the LIB community. This is one of the most important manufacturing operations, because the safety of LIB depends mainly on this process. Two types of conditioning are considered here:

1. Conditioning process in the absence of anode additives is performed in the following way: first, fast charging is done with a rather high current density. Such a charging mode yields SEI on the anode with minimum electrolyte decomposition. After the formation of the SEI film, the cell is kept in its charged state at room or elevated temperature for two or three weeks to grow the SEI film. The voltage of the cells is measured again following the period of storage. Differences in voltage at the start and at the end of storage period are used to sort out the cells. Cells having lower voltage after storage develop a tendency of internal short circuits due to dendrite lithium metal deposition or other metal deposition. Since the cylindrical can is composed of stainless steel, which tolerates gas pressure of up to 20 atm, there would be no need for additives for the graphite anode. However, recent big accidents such as the fire that occurred in the case of cells lacking electrolyte additives or containing less additives has led to increase in the conditioning time.
2. Conditioning process in the case of well-designed additives for anode graphite is performed in a different manner. Since the protective layer film on graphite

is formed by the well-designed additives and not by solvents (common SEI film), the applied charging rate is rather slow (around 1/4C rate). Several charge/discharge cycles are completed at this rate to complete the protective film layer formation. After the completion of several charge/discharge cycles, the cells are kept again at their charged state at room or elevated temperature for several days to grow the protective film.

Figure 7.11(a) shows the appearance of natural graphite. Figure 7.11(b) and (c) shows the SEI film without and with additive (VA), respectively. It is clearly seen that the SEI film formed on the graphite surface is not homogeneous. A large number of minute spots are observed instead. The spots form because of addition of the VA to the electrolyte. Adequately formed SEI film plays a crucial role in the safe operation of the LIB.

Intercalation of Li^+ into graphite in PC-based electrolytes is possible even in the absence of additives, but only at the expense of a reduced PC ratio as shown in Figure 7.12(a). However, in this case a voltage plateau at around 0.6 V appears, lowering the initial coulomb efficiency, and metallic lithium deposition seems to occur. The area of the active sites in an LIB is large. It is the result of the different potentials at each point of the anode. Such nonhomogeneous voltage

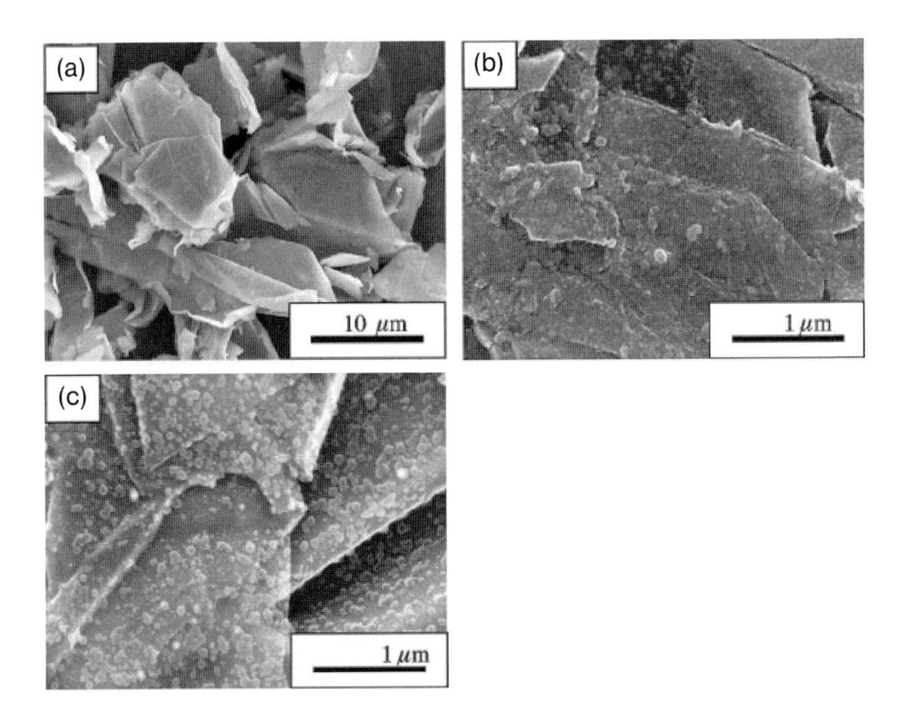

Fig. 7.11 SEM of (a) natural graphite, (b) natural graphite after charge in the absence of VA, and (c) natural graphite after charge in the presence of VA.

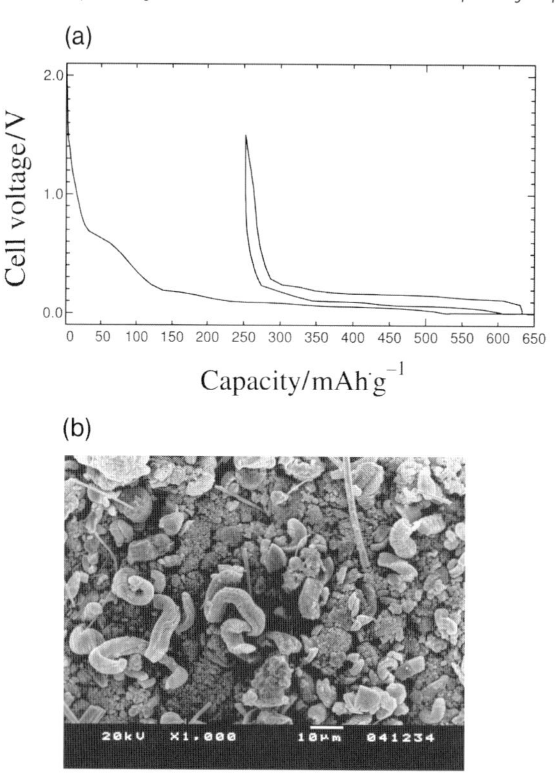

Fig. 7.12 (a) The voltage profile of graphite versus Li/Li^+ in 1 M $LiPF_6$ PC : MEC = 1 : 4. (b) SEM image of graphite electrode cycled versus Li/Li^+ in 1 M $LiPF_6$ PC : MEC = 1 : 4.

distribution facilitates the deposition of metallic lithium because there are points having potential approaching 0 V versus Li/Li^+.

Figure 7.12(b) clearly illustrates that the charge distribution on the graphite surface is not homogeneous. There are some parts of electrode surface that resemble exfoliated/expanded graphite. Li^+ cannot intercalate into the exfoliated graphite. On the other hand, it is a good electric conductor, on which lithium or metal ion deposition may take place.

This would be a typical example of a graphite-based LIB in several tens of cycles. Some graphite would be damaged or exfoliated, and would not accept Li^+ intercalation, but would facilitate Li metal deposition. The conditioning process aims to establish a stable and safety SEI film.

Another important and interesting result revealing the importance of the electrolyte additives is described next [9]. Figure 7.13(a) and (b) shows a graphite anode after the completion of six cycles under the following conditions: voltage window 0.005–2.000 V, current density 0.4 mA cm^{-2} at 0°C. The cell configuration was Li/1 M $LiPF_6$-EC:DMC (1 : 2 by vol.)/graphite. In each cycle, the cells were equilibrated at 0.005 V versus Li/Li^+ for 10 hours. The color of the electrode shown

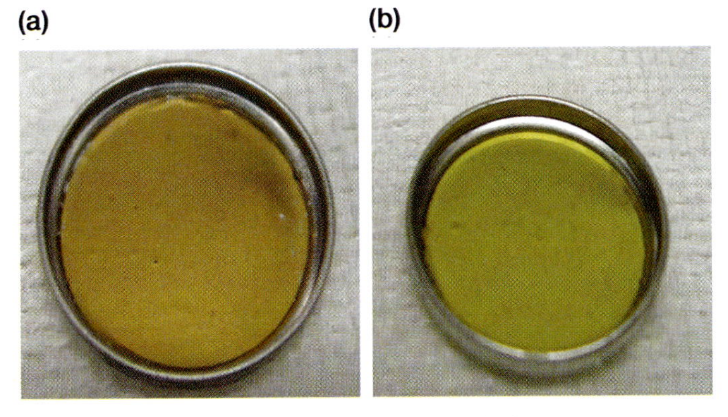

Fig. 7.13 Appearance of the graphite electrode cycled in the (a) absence of electrolyte additives at 0°C versus Li/Li^+ and (b) presence of 0.5 wt % propane sulton at 0°C versus Li/Li^+.

in Figure 7.13(a) is not golden, which is the first-stage color of lithiated graphite (LiC_6) because low-temperature cycling induces Li metal deposition especially in the absence of additives.

When using an appropriate additive (0.5 wt %), for example, PS, the appearance of the same type of electrode under the same conditions is shown in Figure 7.13(b). It has the typical golden color of LiC_6. This means that the additive was decomposed during several charge/discharge cycles and deposited at the active sites of graphite, preventing the formation of metallic Li. Hence, for the LIB industry, additives are the key materials that make it possible to use highly crystalline graphite anodes and suppress the Li metal coexistence with lithiated graphite, thus increasing the safety of LIBs.

References

1. Yoshino, A., Jitsuchika, K. and Nakajima, T. (1985) Jap. Pat. 1989293 (1985/5/10), Assahi Chemical Ind.
2. Yoshitake, H. (1999) Techno-Frontier Symposium Makuhari, Japan.
3. Yoshitake, H. (2000) *Functional Electrolyte in Lithium Ion Batteries* (M. Yoshio, A Kozawa, and N.K. Shinbunsha) Nikkan Kogyou Shinbun, pp. 73–82 (in Japanese).
4. Hamamoto, S. and Hidaka, A., and Abe, K. (1997) US P.6,033,809 (1997/8/22).
5. Fujimoto, M., Takahashi, M., and Nishio, A. (1992) Jap.Pat. 3059832 (1992/7/27).
6. Ishii, Y. Fujita, A., Nishida, T., and Yamada, K. (2001) HitachiKasei Tech. Rep. 36(2001-1)27.
7. Ue, M. (2003) Techno-Frontier Symposium, Makuhari, Japan.
8. Ue, M. (2005) 22nd International Battery Seminar and Exhibit, Fortlauderdale, Fl.
9. Park, G.-J., Nakamura, H., Lee, Y.-S., and Yoshio, M. (2009) *J. Power Sources*, **189**, 602–606.
10. Wang, C., Nakamura, H., Komatsu, H., Yoshio, M., and Yoshitake, H. (1998) *J. Power Sources*, **74**, 142.
11. Yoshitake, H., Abe, K., Kitakura, T., Gong, J.B., Lee, Y.S., Nakamura, H.,

and Yoshio, M. (2003) *Chem. Lett.*, **32**, 134.
12 Abe, K., Yoshitake, H., Kitakura, T., Hattori, T., Wang, H., and Yoshio, M. (2004) *Electrochim. Acta*, **49**, 4613.
13 Nakamura, H., Komatsu, H., and Yoshio, M. (1996) *J. Power Sources*, **63**, 213.
14 Aurbach, D., Daroux, M.L., Faguy, P.W., and Yeager, E. (1987) *J. Electrochem. Soc.*, **134**, 1611.
15 Mao, H. (1999) U.S. Pat. 5,879,834.
16 Abe, K., Takaya, T., Yoshitake, H., Ushigoe, Y., Yoshio, M., and Wang, H. (2004) *Electrochem. Solid-State Lett.*, **7**, A462.
17 Abe, K., Ushigoe, Y., Yoshitake, H., and Yoshio, M. (2006) *J. Power Sources*, **153**, 328–335.
18 Zhang, Q., Noguchi, H., Wang, H., Yoshio, M., Otsuki, M., and Ogino, T. (2005) *Chem. Lett.*, **34**, 1012.

8
Inorganic Additives and Electrode Interface

Shinichi Komaba

8.1
Introduction

Electrolyte additives for lithium-ion batteries are widely studied to improve the battery performance. The active materials used for the positive and negative electrodes as well as the functional electrolytes are required to realize satisfactory higher energy, power densities, and cyclability. In view of recent progress in realizing improved functionality of organic and inorganic electrolyte additives, this chapter reviews the impact of various inorganic ions as electrolyte additives and electrode coatings on the battery performance on the basis of these recent achievements.

In chemical batteries, energy storage and conversion are achieved by the reduction and oxidation of active materials accompanied by electron and ion transfer. The reversible electrochemical reaction enables us to fabricate rechargeable or secondary batteries. Higher energy density depends on the higher electrochemical activity of the electrode materials, and therefore, researchers in the fields of materials science, solid-state chemistry, and electrochemistry are now focusing on highly electroactive materials for negative and positive electrodes. Considerable amount of research on the electrode materials is being carried out in the development of new materials for next generation secondary batteries.

Lithium-ion batteries possess the highest theoretical and practical energy density among the present commercial batteries. For advanced lithium-ion batteries, higher capacity and higher lithium-ion conductivity are required for the active materials and electrolyte, respectively; these essential elements determine the theoretical limit of energy and power densities. The operation voltage of lithium-ion batteries is the highest among practical batteries. This means that the highly oxidative and reductive states appear for the positive and negative electrode, respectively, in fully charged state, so that the stable interface of (negative electrode)/electrolyte and electrolyte/(positive electrode) is very important for long lifetime and adequate safety of the total battery system. To achieve the optimum performance in a practical cell, we need research and development of not only essential components,

Lithium Ion Rechargeable Batteries. Edited by Kazunori Ozawa

ISBN: 978-3-527-31983-1

(negative electrode)/electrolyte/(positive electrode), but also additives that influence the interface. It is here reviewed that electrolyte additives for lithium-ion batteries play an important role in bringing out the capability of the original material. Actually, use of electrolyte additives is one of the most economic and effective methods for the improvement of lithium-ion battery performance.

Several literatures and books [1–3] concerning lithium-ion battery materials have been published already. Recently, Zhang reviewed and discussed organic and inorganic chemical additives and their functionality [4]. In the literature, additives are divided into several categories according to their functions: (i) solid electrolyte interphase (or interface) (SEI) forming improver, (ii) cathode protection agent, (iii) $LiPF_6$ salt stabilizer, (iv) safety protection agent, (v) metallic Li anode improver, and (vi) other agents such as solvation enhancer, Al corrosion inhibitor, and wetting agent. This chapter gives a general description of the function and mechanism of each category of additives.

To improve the negative electrode performance in lithium-based secondary batteries, inorganic additives are known to be effective, some of which are CO_2 [5–7], HF [8, 9], HI [10], AlI_3, MgI_2 [11, 12], $AgPF_6$ [13], and $Cu(CF_3SO_3)_2$ [14]. Among various inorganic electrolyte additives, our group investigates the influence of dissolved metal ions into electrolyte on the performance of lithium-ion batteries. Our early motivation was a quantitative understanding of the influence of a manganese electrolyte additive on the negative electrode because the deterioration of $C/LiMn_2O_4$ was induced by the manganese dissolution from an $LiMn_2O_4$ electrode [15]. We also emphasize the requirement and usefulness of the new additives to suppress the specific deterioration. On the basis of research work on dissolved metal species from positive electrodes [16], it was found that a sodium salt additive as well as a sodium salt coating is effective in enhancing battery performance. This chapter mainly reviews our recent achievements with dissolution of metal-ion additives into an electrolyte and with metal-ion additive coating.

8.2 Transition Metal Ions and Cathode Dissolution

The $C/LiMn_2O_4$ lithium-ion system is one of the most attractive in terms of cost, abundance, toxicity, cyclability, and reversible capacity. However, a serious problem standing in the way of the wider use of the spinel as a cathode is its poor storage performance at high temperature. This issue was addressed by several groups [17–21], and the general consensus was that the structural/chemical instability of the spinel framework resulting in the Mn dissolution was at the origin of the poor performance that was observed at high temperature. The dissolution mechanism and its influence on $LiMn_2O_4$ frame structure have been investigated over the past several years [22–26]. When the manganese species of Mn(II) compounds and/or ions are dissolved from the cathode, the soluble manganese species will be transferred by diffusion and migration toward the opposite anode through the thin

(a)

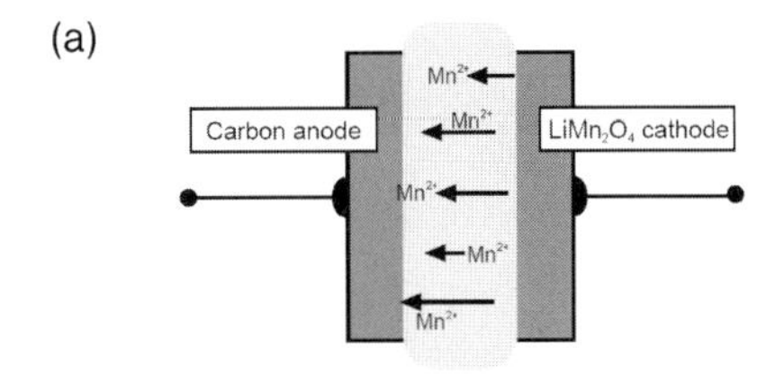

(b)

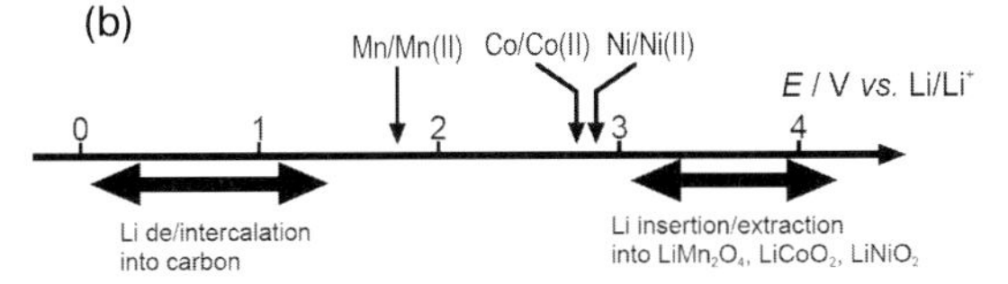

Fig. 8.1 (a) Schematic drawing of C/$LiMn_2O_4$ cell showing that dissolved manganese ions migrate and diffuse toward the carbon anode. (b) Potential relation of the reactions in the lithium-ion battery system, such as Li intercalation/deintercalation into active materials and standard redox potentials manganese, cobalt, and nickel.

separator as illustrated in Figure 8.1. They must be readily deposited on the carbon because of the high deposition potential of Mn/Mn(II) (1.87 V versus Li/Li^+).

In the Li/V_2O_5 secondary cell, the vanadium compounds also dissolved, and were deposited on the Li anode [27]. Furthermore, the influence of the dissolved manganese in primary Li/MnO_2 cells has also been reported [28]. In the case of secondary lithium-ion or metallic lithium cells, several research groups focused on the degradation of the anode, that is, the existence of Mn particle deposits was confirmed on the lithium metal after cycling [29, 30]. A very high impedance was detected for the carbon anode [31]. Amatucci *et al.*, especially, suggested that the most obvious fact would be to have the Mn^{2+} reduction on the carbon anode (essentially some degree of Mn plating) responsible for the large impedance of the anode surface [31], and this fact would be identical to that for the primary Li/MnO_2 cell [28]. A study on the influence of soluble species, which come from the cathode, on the carbon anode is important to understand the degradation mechanism of the battery and to improve the battery system [32]. Here, the influence of Mn(II) electrolyte additive on the carbon is clarified from graphite–Li half-cell tests and compared with those of Co(II) and Ni(II) additives [15, 16].

8.2.1 Mn(II) Ion

Natural graphite anodes usually show a constant reversible capacity of about 300–370 mA h g^{-1} depending on current density, and an irreversible capacity, which is well known to be due to the electroreductive formation of the SEI on graphite, is clearly observed at the first cycle [33, 34]. It is generally accepted that this is due to the reductive decomposition of the electrolyte on the electrode. As a

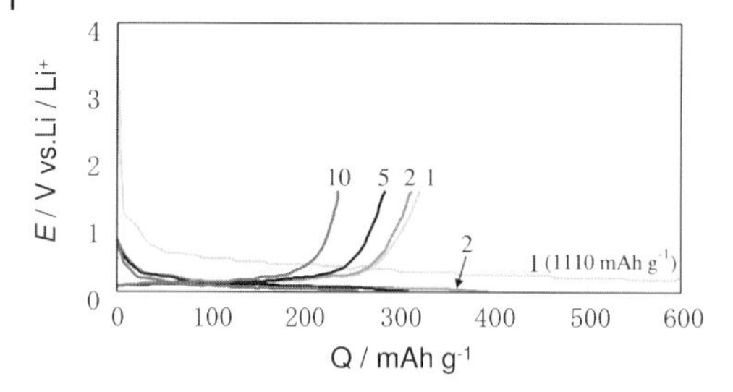

Fig. 8.2 Charge and discharge curves of graphite in $LiClO_4$ EC : PC (2 : 1) electrolyte containing 150 ppm Mn(II) by dissolving manganese perchlorate.

result of this decomposition, an SEI protective film that allows Li^+ ion transfer but prevents electron transfer is formed.

When the graphite electrode is tested in the $Mn(ClO_4)_2$ added electrolyte imitating the Mn dissolution from the Li–Mn spinel into an electrolyte, its battery performance severely suffers from the dissolved manganese. This indicates that the SEI layer did not suppress the reduction of Mn(II). From the galvanostatic tests, the initial charge capacity is remarkably increased by Mn addition, and the reversible capacity is drastically decreased. As shown in Figure 8.2, for 150 ppm Mn(II), the graphite electrode shows initial charge capacity of 1110 mA h g^{-1} from the 0.5-V region, which is thrice as large as the theoretical capacity of LiC_6 though the discharge capacity was about 300 mA h g^{-1}. As reported in [35], the Mn intercalation hardly occurs into graphite by cathodic reduction. So, the huge irreversible capacity is caused by Mn deposition and electrolyte decomposition on the Mn deposit as schematically illustrated in Figure 8.3. The battery performance is degraded by adding a small quantity of manganese. In a practical cell, the concentration of dissolved manganese increases up to several hundred parts per million; therefore, this degradation will occur more severely in a practical cell because of the dependence of Mn(II) concentration and quality. From scanning electron microscopy (SEM) and electron probe micro analyzer (EPMA) analyses, many white particles that consisted of manganese were observed on the entire surface after cycling. This is evidence for the deposition of manganese. In the case of bulk Mn metal electrode, high (electro-)chemical reactivity of electrolyte decomposition below 0.3 V (as shown in Figure 8.4) independent of the SEI was confirmed, suggesting the narrower potential window of the Mn deposits. Consequently, the initial charge capacity becomes very large with Mn deposition owing to the dramatic electrolyte decomposition, and thus the initial coulombic efficiency is very low – less than 30%.

When the Mn dissolution occurs during higher temperature operation (or storage) in a practical cell after cycling, the graphite electrode is already modified with the SEI. In this situation, manganese perchlorate was added immediately after the fifth discharge. Figure 8.5 shows the charge and discharge curves of graphite before and after 150 ppm Mn(II) addition. Though good reversibility is obtained until the fifth cycle, the sixth charge capacity reached 1680 mA h g^{-1}, which is much higher than 1110 mA h g^{-1} for the preaddition as mentioned above. Therefore, the

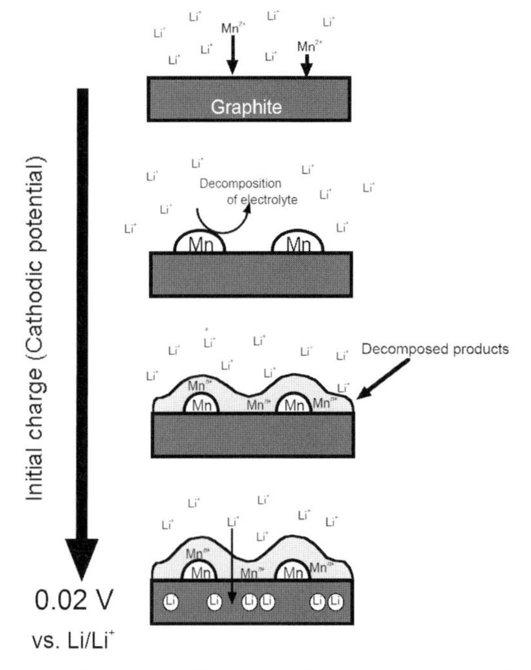

Fig. 8.3 Schematic illustration of the influence of Mn(II) addition in the electrolyte on the initial charge process of the graphite negative electrode.

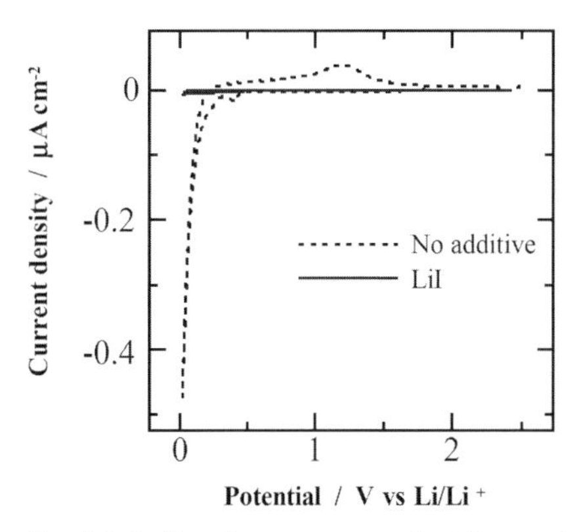

Fig. 8.4 Cyclic voltammograms of a Mn metal electrode at 0.1 mV s^{-1} in a Li-ion electrolyte with (solid line) and without (dashed line) LiI.

SEI layer cannot protect the graphite surface from the dissolved Mn(II), and the decomposition of electrolyte and/or SEI is accelerated by the deposited Mn surface.

From these results, when manganese dissolution occurs in the graphite–$LiMn_2O_4$, the discharge capacity will be decreased more severely by the graphite anode limit; this is because highly charging the graphite to 1680 mA h g^{-1} is not

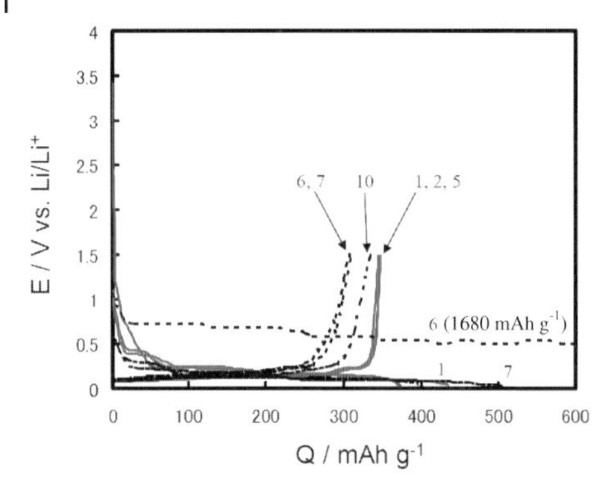

Fig. 8.5 Charge and discharge curves of graphite in $LiClO_4$ EC : DEC (1 : 1) during the initial five cycles, and then adding 150 ppm Mn(II) to the electrolyte before the sixth charge.

possible owing to the capacity balance of graphite and $LiMn_2O_4$ electrodes in a practical cell.

Briefly, the mechanism of degradation of the negative electrode by Mn(II) is as follows: first, manganese is deposited on the graphite surface followed by the drastic decomposition of electrolyte with Mn metal, and therefore, the larger irreversible capacity and the thick deposit on the electrode. Finally, the irreversible reaction with manganese causes the imbalance in the state of charge between negative and positive electrodes, resulting in the remarkable capacity loss of the practical cell.

8.2.2 **Co(II) Ion**

Cobalt dissolution from $LiCoO_2$ occurs in the higher potential region, that is, the overcharge region [36, 37]. A comparison between the influence of Mn(II) and Co(II) should give us a useful information to help understand the difference in the degradation between the C/$LiMn_2O_4$ and C/$LiCoO_2$. The impact of Co(II) ions on the performance of $LiCoO_2$, Li metal, and graphite electrodes is introduced by Markovsky *et al.* [38].

From the charge/discharge test in a 150 ppm Co(II) preadded electrolyte, the initial charge reaction began from 1.6 V versus Li due to cobalt electrodeposition. The deposition potential is higher compared to that of 150 ppm Mn(II) as shown in Figure 8.1, in agreement with the fact that the standard redox potential of Co/Co(II) (2.77 V) is higher than that of Mn/Mn(II). Its capacity of 1770 $mA\,h\,g^{-1}$ is larger than that of 150 ppm Mn(II). Almost all the cobalt ions would be deposited on the graphite during the first charge, and cobalt deposit on the graphite was clearly confirmed by microscopic observation in Figure 8.6. It is thought that the large charge capacity, which is about 5 times larger than the theoretical capacity of LiC_6, is not due to electrochemical intercalation of Co(II) but the metallic cobalt deposition. Furthermore, the electrochemical decomposition of the electrolyte and/or another

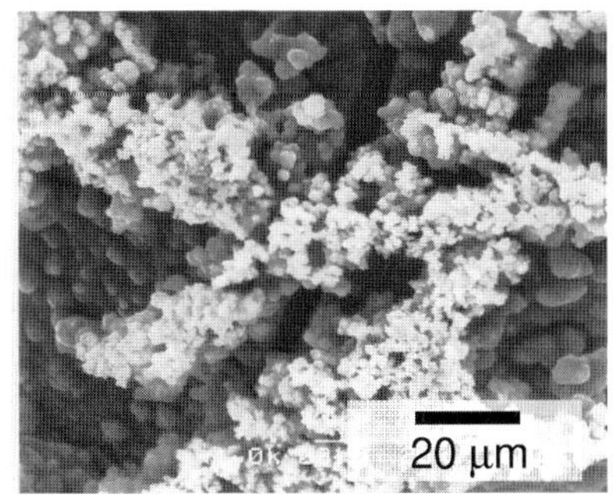

Fig. 8.6 SEM images of graphite electrode surface after 10 cycles in the 150 ppm Co(II) added electrolyte.

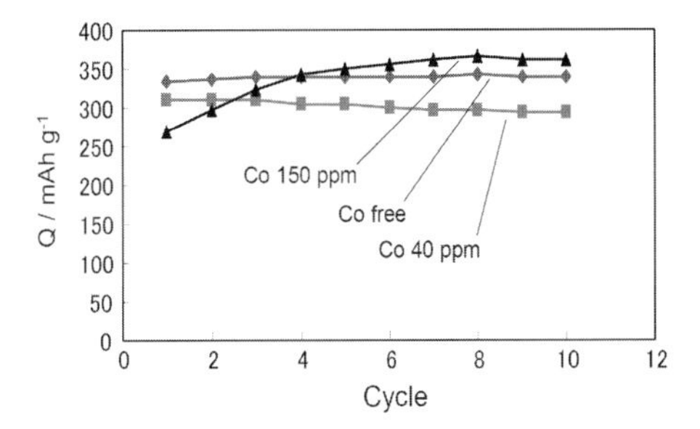

Fig. 8.7 Discharge capacity versus cycle number plots of graphite tested in $LiClO_4$ EC:PC (2:1) containing 0, 40, and 150 ppm Co(II).

electrochemical reaction should be promoted by the electrodeposited Co metal on the graphite.

As is seen in Figure 8.7, the variation in the discharge capacity is quite different from those with the Mn(II) additive. The addition of 150 ppm Co(II) makes the coulombic efficiency quite low during the initial few cycles, and then surprisingly, the discharge capacity and the coulombic efficiency gradually increases up to 10 cycles. At the 10th cycle, the discharge capacity becomes greater than that of the Co-free electrolyte. The discharge reaction appeared in two distinct regions: 0–0.3 and 0.3–1.5 V. As reported in [39], nanosized cobalt oxides, CoO and Co_3O_4, demonstrated a high capacity of 700 mA h g^{-1} between 0.01 and 3 V versus Li as a negative electrode. The mechanism of Li reactivity involves the formation and decomposition of lithia, accompanying the redox reaction of the metal nanoparticles – the so-called "conversion." In this case of 150 ppm cobalt, there is the possibility that the Co deposits on the graphite consist of nanosized particles that are significantly reactive and are covered with a cobalt compound layer, which could be electrochemically active with the reversible absorption of lithium, as schematically drawn in Figure 8.8. The two distinct regions; 0–0.3 and 0.3–1.5 V, should correspond to electrochemical reaction of graphite and cobalt oxide, respectively.

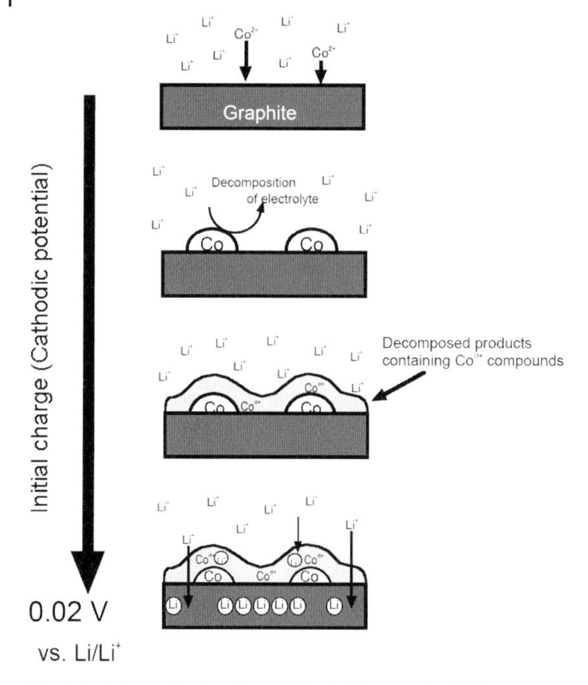

Fig. 8.8 Schematic drawing of the influence of Co(II) addition into the electrolyte during the initial charge of the graphite negative electrode.

These observations confirm that the influence of dissolved Co(II) originates from the deposition of cobalt. After the deposition of Co, since the cobalt compound, which is electroactive with the reversible absorption of lithium, would be formed electrochemically, the additional redox reaction appears at the higher potential. The degradation of lithium deintercalation/intercalation into graphite is prominent in the Mn(II)-containing electrolyte compared to the Co(II)-containing electrolyte [16]. In a practical cell of C/$LiCoO_2$, when cobalt dissolution occurs by overcharging, the influence on the graphite anode would be gradually diminished by the following several cycles.

8.2.3
Ni(II) Ion

The problems of manganese and cobalt dissolution from $LiMn_2O_4$ and $LiCoO_2$, respectively, have been clarified as mentioned above. Though nickel dissolution from the $LiNiO_2$ cathode has not yet been reported to our knowledge, $LiNiO_2$-based material has been used in a practical Li-ion cell. Hence, the interpretation of the influence of the Ni(II) additive is also important.

When the graphite electrode in an Ni(II)-added electrolyte was examined, the initial charge capacity increased because of electrodeposition of metallic Ni. That

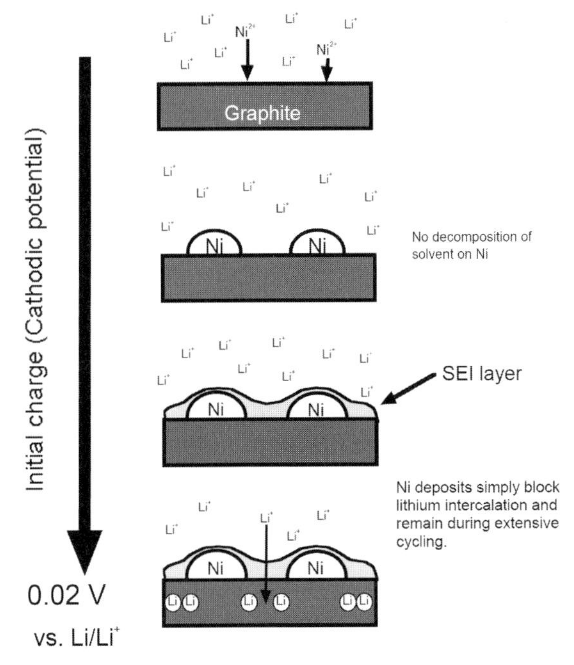

Fig. 8.9 Schematic drawing of the influence of Ni(II) addition into the electrolyte during the initial charge of the graphite negative electrode.

is, the graphite surface was partially covered with nickel deposits *in situ*, which simply hindered the lithium intercalation with no electrolyte decomposition and no lithium alloy formation as illustrated in Figure 8.9, so that their discharge capacities decreased without such a high reductive capacity as in the cases of Co(II) and Mn(II). On the other hand, it was reported that coating with metallic nickel by chemical vapor deposition is effective in decreasing the irreversibility [40].

For the application of Mn-, Co-, and Ni-based oxides, such as spinel $LiMn_2O_4$ and layered $Li(Ni,Mn,Co)O_2$, the above results will play an important role in further understanding the battery performance and suppressing the degradations, the high-temperature performance of the $C/LiMn_2O_4$ cell, and the overcharging performance of $C/LiCoO_2$.

8.3 How to Suppress the Mn(II) Degradation

As mentioned in the above section, the higher temperature degradation of the $C/LiMn_2O_4$ system is mainly caused by the degradation on the negative electrode side. Therefore, it is believed that solving the problem of anode deterioration is very important for development of the entire $C/LiMn_2O_4$ system that is applicable to hybrid and pure electric vehicles.

With regard to spinel type $LiMn_2O_4$ cathode, suppression of Mn(II) dissolution from the spinel is essential for enhanced performance. Recently, battery performance was successfully improved as a consequence of almost no manganese dissolution in fluorine-free electrolyte of 0.7 mol dm^{-3} lithium bis(oxalato)borate (LiBOB) ethylene carbonate (EC): propylene carbonate (PC): dimethylcarbonate (DMC) solution, as described by Amine and coworkers [41]. To our knowledge, from a number of earlier reports, spinel Li–Mn–O free from Mn dissolution is a degradation problem that is yet to be solved. In our opinion, therefore, a protection of the negative electrode from Mn(II) soluble in an electrolyte is also important and useful for understanding and improving battery performance. Here, we review the protective additives of LiI, LiBr, NH_4I [42], and 2-vinylpyridine [43] that are effective in suppressing the Mn(II) reduction (Mn electrodeposition) and the following drastic decomposition of electrolyte on deposited Mn surface.

8.3.1
LiI, LiBr, and NH_4I

The SEI layer does not protect the graphite from the soluble 150 ppm Mn(II), resulting in high irreversibility as described in Figure 8.5. This degradation is successfully suppressed by preaddition of LiI and LiBr into an electrolyte solution, that is, dissolved Mn(II) is electrochemically reduced on the graphite, which brings about large irreversible capacity resulting in quite low efficiency of 20%. The preaddition of iodide or bromide is effective in suppression of the irreversible reaction. Figures 8.10 and 8.11 show the cycle performance of graphite anode in LiI- and LiBr-added electrolytes, respectively. The efficiency is improved from 20 to 50% and 77% by adding a small amount of LiI and LiBr, respectively. Increasing the efficiency at the sixth cycle, the graphite exhibits discharge capacity in LiI- and LiBr-added electrolyte even after the addition of Mn(II). The presence of halogen anions in the electrolyte or at electrode surface influences the SEI-formation process on the graphite surface, and it might form a highly ion-conductive layer [44, 45]. Furthermore, addition of LiI is effective in improving cyclability of Li metal anode as reported by Ishikawa *et al.* [11]. In this case, physical adsorption of the iodide anion on the Li surface inhibits an interfacial reaction between Li and electrolyte.

To clarify the interaction between Mn metal and iodide anions, the voltammetry of metallic manganese electrode was examined in electrolyte solution with and without LiI as shown in Figure 8.4. In general, no alloy formation is known to occur with Li and 3 d metals such as Mn. Obviously, the high (electro-)chemical reactivity of electrolyte decomposition below 0.3 V is promoted on the metallic Mn surface in the LiI-free electrolyte as mentioned above. On the other hand, the LiI addition effectively suppressed the decomposition since reductive/oxidative current scarcely flows in the wide potential region between 0 and 2.5 V as is seen in Figure 8.4. The specific adsorption of iodide on the Mn surface inhibited any electrochemical reactions including the decomposition. Therefore, the electrolyte

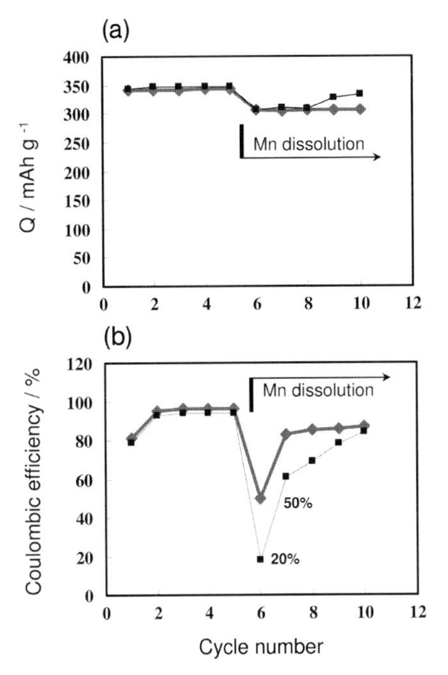

Fig. 8.10 Variation in (a) discharge capacity and (b) coulombic efficiency of a graphite electrode in 1 M $LiClO_4$ EC : DEC containing LiI (as 500 ppm I) (thick line) and no additive (thin line). Manganese perchlorate (150 ppm Mn) was added into the electrolyte before the sixth charge.

decomposition is suppressed by adsorption of iodide and bromide anions on the Mn surface deposited on the graphite and, as a result, the efficiency is improved by LiI and LiBr. When other inorganic and organic compounds containing iodine are added into an electrolyte solution, the efficiency and reversible capacity after Mn addition are also improved similarly.

In general, transition metal ions form ammonia-complex coordinated compounds, and their deposition potential from the complex ions in an aqueous medium is shifted toward the cathodic direction as transition metal ions are stabilized by coordinate bonds with ammonia ligands. As shown in Figure 8.12, when ammonium iodide was dissolved in the electrolyte into which Mn(II) was also added after five cycles, ammonium ions could also suppress Mn deposition by forming an Mn(II)-amine complex with a combination of the above iodide effect as schematically shown in Figure 8.13. The degradation by Mn(II) addition during cycling is more effectively suppressed by NH_4I compared to those for LiI and LiBr. The coulombic efficiency with dissolved Mn(II) is remarkably improved from 20 to 79% by the NH_4I additive (Figure 8.12); furthermore, the charge/discharge curves and discharge capacity are hardly changed in the following cycle, even after Mn(II) addition.

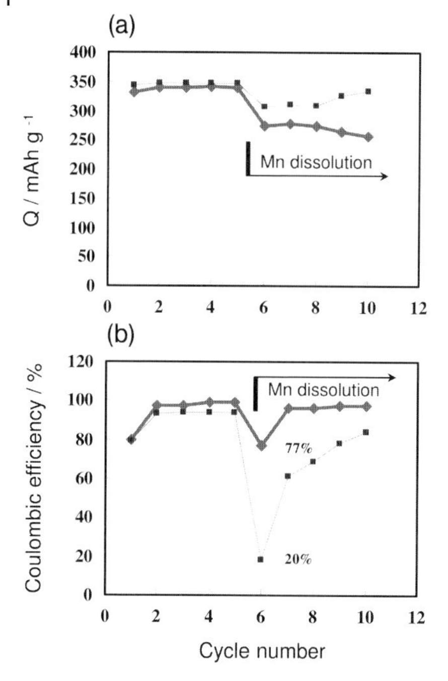

Fig. 8.11 Variation in (a) discharge capacity and (b) coulombic efficiency of a graphite electrode in 1 M $LiClO_4$ EC : DEC containing LiBr (as 500 ppm Br) (thick line) and no additive (thin line). Manganese perchlorate (150 ppm Mn) was added into the electrolyte before the sixth charge.

8.3.2
2-Vinylpyridine

Electrochemical polymerization is among the simplest and most interesting techniques for fabricating a polymer-modified electrode [46, 47]. It is believed that vinyl-type polymers synthesized by electroreduction [48, 49] are advantageous for direct film formation on a negative electrode in lithium-ion cells as pointed out by Besenhard and Winter's group [48]. With the aim of suppressing the degradation of the graphite performance due to dissolved Mn(II), besides LiI, LiBr, and NH_4I [42], 2-vinylpyridine (VP) [50] is also used as an electrolyte additive to effectively inhibit the degradation by dissolved Mn(II). On the other hand, vinylene carbonate (VC) additive hardly eliminates the degradation as described below [43].

Although the electroreductive decomposition of the electrolyte components including the SEI formation on graphite is usually confirmed around 0.8–0.3 V, an additional plateau clearly appears around 0.9–1.0 V in the curves earlier than the electroreductive decomposition in the VP-added electrolytes as seen in Figure 8.14. This is due to the polymerization induced by the electrochemical reduction of VP monomers in Scheme 8.1 as described in previous literatures [48, 49], thus resulting in the formation of poly(2-vinylpyridine) [50].

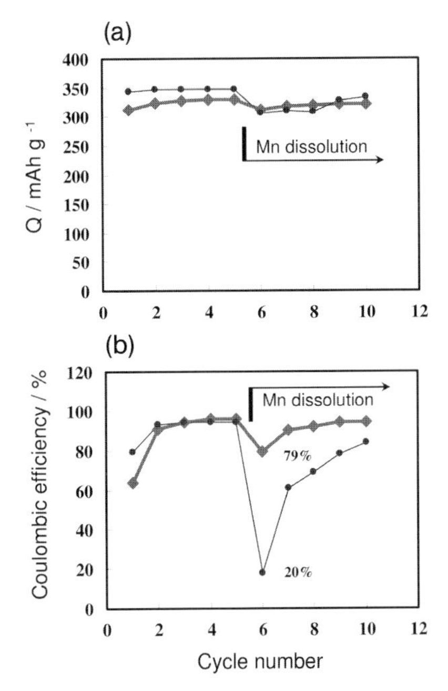

Fig. 8.12 Variation in (a) discharge capacity and (b) coulombic efficiency of a graphite electrode in 1 M $LiClO_4$ EC : DEC containing NH_4I ($[NH_4]$ = 150 ppm, [I] = 1050 ppm) (thick line) and no additive (thin line). Manganese perchlorate (150 ppm Mn) was added into the electrolyte before the sixth charge.

Electroreductive polymerization produces a polymer coating on the graphite during the initial charge. Almost all the monomers in electrolyte seems to be consumed in the first cycle, otherwise, the thick polymer film would hinder subsequent electrochemical film formation at the surface. The electrochemical polymerization occurs at a higher potential than that of the typical SEI formation of 0.8 V, that is, the polymer is deposited before the SEI formation. It is reasonable to think that the surface film formed in the VP-added electrolytes should differ from the typical SEI film, which does not prevent Mn(II) deposition as mentioned above. As described below, this film-forming VP as an electrolyte additive, 0.5 wt%, was capable of suppressing the degradation of the graphite anode, and therefore, improving the performance of graphite/$LiMn_2O_4$ batteries, whereas the addition of VC was hardly effective for the suppression.

Figure 8.15 compares the charge and discharge behaviors in electrolytes with and without 0.5% VP. At the sixth cycle, the charge capacity became significant, about 1600 mA h g^{-1}, with a high irreversibility induced by the electroreduction of Mn(II) because the irreversible reactions simultaneously occurred along with the intercalation in the lower potential range as is shown in Figures 8.3 and 8.5. Surprisingly, the VP addition successfully suppresses these irreversible reactions even though $Mn(ClO_4)_2$ is added into the electrolytes. Furthermore, Figure 8.15

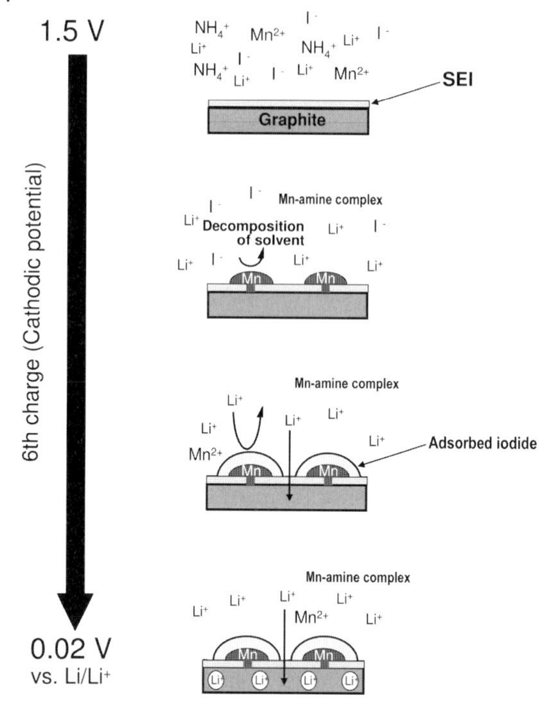

Fig. 8.13 Schematic drawing of the suppression mechanism of the Mn(II) degradation by dissolving NH_4I into the electrolyte.

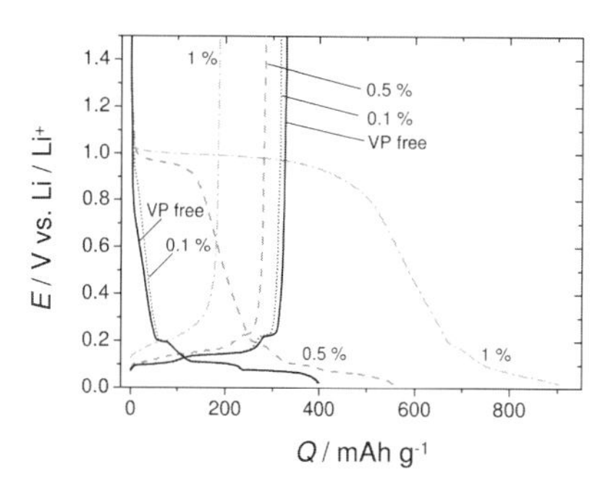

Fig. 8.14 Initial charge and discharge curves of graphite electrodes at $35\,mA\,g^{-1}$ in $1\,mol\,dm^{-3}$ $LiClO_4$ EC : DEC (1 : 1) with 1, 0.5, 0.1 vol% VP and no VP.

confirms that the potential variation and absolute capacity for the VP-added system is very similar to that obtained in the Mn(II)-free electrolyte. It suggests that the stage structures of the lithium-intercalated graphite are successfully formed without any irreversible reactions including the electrochemical plating of Mn metal. Accordingly, the efficiency at the sixth cycle completely recovers from 20 to 98% with maintenance of the high discharge capacity by VP addition as shown

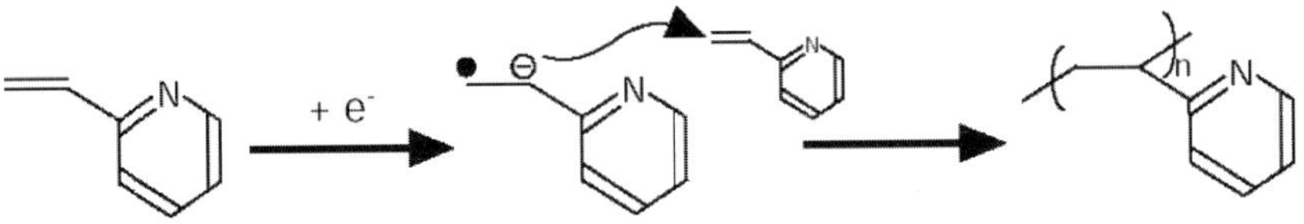

Scheme 8.1 Electropolymerization of VP monomers.

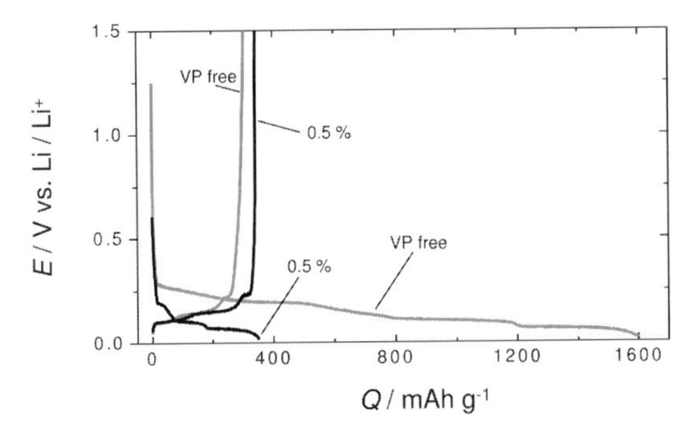

Fig. 8.15 The sixth charge and discharge curves of graphite in 1 mol dm^{-3} $LiClO_4$ EC : DEC with 0.5% VP additive and no additive. Manganese (II) perchlorate ([Mn] = 150 ppm) was added before the sixth charge.

in Figure 8.16. In the cycles subsequent to the fifth, the discharge capacity (~340 mA h g^{-1}) is maintained with satisfactory efficiency.

On the other hand, the VC additive, which has a polymerizable double bond comparable to VP, is well known to positively enhance the performance of the graphite [51–53]. When Mn(II) is added after six cycles in the electrolyte containing VC as shown in Figure 8.17, the charge capacity of graphite anode reaches 717 mA h g^{-1} despite the fact that the capacity without VC was 1680 mA h g^{-1}. Furthermore, the discharge capacity after the Mn(II) addition decreases to about 200 mA h g^{-1}. VC addition makes the capacity retention worse, besides, the efficiency after Mn(II) addition is scarcely improved by the VC addition. On the basis of our observation, it is concluded that VC is hardly capable of depressing the Mn(II) degradation [43]. The SEI film on the carbon anode consists of a heterogeneous mixture of inorganic/organic compounds dependent on the electrolyte additives. It is believed that the functionality of the surface film formed with VC differs from that with VP; the aromatic pyridine ring should modify the functionality of the surface film to suppress the Mn(II) deposition.

As already mentioned, the degradation is due to metallic manganese electrodeposition and also the following electrolyte decomposition promoted on the metallic Mn surface. The irreversibility is almost completely suppressed by the VP addition. From AC impedance measurement, the total cell resistance is significantly reduced by the addition of VP in the electrolyte. The VP addition suppressed the Mn plating and the induced decomposition, so that the entire impedance and resistances estimated from the semicircles become smaller by the protection of

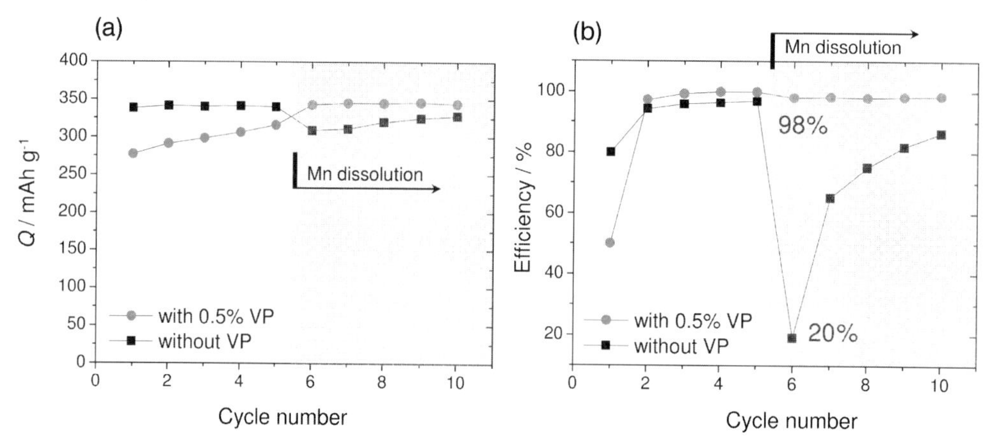

Fig. 8.16 Comparison of (a) discharge capacities and (b) coulombic efficiencies of graphite negative electrodes at 35 mA g^{-1} in 1 mol dm^{-3} $LiClO_4$ EC : DEC (1 : 1) with 0.5% VP additive and without VP additive. Manganese (II) perchlorate ([Mn] = 150 ppm) was added before the sixth cycle.

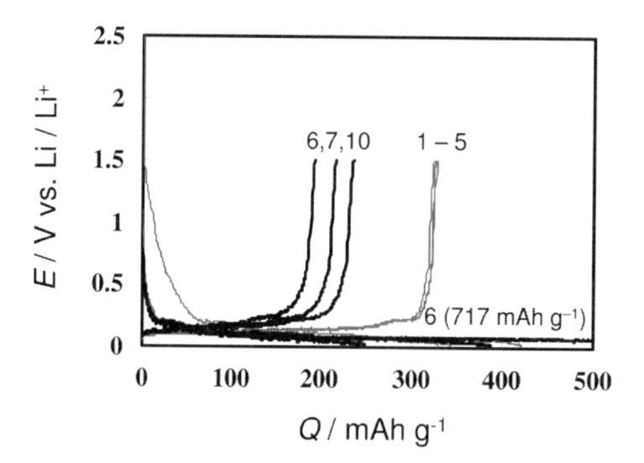

Fig. 8.17 Charge and discharge curves of graphite electrode in 1 mol dm^{-3} $LiClO_4$ EC : DEC (1 : 1) with 6.6 vol% VC additive. Manganese (II) perchlorate ([Mn] = 150 ppm) was added before the sixth cycle.

poly(2-vinylpyridine). As reported previously, the electropolymerized film coating protects steels against corrosion [49], so it is likely that the film formed on graphite protects the lithium-intercalated graphite against the dissolved Mn(II) ions, namely, the protective layer probably possesses a lithium-ion conductivity, but almost no manganese-ion conductivity, and almost no electronic conductivity. Moreover, a stable Mn(II) complex is possibly formed with the pyridine groups of the unreacted monomers, a partly dissolved oligomer, or a polymer in an electrolyte solution similar to the Mn(II)–amine complex as mentioned above. Presumably, manganese deposition is also prevented by the complex formation because the stability of the Mn(II) complex would make the reduction of Mn(II) difficult.

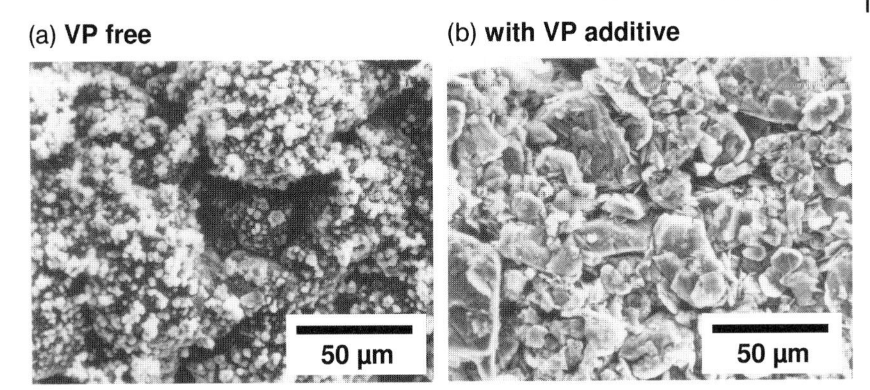

Fig. 8.18 SEM images of graphite electrode surface (a) without and (b) with VP additive after 10 charge–discharge cycles. Manganese (II) perchlorate ([Mn] = 150 ppm) was added before the sixth charge.

Figure 8.18 shows SEM images of graphite electrodes cycled in the VP-free and VP-added electrolytes to which Mn(II) was added after the initial five cycles. For the VP-free electrolyte, graphite particles cannot be observed at all because, with the deposition of many small white particles, an extremely thick deposit covers the graphite electrode. This deposit is formed by the accumulation of metallic Mn and the subsequent significant decomposition products. It is apparent that the degradation of the graphite anode is due to this deposit preventing lithium intercalation into the graphite. Nonetheless, with VP addition, the deposit completely disappears and the flake shape of the pristine graphite is still distinguishable even with the addition of Mn(II). The surface of the graphite particles is smooth and uniform because there are no small white deposits. The fact that the deposit did not appear agrees with the above electrochemical results. Compared to the inorganic additives such as LiI, LiBr, and NH_4I, the effect of the VP additive is much more remarkable owing to the difference in electrode morphology and cycling behavior.

Because the morphologies between the VP-free and -added systems are different (Figure 8.18), XPS was employed to analyze the surface chemicals. As is seen in Figure 8.19, Mn $2p_{3/2}$ peaks are clearly observed at the same binding energy for both cases, but with a different thickness and distribution. In the case of VP addition, the peaks are remarkably weakened and almost disappear after sputtering (note that a single peak at about 640 eV is due to an Auger peak of Ni_{LMM} of the nickel current collector). This proves that the Mn-containing layer becomes thinner by the electropolymerization of VP. Probably, the existence of Mn with the VP additive is due to adsorption and/or complex formation within a thin solid film of poly(2-vinylpyridine). The voltammetry of a manganese metal electrode in an additive-free electrolyte confirmed its high (electro-)chemical reactivity in the electrolyte decomposition as illustrated in Figure 8.4.

The XPS spectra of the electrode surface also confirms that manganese is not metallic (639 eV) and not the dioxide (642 eV), but seems to be divalent or trivalent

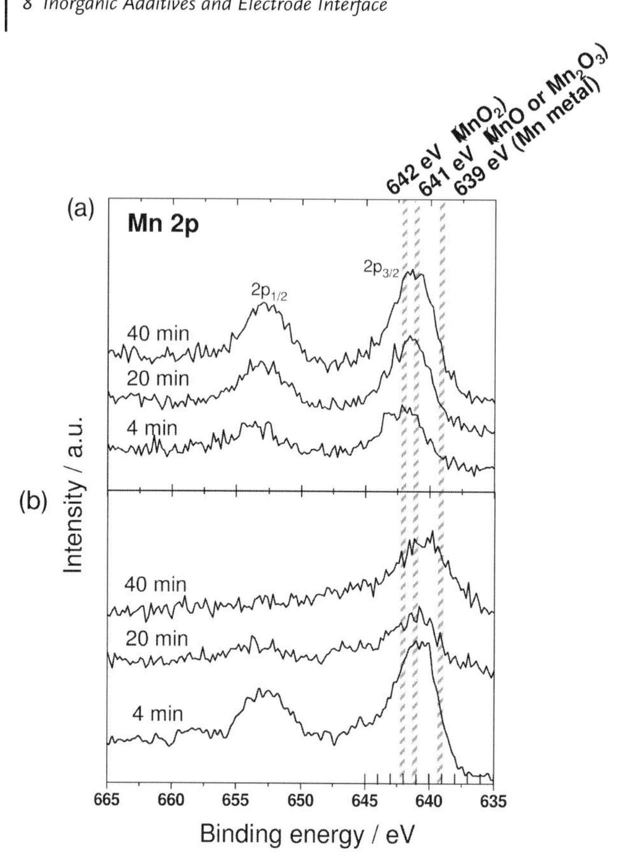

Fig. 8.19 Mn 2p XPS spectra of graphite negative electrode cycled in the electrolytes (a) without and (b) with VP additive after 10 charge and discharge cycles. Mn (150 ppm) was added before sixth charge. The time in minutes indicates Ar^+ ion etching duration.

such as MnO and Mn_2O_3 since all the Mn $2p_{3/2}$ peaks are situated around 641 eV, suggesting an irreversible electrolyte decomposition with Mn as mentioned above. If Mn(II) is dissolved, metallic Mn is first deposited; nevertheless, Mn(II) or Mn(III) compounds, including an organic manganese compound, are readily formed at the surface because of the high reactivity of Mn metal.

Figure 8.20 summarizes a schematic model of the surface of a graphite electrode. For the VP-free electrolyte, when Mn(II) is dissolved in an electrolyte, Mn metal is first deposited on the electrode, followed by a drastic decomposition of the electrolyte by the metal. Finally, a thick layer containing Mn(II)/(III) compounds interferes with the lithium intercalation. When the VP additive is dissolved into the electrolyte, the polymer layer is formed at about 0.9 V before the typical electrolyte reduction at about 0.8 V; therefore, the SEI layer is modified with poly(2-vinylpyridine). This modified layer is capable of eliminating the

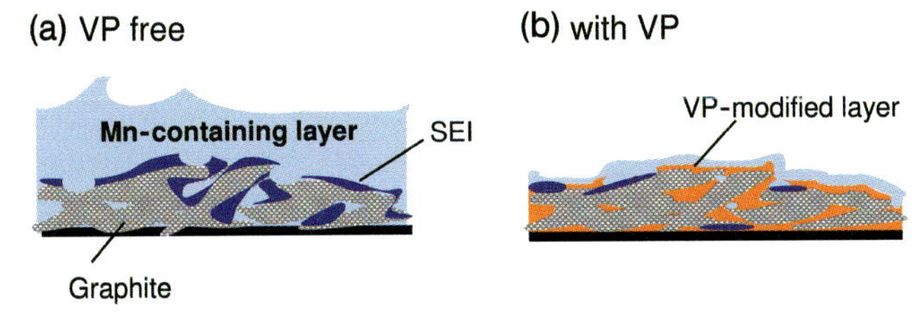

Fig. 8.20 Schematic illustration of graphite electrode in (a) VP-free and (b) VP-added electrolytes after the dissolution of Mn(II) showing that the VP film suppresses the electrodeposition of manganese.

Mn(II) deterioration, because the modified layer successfully blocks the Mn(II) reduction.

For a lithium-ion cell, C/(spinel Li–Mn–O), the degradation of the carbon anode is induced by the electroreduction of Mn(II) dissolved from the spinel; this step can be suppressed by the film-forming additive of 2-vinylpyridine, but VC hardly shows such an effect on the suppression. Although the VP additive has the problem of increasing the initial irreversible capacity and high polarization, it would be solved by precoating the graphite powder with poly(2-vinylpyridine) by a chemical technique prior to cell assembly. Consequently, novel types of additives are essentially important and necessary for suppression of the deterioration.

8.4 Alkali Metal Ions

The inorganic ingredient at the graphite surface is effective for enhancing the battery performance, for example, lithium, sodium, potassium [54], fluorine [55], Mn(II), Co(II), Cu(II), Ni(II), Zn(II), Pb(II), and Ag(I) [56]. The influence of these cations depends on the characteristics and nature of metal elements such as deposition potential, alloy-formation ability with Li, potential window, and so on. These studies have shown that, whatever the cations, the structural modification of the electrode–electrolyte interface is one of the most important factors that determine their anode performances. This section reviews the impact of alkali metal ions – Na^+ and K^+ ions – as electrolyte additives for the enhancement of battery performance, and, additionally, the application of Na_2CO_3- and NaCl-coating for improving the battery performance of graphite.

8.4.1 Na^+ Ion

The electroreducitve intercalation of various kinds of positive ions into graphite was found in aprotic media in the 1970s [57]. Among various carbon materials that are electroactive as intercalation hosts, graphitic derivatives are attractive for battery applications because they exhibit a high specific capacity, low working potential close to that of lithium metal, and superior cycling behavior as the negative electrode [58–60]. Not only lithium but also potassium, rubidium, and alkyl ammonium ions are electrochemically intercalated into carbonaceous materials [57]. However, sodium intercalation into graphite scarcely occurs even under intense or moderate conditions. [61, 62], in spite of its significant advantages over lithium intercalation, notably, a reduction of raw-material cost if sodium-ion batteries with good performance could be developed.

When electrochemical reduction and oxidation of graphite were investigated in an electrolyte containing both lithium and sodium salts, only lithium ions were intercalated into graphite without sodium intercalation and deposition. Nevertheless, both lithium and sodium ions influence the SEI, and the codissolved sodium ions improve the kinetics of lithium intercalation at the electrode interface with no modification of graphite bulk structure and electrolyte conductivity [63]. Therefore, the performance for Li-ion batteries is enhanced by sodium salt additive as mentioned here.

Figure 8.21 shows the initial charge and discharge curves of a graphite electrode in $LiClO_4$ and $NaClO_4$ solutions. In the sodium-free electrolyte, the natural graphite shows a reversible capacity of about 320 mA h g^{-1} due to the lithium intercalation, and an irreversible reduction is observed due to electroreductive formation of the SEI on graphite around 0.8–0.3 V [63, 64], that is, the typical negative electrode performance of graphite in lithium-ion cells. By dissolving 0.22 mol dm^{-3} $NaClO_4$, higher electrochemical performance is achieved for the graphite electrode as shown in Figure 8.21, namely, the irreversible reduction is obviously suppressed, and a higher reversible capacity is obtained. In this case, reversible lithium intercalation predominantly occurs with no side reactions such as sodium deposition and intercalation as supported by *ex situ* X-ray diffraction (XRD), suggesting that the SEI layer depressed electron transfer between graphite and sodium ions because of the selective permeability of lithium ions. There still remains the possibility of slight sodium intercalation into the most superficial edge part of graphite, which was difficult to detect with laboratory XRD or electrochemical techniques. The extremely small amount of sodium can be reversibly inserted into graphite in $NaClO_4$ EC [65] or $NaPF_6$ EC : DEC solutions [66]. The addition of 0.44 mol dm^{-3} $NaClO_4$ also gave us the comparative electroactivity; however, one can notice that not only the whole reductive but the oxidative potential curves are also shifted 40–50 mV toward the same negative direction, indicating that this is not due to a polarization. The electrolyte conductivity is hardly influenced by 0.22 mol dm^{-3} $NaClO_4$. To suppress the irreversibility and accomplish higher electroactivity, the optimum concentration of sodium perchlorate was about 0.22 mol dm^{-3}.

Figure 8.22 represents cyclic voltammograms of graphite electrode tested in Na-free and 0.22 mol dm^{-3} $NaClO_4$-added electrolytes. All redox couples in the Na-free electrolyte, of course, correspond to the formation of stage structures of lithium-intercalated graphite, i.e., stages I, II, III, IV, and diluted I. In spite of sufficient concentration of sodium salt, additional electroreduction including background current is hardly observed at potentials lower than the Na deposition

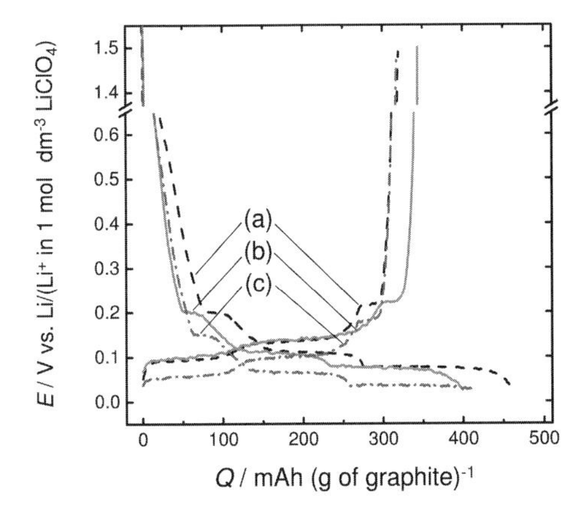

Fig. 8.21 Charge and discharge curves of graphite in (a) 0, (b) 0.22, and (c) 0.44 mol dm^{-3} $NaClO_4$ and 1 mol dm^{-3} $LiClO_4$ EC : DEC.

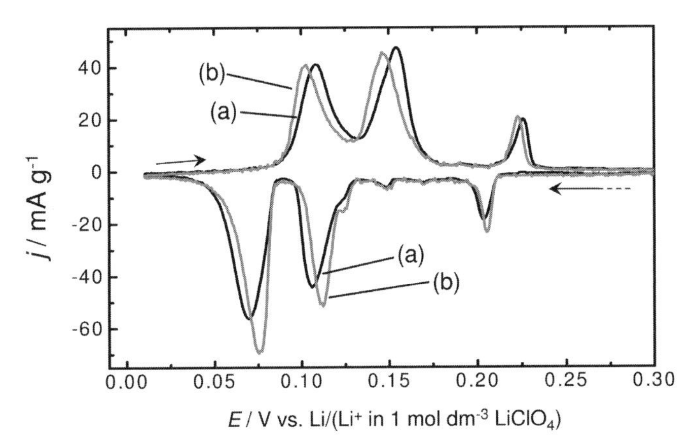

Fig. 8.22 Cyclic voltammogram of graphite electrodes in $LiClO_4$ EC : DEC (a) with and (b) without $NaClO_4$ at 0.02 mV s^{-1}.

potential in the voltammogram, so that metal plating of Na is not observed, while lithium metal deposition at intercalated graphite electrode is observed with small polarization. This implies that the SEI layer depressed electron transfer between graphite and sodium ions because of the selective permeability of lithium ions, that is, the SEI is not a sodium-ion conductor. When adding $NaClO_4$, the basic redox response is quite similar to that of lithium intercalation. However, each current peak becomes sharp and intense with lower polarization. From the viewpoint of lithium-ion battery application, the performance of graphite will successfully improve with the use of Na^+ ions as an electrolyte additive.

Figure 8.23 exhibits the variation in discharge capacities of graphite at constant current. Capacity retention tested at both current densities was improved by adding sodium salt, showing that the Na^+ additive clearly enhances the cycle performances.

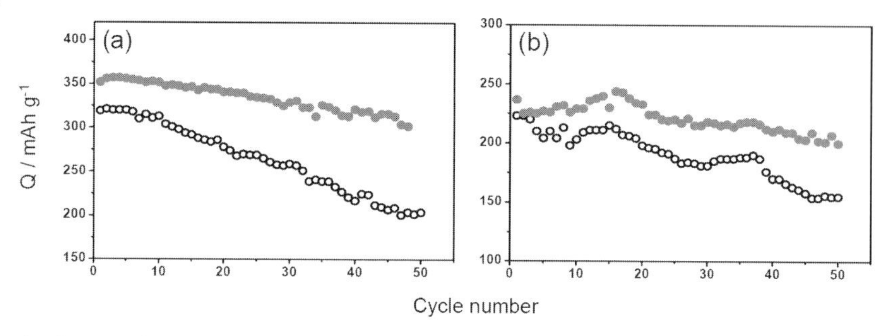

Fig. 8.23 Discharge capacity versus cycle number plots of graphite electrode with (filled plots) and without (open plots) Na^+ at (a) 17 and (b) 350 mA g^{-1}.

Since the potential variation at the subsequent cycles agreed with the potential plateaus due to lithium intercalation, the redox over 50 cycles would be due to the reversible lithium intercalation with no sodium plating and intercalation. This reminds us of the influences of other alkaline metal ions: the addition of KPF_6 exhibited a negative effect (see below), though K_2CO_3 addition enhanced the performance of natural graphite [67]. These facts suggest that the alkali salt additives possess combined effects of not only cation but also anion species, concentrations, solvents, and so on, to vary the whole electrochemistry of electrode, electrolyte, and interface.

To discuss the interface structure, AC impedance measurements were carried out in Na-free and Na-added (0.22 mol dm^{-3} $NaClO_4$) electrolytes after cycling at 350 mA g^{-1}. Typical Cole–Cole plots obtained at the 50th cycles are superimposed in Figure 8.24. As expected, the impedance loci after 50 cycles are quite different from Figure 8.23(b). At least two semicircles and an inclined line appear in each locus as seen in Figure 8.24. The first semicircle appearing at the high-frequency region is attributed to the impedance relative to the presence of an SEI layer at the graphite–electrolyte interface; the second semicircle appearing at the middle-frequency region is attributed to the impedance of the charge-transfer process. Finally, the straight line in the low-frequency region is attributed to the Warburg diffusion associated with the finite lithium diffusion in the graphite lattice. This figure indicates that the resistance of both the SEI layer (high-frequency region) and the charge transfer (middle-frequency range) in an Na^+-added electrolyte are approximately half of those observed in a Na^+-free electrolyte. The dissolved $NaClO_4$ in the 1 mol dm^{-3} $LiClO_4$ electrolyte successfully modifies the properties of the SEI film; as a result, the improvement in the electrochemical performance is attributed to the interface modification by the Na salt additive.

Figure 8.25 compares the electrode morphologies after the 50th cycle in $LiClO_4$ solution without and with 0.22 mol dm^{-3} $NaClO_4$. White particles in submicrometer size on the surface of the graphite flakes are observed in the SEM photos in the case of sodium-free solution. On the other hand, for the Na^+ additive system, the deposits hardly appear at the surface, and the electrode morphology is composed of the flakelike shape of pristine graphite even after 50 cycles. The deposits are formed

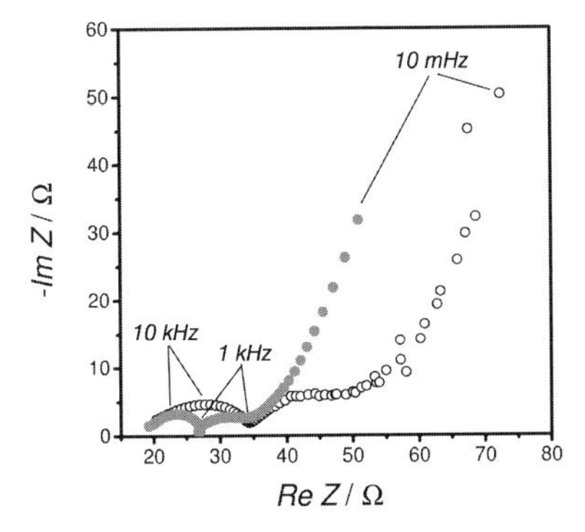

Fig. 8.24 Cole–Cole plots of graphite electrodes with (filled) and without Na^+ (open) additive after the 50th reduction.

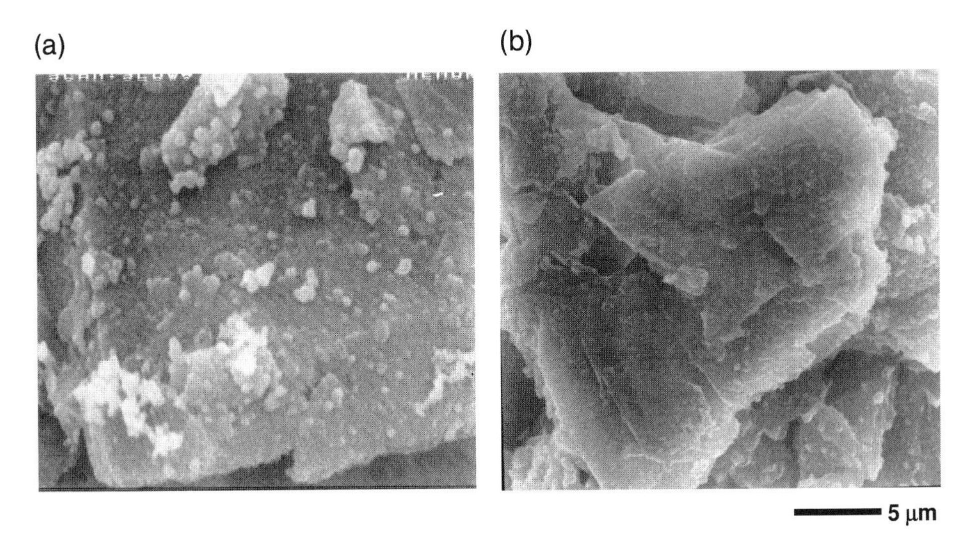

Fig. 8.25 SEM images of graphite electrodes tested in (a) Na^+-free and (b) with Na^+-added electrolytes after the 50th discharge.

by the irreversible decomposition in the first cycle and their number increased during successive cycles.

Furthermore, from TEM photos in Figure 8.26, the rough and disordered layer is also confirmed on the surface for the Na-free system. However, for the Na-additive system, the graphite surface is covered with the homogeneous and uniform layer. These surface layers should relate to the SEI, and it is likely that this uniform interface including the SEI layer contributed to the enhancement of the electrochemical performances. The surface morphologies in both micrometer and nanometer scales are influenced by the presence of sodium, as shown in Figures 8.25 and 8.26. It is believed that the uniform interface contributes to the enhancement in the performance as a negative electrode.

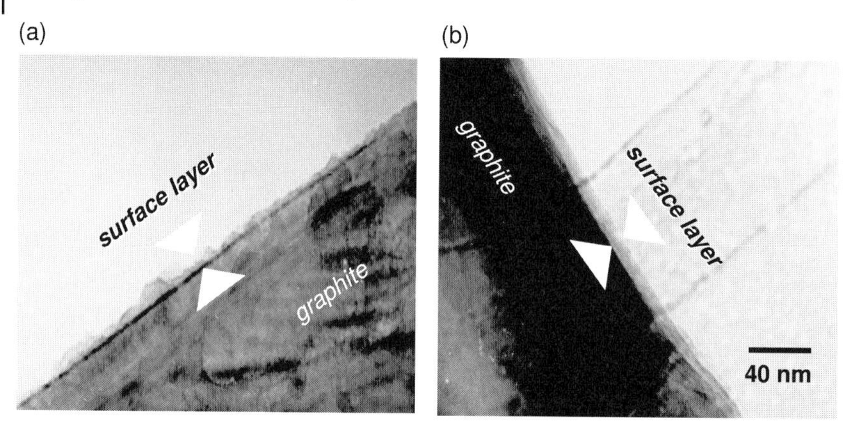

Fig. 8.26 TEM images of graphite electrodes tested in (a) Na^+-free and (b) with Na^+-added electrolytes after the 50th discharge.

The graphite electrode surface, i.e., the SEI layer after cycling, was analyzed by TOF-SIMS as shown in Figure 8.27. Clearly, the definite amount of sodium ingredient ($m/z = 23$) is confirmed at the surface cycled in the Na-additive electrolyte. As this element is not detectable in the bulk of the material by XRD, sodium is present only in the surface layer and uniformly distributed in the interface. This sodium component resulted from entrapment in the SEI layer during the irreversible reduction similar to the lithium ingredient ($m/z = 6$ and 7). As described in [68], the lithium-ion conductivity in solid oxide electrolyte is enhanced by the larger interstitial space for the diffusion path introduced by the partial substitute of the larger ions for parent element in the oxide framework. On the basis of earlier reports, the incorporation of sodium should result in the partial substitution for lithium in the SEI layer that enlarged the interstitial space, resulting in the highly uniform Li conduction through the entire SEI layer. Probably, the desolvation of

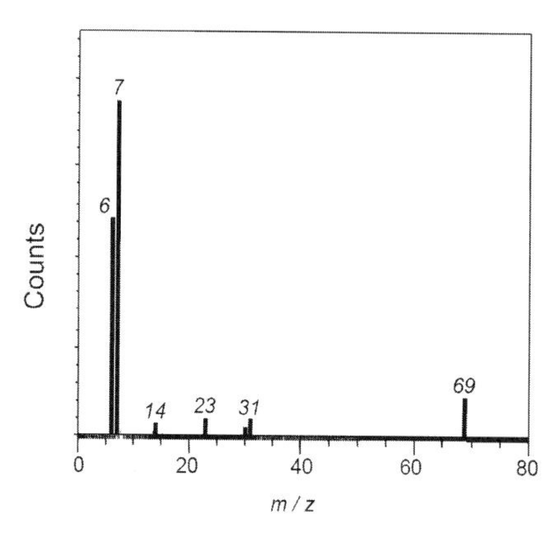

Fig. 8.27 TOF-SIMS positive ion spectrum of graphite after 100 charge/discharge cycles in Na^+-added electrolyte.

Li^+ ions during intercalation is made smoother and easier leading to less decomposition of solvent molecules at the sodium-containing SEI. Surprisingly, it was found that the intensity ratio of $^6Li/^7Li$ in TOF-SIMS results was much higher than that of the $^6Li/^7Li$ isotopic abundance, 0.1, which was also observed for the Na-free one. It seems that the SEI layer may possess an isotopic effect such as ^6Li enrichment.

From the surface–interface analyses, it was found that the improvement of the electrochemical performances is attributed to the interface modification by Na salt additive, that is, the interfacial kinetics rate is higher in presence of a sodium additive. These points explain the improvement in the electrochemical behavior of graphite in the codissolved electrolyte.

To verify the Na effect of the entrapment in the SEI, the first reduction and oxidation of graphite was carried out in the Na-added electrolyte and then the graphite electrode was cycled in the Na-free electrolyte after the second cycle by transferring the electrode from the Na-added cell to the Na-free cell as shown in Figure 8.28. When a sodium salt is contained in the SEI at the first cycle, the Na effect is maintained even in the Na-free electrolyte, and when the graphite is cycled once in an Na-free one, the capacity variation is rather the same as the result in the Na-free electrolyte. This proved that the coexistence of Na after the second cycle, i.e., SEI formation, has almost no effect on the electrochemistry. The Na additive is required only during the first cycle, suggesting that the SEI modification is almost complete during the cycle. Furthermore, the charge and discharge of $LiMn_2O_4$ were confirmed to be independent of the sodium salt addition, supporting no remarkable influence on reversible lithium intercalation of $LiMn_2O_4$, which is important in a practical cell.

It is summarized that sodium addition is found to be much more effective in easily enhancing the performance of a graphite anode and the solvation/desolvation process of Li ions, that is, not only in the suppression of the initial irreversibility

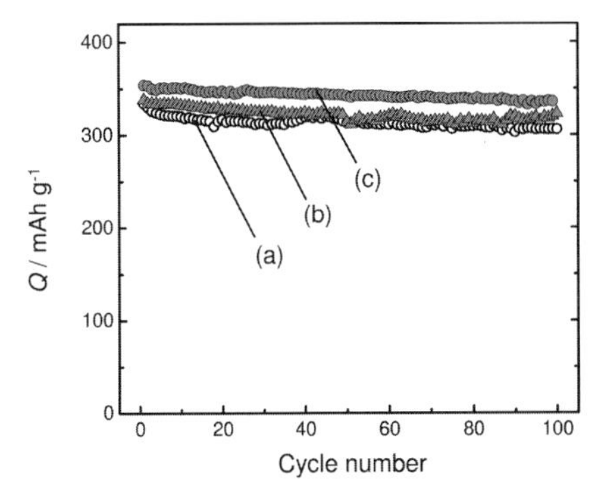

Fig. 8.28 Cycle number versus discharge capacity plots of graphite cycled at 350 mA g^{-1}; (a) all cycles in an Na^+ free electrolyte, (b) the first cycle in an Na^+ free and the subsequent cycles to the first in an Na^+ added electrolyte, and (c) the first cycle in an Na^+ added and the subsequent cycles in an Na^+ free electrolyte.

but also in the better retention of discharge capacity. This enhancement comes from changing the interface structure including the SEI film. We believed that the addition has high potential for application in practical lithium-ion batteries since its method is simple and effective.

8.4.2
K^+ Ion

As the existence of surface layer plays an important role in reversible lithium intercalation into interspace between graphene layers as mentioned, chemical modification of graphite surface including an SEI film has attracted wide attention to improve battery performances. As reported by Tossici *et al.*, potassium intercalated graphite (KC8) prepared by chemical reaction could be employed as a high-rate negative electrode for rechargeable lithium-ion batteries [69–71] because of the formation of an insoluble protective $KClO_4$ layer on the surface [72]. Our group previously studied the performance of a graphite electrode in a PF_6^- based electrolyte containing both lithium and potassium ions [73]. Recently, K_2CO_3 addition enhanced the performance of natural graphite, suggesting that the alkali salt additives possess the combined effects of cation and anion species [67]. This section presents a comparative study on the influences of Na^+ and K^+ ions, which have different electrochemical reactivities, as electrolyte additives on a graphite anode.

Figure 8.29 shows charge and discharge curves of a graphite electrode examined in $LiPF_6$ EC : DEC solution without and with 0.2 mol dm^{-3} KPF_6. In an additive-free electrolyte, the graphite electrode shows a reversible capacity of 339 mA h g^{-1}, and an irreversible capacity of 65 mA h g^{-1} is observed due to the electroreductive formation of the SEI [64]. In case of a K^+ added electrolyte, the irreversible capacity increased up to 76 mA h g^{-1}, and a lower reversible capacity of 322 mA h g^{-1} is obtained.

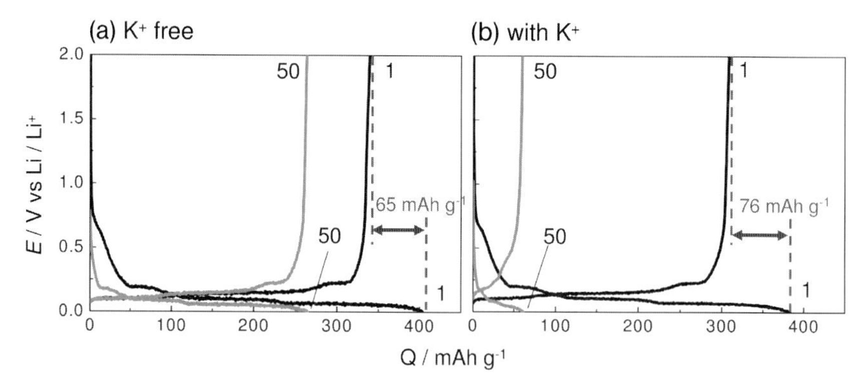

Fig. 8.29 Charge and discharge curves of graphite negative electrodes in (a) potassium free and (b) 0.2 mol dm^{-3} KPF_6 added EC : DEC (1 : 1) electrolyte containing 1 mol dm^{-3} $LiPF_6$ at 70 mA g^{-1}.

The variation in discharge capacities of a graphite negative electrode with and without K^+ addition during cycling is different as plotted in Figure 8.30. At the first cycle examined at 70 and 350 mA g^{-1}, the reversible capacities for a potassium-ion-added electrolyte are similar to those of the additive-free one. However, the discharge capacity faded more rapidly by dissolving KPF_6. The addition of K^+ ions into an electrolyte is unfavorable to battery performance of a graphite anode, depending on the charge/discharge rate. One can note the negative effect of K^+ ion addition despite the positive effect of a sodium additive.

Figure 8.31 shows ex situ XRD patterns of graphite electrodes charged to 2.0, 0.2, 0.1, and 0.0 V in additive-free and potassium-added electrolytes followed by

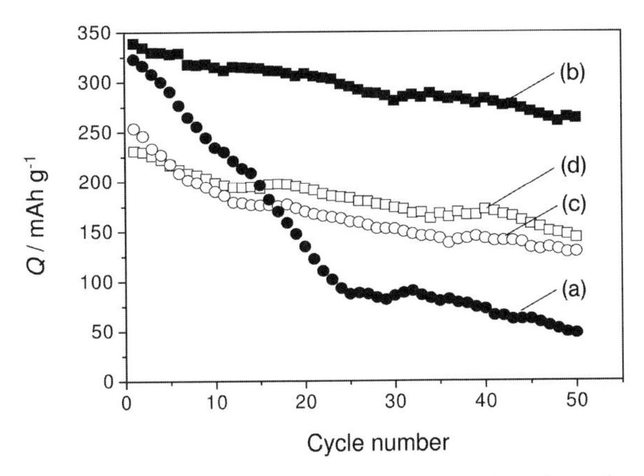

Fig. 8.30 Discharge capacities versus cycle number plots of graphite electrodes (a) with and (b) without KPF_6 at 70 mA g^{-1}, and (c) with and (d) without KPF_6 at 350 mA g^{-1}. Electrolyte: 1 mol dm^{-3} $LiPF_6$ EC : DEC (1 : 1).

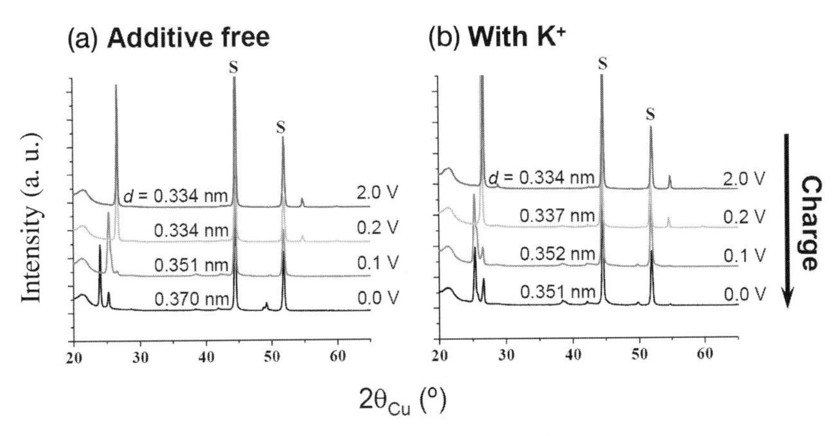

Fig. 8.31 *Ex situ* XRD patterns of graphite during the initial charge in $LiPF_6$ electrolytes (a) with and (b) without KPF_6. Galvanostatic charging at 70 mA g^{-1} was stopped at 2.0, 0.2, 0.1, and 0.0 V versus Li. S: current corrector (Ni).

equilibrating the electrode potential for several hours in a cell. In the diffractograms, *d* spacing values estimated from the strongest diffraction peaks are inserted in each pattern of the figure. Since 00*l* peaks shift toward lower angle, electrochemical intercalation occurs in the space between graphene layers. However, the variation in *d* values of the 00*l* peaks in an additive-free electrolyte is different from that in a K^+ added one. When the electrode is charged to 0.2 V, *d* values is slightly increased by adding K^+ ions, that is, the difference is 3 pm (= 0.337–0.334 nm). As described by Tossici *et al.* [69, 70], small (residual) amount of potassium derived from KC8 in a lithium-containing electrolyte expands the graphite structure even if potassium ions are removed from KC8 compared with that of pure graphite. It is likely that the polarization is influenced by the slight interlayer expansion, which favors the simultaneous lithium intercalation process accompanied by enhancement of its kinetics. In consideration of the low expansion (3 pm) compared to the K-deintercalated graphite (6 pm [70]), a graphite electrode predominantly undergoes lithium intercalation in a potassium-added electrolyte at >0.1 V because of the larger size and the lower concentration of potassium ions. The fully charged electrodes possess different *d* spacings. Values of *d* spacing increases from 0.351 to 0.370 nm in an additive-free system, which is close to $d002 = 0.372$ nm for LiC_6 [61, 74], however, almost constant *d* spacing is observed in K^+ added system suggesting no intercalation between 0.1 and 0.0 V during the initial charge. Electroplating of metallic potassium should occur instead of lithium intercalation, thus metallic potassium might obstruct graphite surface to prevent lithium intercalation into graphite at less than 0.1 V. These results suggest that potassium plating cannot be prevented by the SEI layer formed between 0.3 and 0.8 V in a KPF_6 dissolved electrolyte, and the graphite electrode suffers from deposition of potassium during every charging. Indeed, the surface of electrode after successive cycles was greatly different, depending on potassium addition [73].

The Na^+ additive clearly enhances the cycle performances of graphite, and a sodium component is detected at the surface that is similar to the carbonaceous material prepared from molten carbonates [54]. On the other hand, the K^+ additive has a disappointing influence on the graphite electrode performance though the intercalated potassium somewhat expands graphite structure along *c*-axis. In summary, the addition of alkaline metal ions such as Na^+ and K^+ has different influences on the electrochemistry of graphite electrode in lithium salt containing electrolytes, which is due to their different electrochemical behavior at the surface. As known generally, sodium intercalation into graphite occurs under very limited conditions compared to that of other alkali metals [62]. The enhancement by Na^+ addition is considered as one of the particular behaviors of Na^+, that is, no sodium intercalation and no electrodeposition of Na metal but positive modification of SEI layer. However, the potassium component in an electrolyte does take part in electrochemical deposition/dissolution and intercalation resulting in negative modification of a graphite anode.

8.5 Alkali Salt Coating

Electrolyte additives exhibits a significant effect on modifying the SEI formation during the first charge (electroreduction) process. With the aim of enhancing the negative electrode performance in lithium-based rechargeable batteries, various organic and inorganic additives dissolved into an electrolyte solution are effective as already mentioned. The impact of Na^+ ions as an electrolyte additive on the enhancement of battery performance is attributed to the entrapment of Na^+ ions in the SEI layer [16, 63]. Here, we review the Na_2CO_3- and NaCl-coating that improves the battery performance of graphite as found from our studies on the effect of additives.

Coating with water-soluble and air-stable chemicals such as Na_2CO_3 and NaCl can be completed with only aqueous solutions in air at room temperature, so that the pretreatment for coating graphite powders is significantly simple and easy compared to those of oxide coating [74, 75]. In the case of conventional additives for lithium-ion, we have to pay attention to the influence on the opposite both negative and positive electrodes. On the other hand, the insoluble coating method should enable the graphite surface to be modified with a minimum amount of sodium salts that should not have any influence on the positive electrode.

As reported by Zhang *et al.* [76], Li_2CO_3 coating is effective to improve the electrochemical performance of a graphite anode. From our investigation, a battery performance of a graphite electrode is also successfully enhanced by pretreatment of graphite powders with NaCl and Na_2CO_3 aqueous solution. A surface analysis with TOF-SIMS confirmed the existence of the corresponding alkali components in all cases for Li_2CO_3, Na_2CO_3, and K_2CO_3. We believe that the active part such as the edge plane of graphite is preferably modified by them because the edge plane would successfully serve as adsorbing and/or nucleating seed for their coating during drying. This improvement is due to the participation of Na_2CO_3 in the SEI formation. We believe that it results from the cooperative outcome of the positive effects of both the sodium [63] and the carbonate ions [6].

Figures 8.32 and 8.33 show the initial charge and discharge curves of Na_2CO_3 and NaCl coated graphite electrodes. Obviously, the reversible capacity increases

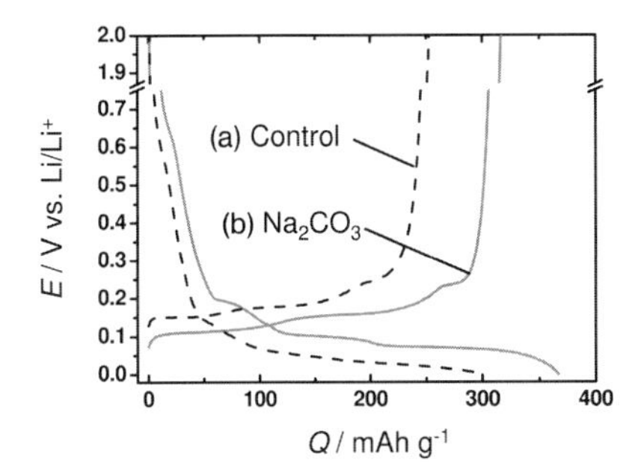

Fig. 8.32 Initial charge and discharge curves of graphite (a) with no treatment and (b) treated with Na_2CO_3 aqueous solution in $LiClO_4$ EC : DEC electrolyte.

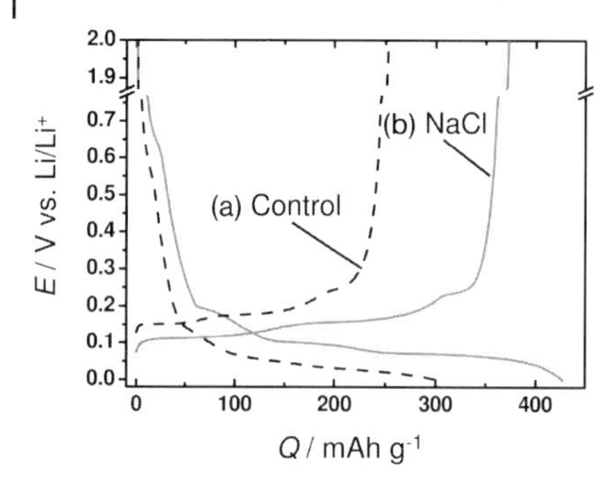

Fig. 8.33 Initial charge and discharge curves of graphite (a) with no treatment and (b) treated with NaCl aqueous solution in $LiClO_4$ EC : DEC electrolyte.

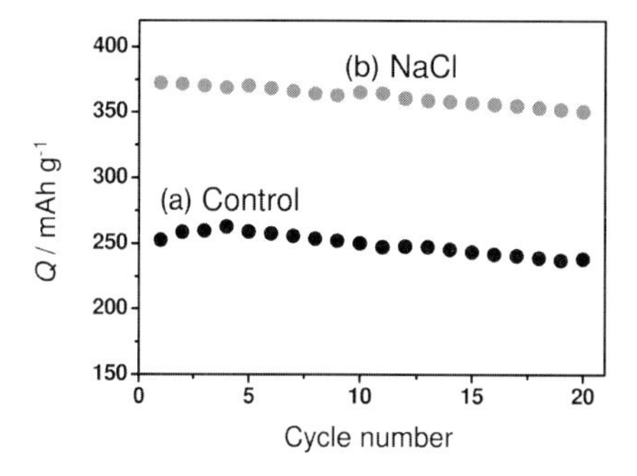

Fig. 8.34 Discharge capacity versus cycle number plots of (a) control and (b) NaCl-coating graphite electrodes under galvanostatic condition at 175 mA g^{-1}.

with no significant change in the potential variation suggesting that the SEI layer is modified with the Na_2CO_3 and NaCl, resulting in faster kinetics for the lithium intercalation into graphite. Modification of graphite surface with some halogens is known to improve the battery performance of lithium-based anodes [8–10, 42, 55]. The improvement is attributed to the merged effects of the sodium/chloride ions on the interface. The NaCl-coating demonstrates a capacity higher than 350 mA h g^{-1} over 20 cycles as shown in Figure 8.34, whereas the control shows about 250 mA h g^{-1}. Thus, we obtain the superior cyclability for the NaCl-coating to the pristine and alkali carbonate modified graphite.

The dependences of efficiency and discharge (delithiation) capacity on the NaCl concentration are shown in Figure 8.35. The optimum concentration improves performance, that is, the optimum concentration gives the appropriate SEI layer for high-rate performance with higher coulombic efficiency. To the best of our knowledge, this is the first application of sodium chloride and carbonate in lithium-ion batteries.

If chloride anions are added into an electrolyte to improve the performance of graphite, the battery suffers from oxidation of the anions at the cathode.

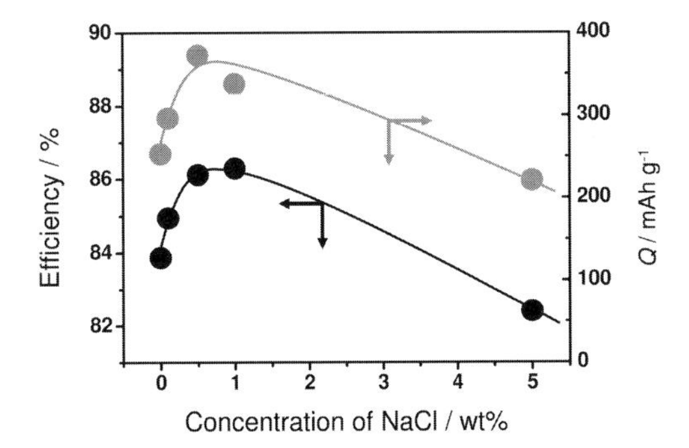

Fig. 8.35 Dependences of the initial coulombic efficiency and initial discharge capacity of graphite electrodes on NaCl concentration used for the pretreatment of graphite powders.

Furthermore, when we adopt inorganic additives in electrolytes for lithium-ion batteries, they must be soluble in organic solvents and compatible with both anode and cathode. However, this pretreatment method makes possible the use of the minimal amount of insoluble inorganics in electrolyte solutions to modify the electrode interface; besides, one need not take into account the influence of the coatings on the opposite electrode at all. Because of the advantages of this cheap, environmentally benign, and easy process, it has high potential for application in practical lithium-ion batteries.

8.6 Summary

We here reviewed the effects of metal-ion additive and salt coatings on graphite anode performance on the basis of our recent reports. In our early motivation, we studied the influence of manganese dissolution on the negative electrode because the deterioration of $C/LiMn_2O_4$ was induced by the manganese dissolution from an $LiMn_2O_4$ electrode. Next, we emphasized the requirement and usefulness of the new additives to suppress the manganese deterioration. On the basis of systematic research on dissolved metal species including transition metal and alkali metal ions as reviewed here, we suggest that sodium salt additives as well as sodium salt coatings are effective in enhancing battery performance.

Acknowledgments

This study was partly supported by the program *"Development of Rechargeable Lithium Battery with High Energy/Power Density for Vehicle Power Sources"* of the Industrial Technology Research Grant Program from the New Energy and Industrial Technology Development Organization (NEDO) of Japan, Yazaki Memorial

Foundation for Science and Technology, and a research grant from Iwatani Naoji Foundation.

References

1 Mizushima, K., Jones, P.C., Wisemanm, P.J., and Goodenough, J.B. (1980) *Mat. Res. Bull.*, **15**, 783.

2 Tarascon, J.M. and Armand, M. (2001) *Nature*, **414**, 359.

3 van Schalkwijk, W.A. and Scrosati, B. (eds) (2002) *Advances in Lithium-Ion Batteries*, Kluwer Academic/Plenum Publishers, New York.

4 Zhang, S.S. (2006) *J. Power Sources*, **162**, 1379.

5 Aurbach, D., Ein-Eli, Y., Markovsky, B., Zaban, A., Luski, S., Carmeli, Y., and Yamin, H. (1995) *J. Electrochem. Soc.*, **142**, 2882.

6 Osaka, T., Momma, T., Tajima, T., and Matsumoto, Y. (1995) *J. Electrochem. Soc.*, **142**, 1057.

7 Osaka, T., Komaba, S., Uchida, Y., Kitahara, M., Momma, Y., and Eda, N. (1999) *Electrochem. Solid-State Lett.*, **2**, 215.

8 Shiraishi, S., Kanamura, K., and Takehara, Z. (1999) *J. Electrochem. Soc.*, **146**, 1633.

9 Shiraishi, S., Kanamura, K., and Takehara, Z. (1999) *J. Appl. Electrochem.*, **29**, 867.

10 Shiraishi, S., Kanamura, K., and Takehara, Z. (1999) *J. Appl. Electrochem.*, **25**, 584.

11 Ishikawa, M., Yoshitake, S., Morita, M., and Matsuda, Y. (1994) *J. Electrochem. Soc.*, **141**, L159.

12 Ishikawa, M., Machino, S., and Morita, M. (1999) *J. Power Sources*, **473**, 279.

13 Wu, M.S., Lin, J.C., and Chiang, P.C.J. (2004) *Electrochem. Solid-State Lett.*, **7**, A206.

14 Wu, M.S., Chiang, P.C.J., Lin, J.C., and Lee, J.T. (2004) *Electrochim. Acta*, **49**, 4379.

15 Kumagai, N., Komaba, S., Kataoka, Y., and Koyanagi, M. (2000) *Chem. Lett.*, 1154.

16 Komaba, S., Kumagai, N., and Kataoka, Y. (2002) *Electrochim. Acta*, **47**, 1229.

17 Pistoia, G., Antonini, A., Rosati, R., and Zane, D. (1997) *Electrochim. Acta*, **41**, 2683.

18 Xia, Y. and Yoshio, M. (1997) *J. Power Sources*, **66**, 129.

19 Xia, Y., Zhou, Y., and Yoshio, M. (1997) *J. Electrochem. Soc.*, **144**, 2593.

20 Komaba, S., Kumagai, N., Sasaki, T., and Miki, Y. (2001) *Electrochemistry*, **69**, 784.

21 Myung, S.T., Komaba, S., and Kumagai, N. (2001) *J. Electrochem. Soc.*, **148**, A482.

22 Terada, Y., Nishiwaki, Y., Nakai, I., and Nishizawa, F. (2001) *J. Power Sources*, **97,98**, 420.

23 Aoshima, T., Okahara, K., Kiyohara, C., and Shizuka, K. (2001) *J. Power Sources*, **97,98**, 377.

24 Saitoh, M., Yoshida, S., Yamane, H., Sano, M., Fujita, M., Kifune, K., and Kubota, Y. (2003) *J. Power Sources*, **122**, 162.

25 Wang, L.F., Ou, C.C., Striebel, K.A., and Chem, J.S. (2003) *J. Electrochem. Soc*, **150**, A905.

26 Akimoto, J., Takahashi, Y., and Kijima, N. (2005) *Electrochem. Solid-State Lett.*, **8**, A361.

27 Arakawa, M., Nemoto, Y., Tobishima, S., Ichimura, M., and Yamaki, J. (1993) *J. Power Sources*, **43,44**, 517.

28 Okazaki, R., Takada, K., Oguro, S., Kondo, S., and Matsuda, Y. (1994) *Denki Kagaku (presently Electrochem).*, **62**, 1074.

29 Blyr, A., Sigala, C., Amatucci, G., Guyomard, D., Chabre, Y., and Tarascon, J.M. (1998) *J. Electrochem. Soc.*, **145**, 194.

30 Cho, J. and Thackeray, M.M. (1999) *J. Electrochem. Soc.*, **146**, 3577.

31 Amatucci, G., Du Pasquier, A., Blyr, A., Zheng, T., and Tarascon, J.M. (1999) *Electrochim. Acta*, **45**, 255.

32 Tsunekawa, H., Tanimoto, S., Marubayashi, R., Fujita, M., Kifunc, K., and Sano, M. (2002) *J. Electrochem. Soc.*, **149**, A1326.
33 Novak, P., Joho, F., Imhof, R., Panitz, J.C., and Haas, O. (1999) *J. Power Sources*, **81,82**, 212.
34 Aurbach, D., Markovsky, B., Weissman, I., Levi, E., and Ein-Eli, Y. (1999) *Electrochim. Acta*, **45**, 67.
35 Maeda, Y. and Touzain, Ph. (1988) *Electrochim. Acta*, **33**, 1493.
36 Amatucci, G., Tarascon, J.M., and Klein, L.C. (1996) *Solid State Ionics*, **83**, 167.
37 Myung, S.T., Kumagai, N., Komaba, S., and Chung, H.T. (2001) *Solid State Ionics*, **139**, 47.
38 Markovsky, B., Rodlin, A., Salitra, G., Talyosef, Y., Aurbach, D., and Kim, H.J. (2004) *J. Electrochem. Soc.*, **151**, A1068.
39 Poizot, P., Laruelle, S., Grugeon, S., Dupont, L., and Tarascon, J.M. (2000) *Nature*, **407**, 496.
40 Sandu, I., Brousse, T., and Schleich, D.M. (2003) *Ionics*, **9**, 329.
41 Amine, K., Liu, J., Kang, S., Balharouak, I., Hyung, Y., Vissers, D., and Henriksen, G. (2004) *J. Power Sources*, **129**, 14.
42 Komaba, S., Kaplan, B., Ohtsuka, T., Kataoka, Y., Kumagai, N., and Groult, H. (2003) *J. Power Sources*, **119,121**, 378.
43 Komaba, S., Itabashi, T., Ohtsuka, T., Groult, H., Kumagai, N., Kaplan, B., and Yashiro, H. (2005) *J. Electrochem. Soc.*, **152**, A937.
44 Shahi, K., Wagner, J.B., and Owens, B.B. (1983) in *Lithium Batteries*, (eds J.P. Gabano), Academic Press, London, p. 418.
45 Wang, H. and Yoshio, M. (2001) *J. Power Sources*, **101**, 35.
46 Komaba, S. and Osaka, T. (1998) *J. Electroanal. Chem.*, **453**, 19–23.
47 Komaba, S., Fujihana, K., Osaka, T., Aiki, S., and Nakamura, S. (1998) *J. Electrochem. Soc.*, **145**, 1126.
48 Moller, K.C., Santner, H.J., Kern, W., Yamaguchi, S., Besenhard, J.O., and Winter, M. (2003) *J. Power Sources*, **119,121**, 561.
49 Sekine, I., Kohara, K., Sugiyama, T., and Yuasa, M. (1992) *J. Electrochem. Soc.*, **139**, 3090.
50 Komaba, S., Ohtsuka, T., Kaplan, B., Itabashi, T., Kumagai, N., and Groult, H. (2002) *Chem. Lett.*, 1236.
51 Sanyo Co. (1993) *Japan Kokai Tokkyo Koho*, 3066126.
52 Jehoulet, C., Biensan, P., Bodet, J.M., Broussely, M., Moteau, C., and Tessier-Lescourret, C. (1997) Abstract No. 135, The Electrochemical Society and International Society of Electrochemistry Meeting Abstracts, Vol. 97-(2), Paris, France, Aug. 31-Spt. 5, p. 153.
53 Zhang, X., Kostecki, R., Richardson, T.J., Pugh, J.K., and Ross, P.N. Jr (2001) *J. Electrochem. Soc.*, **148**, A1341.
54 Groult, H., Kaplan, B., Komaba, S., Kumagai, N., Gupta, V., Nakajima, T. and Simon, B. (2003) *J. Electrochem. Soc.*, **150**, G67.
55 Groult, H., Nakajima, T., Perrigaud, L., Ohzawa, Y., Yashiro, H., Komaba, S., and Kumagai, N. (2005) *J. Fluorine Chem.*, **126**, 1111.
56 Komaba, S., Kataoka, Y., Kumagai, N., Ohtsuka, T., and Kumagai, N. (2001) Joint International Meeting (The 200th Meeting of ECS, The 52nd Meeting of ISE), No. 239, San Francisco, Sep. 6, 2001.
57 Besenhard, J.O. and Fritz, H.P. (1974) *J. Electroanal. Chem.*, **53**, 329.
58 Fong, F., von Sacken, U., and Dahn, J.R. (1990) *J. Electrochem. Soc*, **137**, 2009.
59 Sumiya, K., Suzuki, J., Takasu, R., Sekine, K., and Takamura, T. (1990) *J. Electroanal. Chem.*, **150**, 462.
60 Kaplan, B., Groult, H., Barhoun, A., Lantelme, F., Nakajima, T., Gupta, V., Komaba, S., and Kumagai, N. (2002) *J. Electrochem. Soc.*, **149**, D72.
61 Ge, P. and Fouletier, M. (1988) *Solid State Ionics*, **28–30**, 1172.
62 Mizutani, Y., Abe, T., Inaba, M., and Ogumi, Z. (2002) *Syn. Met.*, **125**, 153.
63 Komaba, S., Itabashi, T., Watanabe, M., Groult, H., and Kumagai, N. (2007) *J. Electrochem. Soc*, **154**, A322.

64 Besenhard, J.O., Winter, M., Yang, J., and Biberacher, W. (1995) *J. Power Sources*, **54**, 228.
65 Thomas, P. and Billlaud, D. (2000) *Electrochim. Acta*, **46**, 39.
66 Stevens, D.A. and Dahn, J.R. (2001) *J. Electrocem. Soc.*, **148**, A803.
67 Zheng, H., Fu, Y., Zhang, H., Abe, T., and Ogumi, Z. (2006) *Electrochem. Solid-State Lett.*, **9**, A115.
68 Tamura, S., Mori, A., and Imanaka, N. (2004) *Solid State Ionics*, **175**, 467.
69 Tossici, R., Berrettoni, M., Nalimova, V., Marassi, R., and Scrosati, B. (1997) *J. Electrochem. Soc.*, **144**, 186.
70 Tossici, R., Berrettoni, M., Rosolen, M., Marassi, R., and Scrosati, B. (1996) *J. Electrochem. Soc.*, **143**, L64.
71 Tossici, R., Antoine, L., Janot, R., Guerard, D., and Marassi, R. (2004) *J. Phys. Chem. Solid.*, **65**, 205.
72 Tossici, R., Croce, F., Scrosati, B., and Marassi, R. (1999) *J. Electroanal. Chem.*, **474**, 107.
73 Komaba, S., Itabashi, T., Kimura, T., Groult, H., and Kumagai, N. (2005) *J. Power Sources*, **146**, 166.
74 Kim, S.S., Kadoma, Y., Ikuta, H., Uchimoto, Y., and Wakihara, M. (2001) *Electrochemistry*, **69**, 830.
75 Kottegoda, I.R.M., Kadoma, Y., Ikuta, H., Uchimoto, Y., and Wakihara, M. (2002) *Electrochem. Solid-State Lett.*, **5**, A275.
76 Zhang, S.S., Xu, K., and Jow, T.R. (2003) *Electrochem. Commun.*, **5**, 979.

9
Characterization of Solid Polymer Electrolytes and Fabrication of all Solid-State Lithium Polymer Secondary Batteries

Masataka Wakihara, Masanobu Nakayama, and Yuki Kato

This chapter deals with the lithium conducting polymer electrolytes without volatile organic solvents: (i) their characterization and (ii) the fabrication of all-solid-state lithium polymer secondary batteries (LPBs). To date, most of efforts have been devoted to developing fast lithium-ionconductors and analyzing their conduction mechanism. The first part of this chapter attempts to introduce briefly these studies briefly, especially those on the polymer electrolytes whose ionic conductivity was enhanced by adding plasticizer. On the other hand, the attempts to fabricate the all-solid-state LPB are limited, because of technical difficulty in forming good electrode (inorganic material)/electrolyte(polymer) interfaces. The latter part of this chapter focuses on the recent reports on the fabrication of the all-solid-state LPB and discusses the practical problems observed in these batteries.

9.1
Molecular Design and Characterization of Polymer Electrolytes with Li Salts

9.1.1
Introduction

Polymer electrolytes consist of a polymer matrix and lithium salts. Besides the safety of the polymer compared with organic solvents, other interesting properties, such as flexibility, easier manipulation, and high-temperature use, give us possibilities for the advanced lithium polymer secondary batteries (LPBs). Polymer electrolytes without any volatile organic solvents are often called *solid polymer electrolytes* (*SPEs*), and this section focuses mainly on these. On the other hand, polymer electrolytes containing a large amount of organic solvents, which have been commercialized recently, are called *gel polymer electrolytes*, [1]. In the case of the gel polymer electrolytes, the polymer matrix plays the role of only the "holder" for major part of the organic solvents.

Ever since the studies by Wright *et al.* and Armand *et al.* [2, 3] on their lithium conductivity, poly(ethylene oxide) (PEO) and its derivatives have been known to be typically representative of the polymer matrix for the SPE. This PEO-based polymer can dissolve Li salts, i.e., the polymer plays the role of a

Lithium Ion Rechargeable Batteries. Edited by Kazunori Ozawa

ISBN: 978-3-527-31983-1

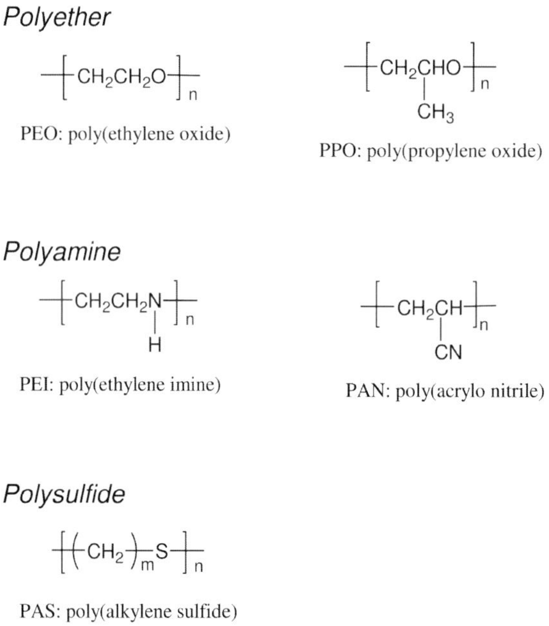

Polyether

$-[CH_2CH_2O]_n-$

PEO: poly(ethylene oxide)

$-[CH_2CH(CH_3)O]_n-$

PPO: poly(propylene oxide)

Polyamine

$-[CH_2CH_2N(H)]_n-$

PEI: poly(ethylene imine)

$-[CH_2CH(CN)]_n-$

PAN: poly(acrylo nitrile)

Polysulfide

$-[(CH_2)_m S]_n-$

PAS: poly(alkylene sulfide)

Fig. 9.1 Basic unit of polymer matrix chains for polymer electrolytes.

solvent, since heteroatoms (oxygen atom, –O–) of polymer chains acting as electron donors coordinate Li ions. Similar effects were observed in nitrogen atoms of imide group (–NH–) and thiol sulphur atoms (–S–). Basic polymer chains for polymer electrolytes are summarized in Figure 9.1. For the solvation of lithium salts in polymers, the solvation Gibbs energy should be larger than the lattice Gibbs energy of lithium salts. The solvation Gibbs energy depends on the ion-solvent interactions, which include electrostatic interactions and electron pair donor–acceptor (nonelectrostatic) interactions. The order of the power of the above electron donors for lithium ions, which are represented using a scale called *donor number*, follows the relative values of the negative charge on the heteroatoms [4]:

$$-\mathrm{O}- > \dot{\mathrm{N}}- > -\ddot{\mathrm{S}}-$$

In high molecular weight polymers, the conformation of polymer chains as well as the donor power is also important for the solvation. In low molecular weight solvents, e.g., propylene carbonate (PC) and dimethyl carbonate (DMC) used as conventional liquid electrolytes, solvation of lithium ions depends mainly on the number of molecules that may pack around a lithium ion. On the other hand, lithium ions are more likely to be coordinated by heteroatoms on the same chain with the possibility of some coordination by neighboring atoms, and therefore, the chain must wrap around the lithium ion without excessive strain. Taking polyethers as an example, it was found that (i) $-(CH_2CH_2O)_n-$ provides just the right space for maximum solvation and (ii) $-(CH_2O)_n-$ and $-(CH_2CH_2CH_2O)_n-$ are much weaker solvents [4, 5]. Accordingly, $-(CH_2CH_2O)_n-$, i.e., PEO would be the most favorable basic unit for polymer electrolytes, and these PEO chains would form a helix coil structure, in which lithium ions are solvated stably [6–8]. From this point of view, studies on PEO-based matrices have been the "mainstream" for polymer electrolytes [9–11].

Primitive polymer electrolytes using linear PEO with high molecular weights (>100 000) showed relatively high ionic conductivity ($>10^{-5}$ S cm^{-1}) over 60 °C, while the ionic conductivity showed a sudden decrease below 60 °C because of crystallization of the polymer chain [3–5]. In the late 1970s, it was suggested that conductivity resulted from the hopping of lithium ions between vacant sites inside polymer chains, corresponding to the hopping mechanism of the conventional solid-state ionic conductors [12]. However, the studies that followed revealed that it is not the crystalline but the amorphous phase in polymer electrolytes that is mainly responsible for ion transport [13–16]. Shriver *et al.* suggested that, in the amorphous phase, lithium ions move along with segmental motion of polymer chains [17]. Therefore, the studies in the late 1980s and the early 1990s focused on suppressing the crystallization of polymer chains to prevent the decrease in ionic conductivities around room temperature. In this context, it was found that poly(propylene oxide) (PPO), an analog to PEO (see Figure 9.1), forms amorphous complexes with lithium salts; this has been the focus of a large number of studies [18–20]. Although PPO-based electrolytes exhibit higher ionic conductivity than PEO-based ones around room temperature under a certain degree of lithium salt concentration, PPO cannot generate a high concentration of carrier ions. Thus, the attempts to improve SPEs shifted to the suppressing the crystallization of PEO-based electrolytes [5].

Cross-linking of PEO with low molecular weights or derivatives was very useful in suppressing the crystallization of polymer chains and also to retain mechanical strength (Figure 9.2). Both radiation [21–23] and chemical [24–27] cross-linking have been used extensively to produce amorphous and mechanically stable matrices. Radiation cross-linking has the additional practical advantage that polymer electrolyte films can be fashioned to the desired thickness or shape for a device because polymer matrices can be incorporated into a cell of the device prior to cross-linking [21–23]. With regard to chemical cross-linking, the urethane bond [24–27], ester bond [27], siloxane bond [28, 29], etc., were utilized. Furthermore, to get higher ionic mobilities, studies to increase the segmental motion of polymer chains have been carried out, for example, by (i) synthesis of "comblike" polymers [30–33], (ii) adding ceramic fillers [34–52], and iii) introducing plasticizers, as described later in this chapter.

Molecular design of lithium salts with lower lattice energies as well as matrix polymers is also important for obtaining higher ionic conductivity. From the time that early studies on polymer electrolytes began, attempts have been made to use $LiClO_4$ and $LiCF_3SO_3$ and such typical salts for conventional electrolyte solutions like $LiPF_6$ and $LiBF_4$ [9–11]. However, their solubilities are not so high in polymer electrolytes because their lattice energies are not low enough in polymer matrices with low permittivity, for example, the permittivity ε_r of PEO is around 8, whereas

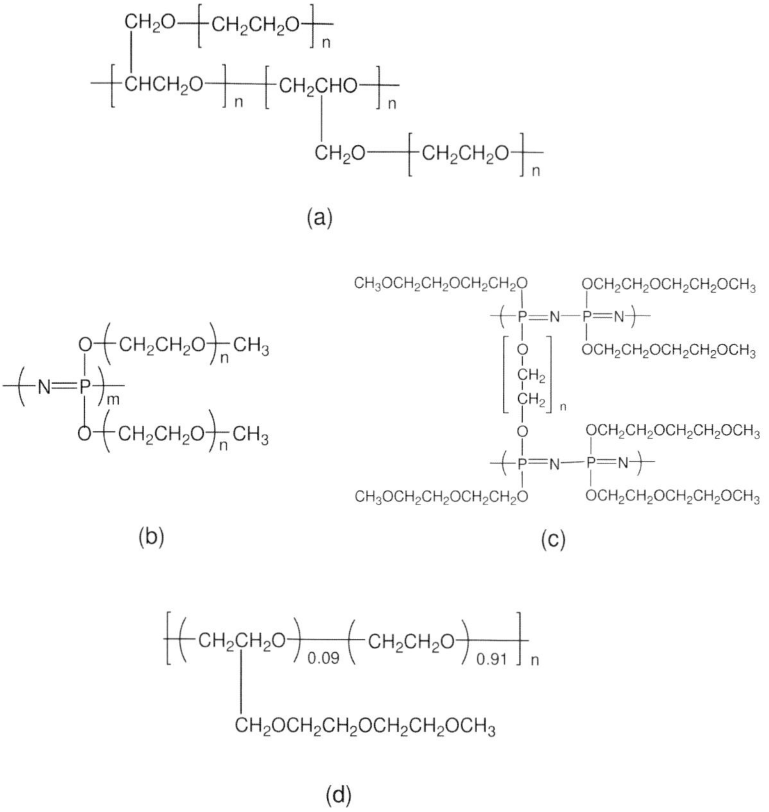

Fig. 9.2 Schematic image of polymer matrix: (a) cross-linked PEO (often radiation is used) [21–23]; (b) typical "comblike" polymer, MEEP (poly[bis-2-(2-methoxyethoxy) ethoxy phosphazenel]) [24–26]; (c) cross-linking MEEP with polyether [24–26]; (d) P(EO/MEEGE (2-(2-methoxyethoxy) ethyl glycidyl ether)) [27].

ε_r of PC is 65 (the permittivity is related to electrostatic interactions with matrices). To date, lithium bis-trifluoromethanesulfonimide ($LiN(CF_3SO_2)_2$) (often termed LiTFSI), with its lower lattice energy, has been the most popular of lithium salts from the time it was first reported by Armand *et al.* [53]. Later studies presented new types of lithium salts, with lower lattice energies, such as $LiC(CF_3SO_2)_3$ [53], $LiN(SO_2CF_2CF_3)_2$ (LiBETI) [54, 55], etc., some of which contributed high ionic conductivity of over 10^{-5} S cm^{-1} in PEO.

In the preceding text, the molecular design strategies of SPEs based on PEO and Li salts were introduced briefly. In the following section, our attention focuses on the most promising approach, that is, the introduction of plasticizer into the SPE, since this method does not require special molecular design for the matrix polymer, such as was seen in the "comb-polymer" approach. The preparation of SPE films and their fundamental physical properties are described thereafter.

9.1.2
Solid Polymer Electrolytes with Plasticizers

Typical examples of plasticizers for SPEs are organic solvents, such as PC and ethylene carbonate (EC); SPEs with these plasticizers are often called *gel polymer electrolytes*. In these gel polymer electrolytes, organic solvents are wholly responsible for ionic transport. (For practical purposes, organic solvents are sometimes restricted to 80–90 wt %). However, the safety problem associated with the use of volatile organic solvents essentially remains.

The use of a small quantity of non- or low-volatile oligomer as plasticizer would reduce such safety issues to a level sufficient for practical use. Abraham *et al.* [56] proposed the addition of poly(ethylene glycol) (PEG) with low molecular weights as plasticizer for SPEs, and mentioned the system as having "PEO-like" polymer electrolytes. They also reported that these electrolytes show high ionic conductivities without loss of high thermal stability. Similar systems were reported by Morita *et al.* [57] and Ito *et al.* [58]. For other approaches, phosphate solvents were suggested as noninflammable plasticizers for polymer electrolytes [59–61]. Kasuya *et al.* [60] reported the safety of the phosphate solvents in coin-type cells. In these SPEs, the plasticizers partially support the enhancement in ionic conductivity of the electrolyte, since liquid-type plasticizers increase in ionic mobility by suppressing the crystallization or increasing the segmental motion of polymer chains [60, 61]. However, dissociation of lithium salts is lower than in gel polymer electrolytes because of the low permittivity of the plasticizer.

Kato *et al.* [62] proposed borate derivatives with poly(ethylene glycol) (B-PEG), a new type of plasticizer. Lewis acidity of B-PEG would enhance dissociation of lithium salts, leading to high carrier-ion concentration. At the same time, the function of plasticizer would increase in ionic mobility. It is worth noting that the B-PEG has already been used as a main component of the fluid for the brake system in automobiles because of its high boiling point and good fluidity even at low temperature. Consequntly, higher ionic conductivity at low temperature and better thermal stability was expected. Later, Masuda *et al.* [63] expanded this idea by developing a new plasticizer of an aluminate derivative with poly(ethylene glycol) (Al-PEG) to enhance Lewis acidity. Hereinafter, the preparation of SPE films using these plasticizers and their physical properties are presented in detail.

9.1.3
Preparation of SPE Films with B-PEG and Al-PEG Plasticizers

The B-PEG and Al-PEG were synthesized by a relatively simple reaction between methoxy poly(ethylene glycol) $CH_3O(CH_2CH_2O)_nH$ (*n* is variable, 3–12 or larger) and B- or Al-containing compounds, boric acid anhydride (B_2O_3) or aluminum isopropoxide (Figure 9.3). The starting mixtures were dissolved in toluene under inert gas atmosphere and refluxed around 100 °C to eliminate water or isopropanol generated from this reaction. The B-PEG and Al-PEG obtained are sensitive to water

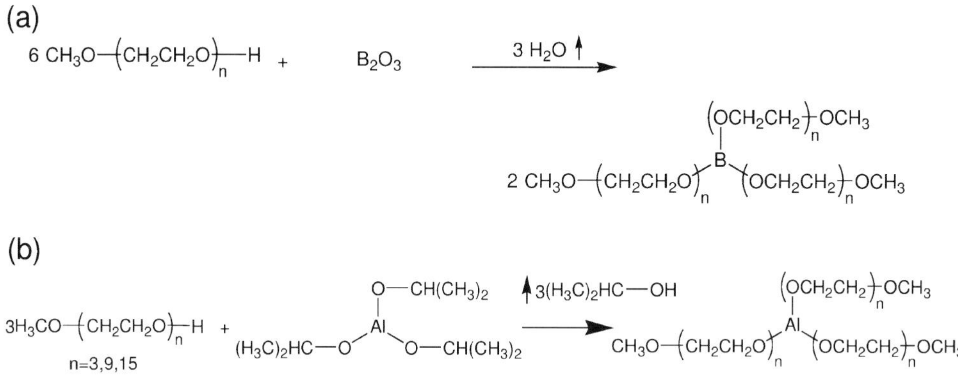

(a)

$$6\ CH_3O{-}(CH_2CH_2O)_n{-}H + B_2O_3 \xrightarrow{3\ H_2O\uparrow} 2\ CH_3O{-}(CH_2CH_2O)_n{-}B[(OCH_2CH_2)_n{-}OCH_3]_2$$

(b)

$$3H_3CO{-}(CH_2CH_2O)_n{-}H + Al[O{-}CH(CH_3)_2]_3 \xrightarrow{\uparrow 3(H_3C)_2HC{-}OH} CH_3O{-}(CH_2CH_2O)_n{-}Al[(OCH_2CH_2)_n{-}OCH_3]_2$$

n=3,9,15

Fig. 9.3 Scheme of the reaction for (a) B-PEG and (b) Al-PEG.

in the moisture because of hydrolysis, thus requiring handling of the compound under water-eliminated conditions by using an Ar-filled glove box [62, 63].

These plasticizers prepared as above can be easily added to various types of polymer matrices that have been developed, for example, the lithium-ion conductor. The procedure using radical polymerization for the preparation of SPEs is usually as follows: (i) appropriate amounts of monomers or oligomers for polymer matrix, plasticizers, and lithium salts are blended and dissolved as homogeneous viscous solution; (ii) the blended solution is poured into a pool or plate made from nonadhesive material, such as Teflon; (iii) a film-shaping treatment is performed, such as the doctor-blade method, as the need arises; and (iv) the monomers or oligomers are polymerized by heating, UV irradiation, and so on. For the above process, no major change is needed for the polymer film preparation using plasticizers. For example, Kato and Wakihara *et al.* prepared the SPE containing B-PEG using the above method. They chose a copolymer of two types of poly(ethylene glycol) methacrylate (PEGMA) – PEG-dimethacrylate (PDE600) and PEG-monomethacrylate (PME4000) – as shown in Figure 9.4 as the matrix polymer. $LiCF_3SO_3$ $LiClO_4$ or $LiN(CF_3SO_2)_2$ was used for the lithium salt. The details are described in, for example, Ref. [62].

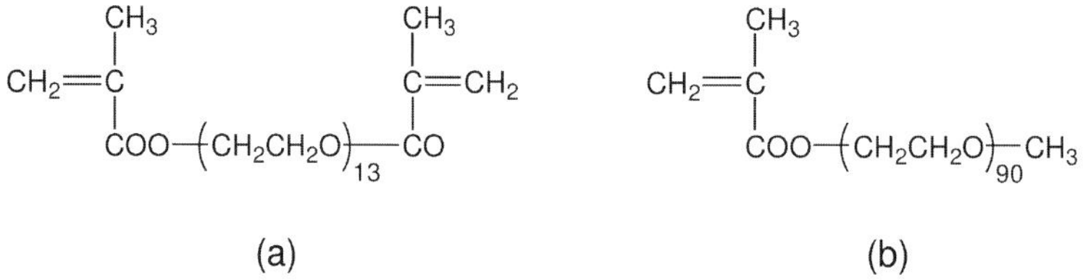

$$CH_2{=}C(CH_3){-}COO{-}(CH_2CH_2O)_{13}{-}CO{-}C(CH_3){=}CH_2$$

(a)

$$CH_2{=}C(CH_3){-}COO{-}(CH_2CH_2O)_{90}{-}CH_3$$

(b)

Fig. 9.4 Molecular structures of poly(ethylene glycol)-methacrylates (PEGMA): (a) poly(ethylene glycol) (600)-dimethacrylate (PDE600), (b) poly(ethylene glycol) (4000)-monomethacrylate (PME4000).

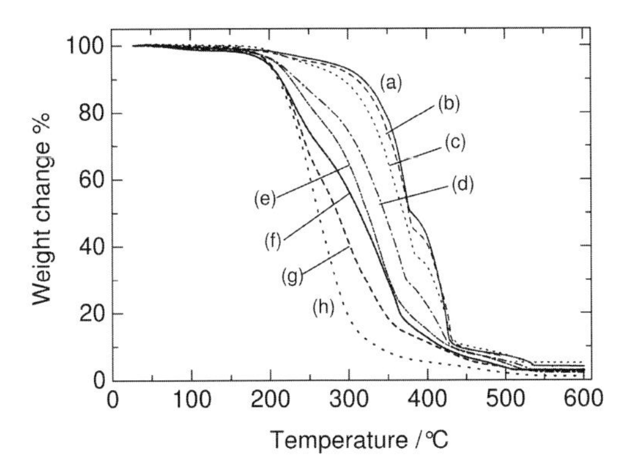

Fig. 9.5 Thermogravimetric (TG) curves of polymer electrolytes, PDE600 : PME4000 : PEG-borate ester($n = 12$) in a weight ratio of 1 : 1: 5 containing $LiN(CF_3SO_2)_2$ with various concentrations. Molar ratio of lithium atom to ether oxygen atoms (Li/EO): (a) 1/6, (b) 1/8, (c) 1/12, (d) 1/16, (e) 1/24, (f) 1/32, (g) 1/64; (h) without $LiN(CF_3SO_2)_2$.

9.1.4 Evaluation of SPE Films with B-PEG Plasticizers

Properties, such as thermal stability, electrochemical stability, and so on, are important factors for the realization of the practical use of LPB as well as ionic conduction property. This subsection introduces these evaluation methods for SPEs and their results in the case of the previous studies on the SPE with B-PEG plasticizers [62].

The thermal stability of materials is an indicator of the possible working range of temperatures for these materials. Since the LPB is suitable technology for large-scale batteries, safety measures against accidents are required, and higher temperature environmental use may also be demanded, in comparison with portable devices. Thermogravimetric (TG) analysis during heating samples is often useful as a method of investigation of stability. Figure 9.5 shows TG curves of the polymer electrolyte samples, PDE600 : PME4000 : B-PEG ester ($n = 12$) in a weight ratio of 1 : 1 : 5, containing $LiN(CF_3SO_2)_2$ with various concentrations. An increase in thermal stability was observed with increasing concentration of $LiN(CF_3SO_2)_2$ in the samples. Generally, lithium ions interact with ether oxygen atoms in PEO-based polymer electrolytes [1, 5], and therefore, the lithium ions would play a role of electrostatic cross-linking points of EO (ethylene oxide) chains in the polymer electrolytes. Accordingly, an increase in the quasi cross-linking points by the lithium ions with ether oxygen atoms should lead to the increase in thermal stability of the polymer electrolyte samples. A similar tendency was reported with regard to glass transition temperatures of SPEs (a discussion about glass transition of SPEs is given later in this section). Figure 9.6 represents TG curves of the SPEs containing $LiN(CF_3SO_2)_2$ (Li/EO = 1/8 in molar ratio) and added with the B-PEG whose EO chain length is $n = 3$, 6, or 12. As can be

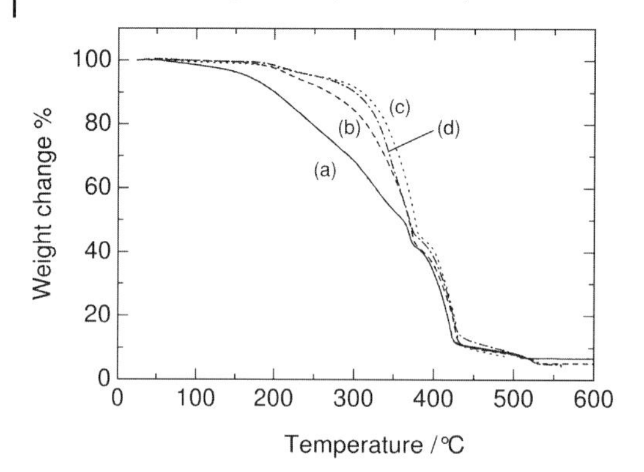

Fig. 9.6 Thermogravimetric (TG) curves of polymer electrolytes, PDE600 : PME4000 : PEG-borate ester (1 : 1 : 5) containing $LiN(CF_3SO_2)_2$ (Li/EO = 1/8). EO chain length of PEG-borate ester: (a) $n = 3$, (b) $n = 6$, (c) $n = 12$; (d) without PEG-borate ester.

seen in the figure, SPE without B-PEG was observed to be thermally stable up to about 200 °C. One can expect the decrease in the thermal stability for SPEs with B-PEG due to the smaller molecular weight of B-PEG plasticizers than that of the matrix polymer. However, the thermal stability of the SPE was comparable to that without the B-PEG, when the B-PEG whose EO chain length is $n = 12$ was added as plasticizer for the matrix polymer. On the other hand, the polymer electrolyte containing the B-PEG ($n = 6$) was observed to be thermally stable up to almost 200 °C, and in the case of the sample containing the B-PEG ($n = 3$), a weight loss was observed above 150 °C. As a consequence, the polymer electrolytes containing the B-PEG whose EO chain length is $n = 6$ or 12 have excellent thermal stability, over 200 °C, making them desirable for the development of safe LPB. A similar tendency was also observed in the SPE containing the Al-PEG [63].

Another aspect of the study of thermal properties is to measure crystallization and glass transition temperatures of polymer chains, since they relate to the ionic conduction mechanism. Figure 9.7 shows the DSC (differential scanning calorimetry) curves of the SPEs containing $LiN(CF_3SO_2)_2$ (Li/EO = 1/8), to which the PEG-borate ester whose EO chain length is $n = 3$, 6 or 12 has been added. Although in a DSC curve PEO-based polymer electrolytes containing a crystalline phase of polymer chains often show an endothermic peak due to melting of the phase [64–67], such a peak was not observed in the curves of the samples. This result indicates that the polymer electrolyte samples do not contain any crystalline phases, or that the electrolytes exist solely in an amorphous phase. Since the amorphous phase is mostly owing to the ionic conduction as mentioned above, the addition of plasticizer is advantageous for designing SPE materials. Meanwhile, a heat capacity change due to glass transition of polymer chains was observed around −50 °C in each of the DSC curves. The glass transition temperature, T_g, is defined as the midpoint of the heat capacity change, and the temperatures of these SPEs

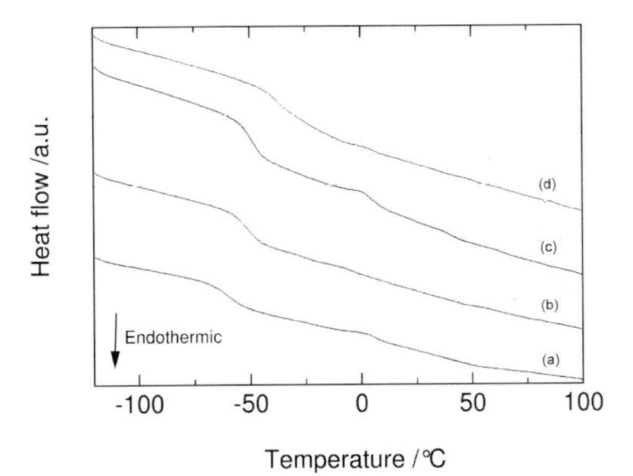

Fig. 9.7 DSC curves of polymer electrolytes, PDE600 : PME4000 : PEG-borate ester (1 : 1 : 5) containing $LiN(CF_3SO_2)_2$ (Li/EO = 1/8). EO chain length of PEG-borate ester: (a) $n = 3$, (b) $n = 6$, (c) $n = 12$; (d) without PEG-borate ester.

Table 9.1 Glass transition temperatures T_g of PDE600 : PME4000 : PEG-borate ester whose EO chain length is $n = 3$, 6, 12 (1 : 1 : 5 in weight ratio) + $LiN(CF_3SO_2)_2$ (Li/EO = 1/8)

EO chain length (n)	T_g (°C)
none additive	−36.3
3	−58.4
6	−49.1
12	−47.6

are summarized in Table 9.1. As seen in the results, T_g of the polymer electrolytes decreased with the addition of the B-PEG plasticizers. In addition, the shorter the EO chain length of the B-PEG (and Al-PEG), the lower was the T_g of the polymer electrolyte sample. Generally, T_g is correlated with flexibility of polymer chains, so that T_g is correlated with segmental motion of polymer chains, which is the driving force of ion conduction. Therefore, this result indicates that addition of the B-PEG or Al-PEG into the polymer electrolytes induces an increase in the mobility of the polymer chains, probably due to the plasticization effect of the ester on the matrix of the electrolytes.

The influence of the selection of lithium salts, $LiClO_4$ or $LiCF_3SO_3$ as well as $LiN(CF_3SO_2)_2$, on the thermal properties of the polymer electrolytes was also investigated. DSC curves of the polymer electrolytes, containing one of the lithium salts (Li/EO = 1/24) with the B-PEG whose EO chain length is $n = 3$ added to it, are shown in Figure 9.8. Each DSC curve showed the crystallization and the melting of the crystalline phase of EO chains at around −20 and 25 °C, respectively. Glass transition was observed in each DSC curve at around −70 °C.

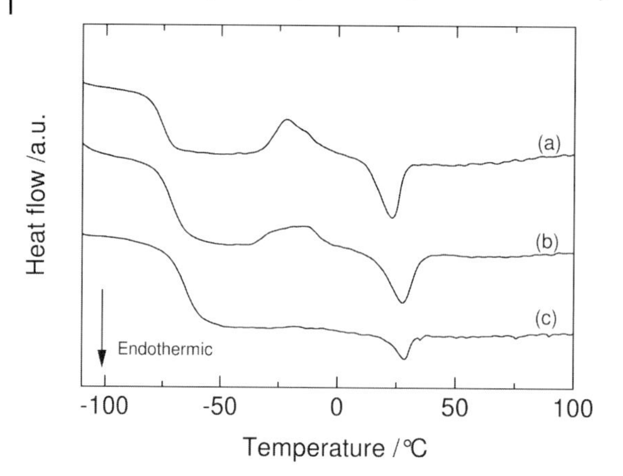

Fig. 9.8 DSC curves of polymer electrolytes, PDE600 : PME4000 : PEG-borate ester ($n = 3$) (1 : 1 : 5) containing lithium salt (Li/EO = 1/24). Lithium salt for the polymer electrolytes: (a) $LiN(CF_3SO_2)_2$, (b) $LiCF_3SO_3$, (c) $LiClO_4$.

Table 9.2 Glass transition temperatures T_g of PDE600 : PME4000 : PEG-borate ester ($n = 3$) (1 : 1 : 5 in weight ratio) with $LiN(CF_3SO_2)_2$, $LiCF_3SO_3$, or $LiClO_4$ (Li/EO = 1/24)

Lithium salt	T_g (°C)
$LiN(CF_3SO_2)_2$	−74.0
$LiCF_3SO_3$	−68.5
$LiClO_4$	−65.5

Table 9.2 summarizes T_g of the polymer electrolyte samples. As shown in the result, T_g of the polymer electrolyte differed with the lithium salt used, and the temperature of the sample with $LiN(CF_3SO_2)_2$ was lower than that of the samples with $LiCF_3SO_3$ or $LiClO_4$. Therefore, the mobility of polymer chains in the polymer electrolytes with $LiN(CF_3SO_2)_2$ should be higher than those of the electrolytes with $LiCF_3SO_3$ or $LiClO_4$; furthermore, the order for mobility levels is expected to be $LiN(CF_3SO_2)_2 > LiCF_3SO_3 > LiClO_4$. This order is probably due to the size of the anions leading to plasticization of the polymer chains. Similar tendencies were also reported by some research groups [68, 69].

As described above, the studies of the thermal property of SPEs give us useful information related to their safety and ion mobility. The results for SPEs with B-PEG or Al-PEG as plasticizer indicate that it is difficult to satisfy the best properties with regard to both safety and ionic conductivity, since elongation of the EO chain in B- and Al-PEG leads to higher safety but lower conductivity, and vice versa. Thus, optimization is necessary to construct a practical LPB.

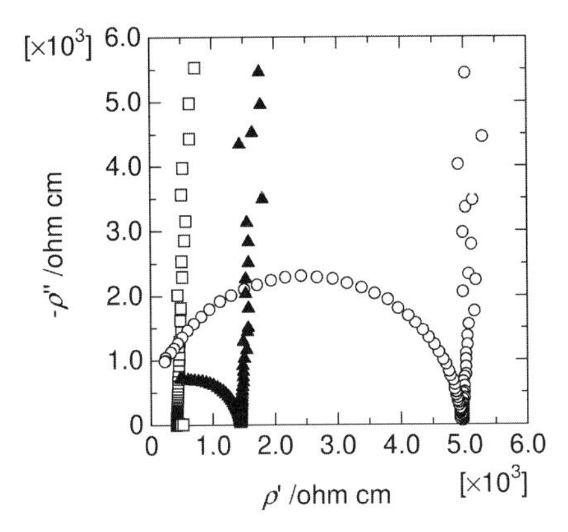

Fig. 9.9 Typical examples of Cole–Cole plots for impedance measurement of SPE; The data was obtained using SPE synthesized from PDE600 : PME4000 : B-PEG ($n = 12$) (1 : 1 : 5) containing $LiN(CF_3SO_2)_2$ (Li/EO = 1/8) at various temperatures: ○ at 40 °C, ▲ at 60 °C, □ at 90 °C.

9.1.5 Ionic Conductivity of SPE Films with B-PEG Plasticizers

Ionic-conduction phenomena have been investigated mostly by alternating current (AC) impedance techniques, which can separate the various impedance components. The cell construction for the impedance measurements is typically a symmetrical arrangement, such as stainless|SPE|stainless. These cells give impedance data that are often represented by Cole–Cole plots (often also called *Nyquist plots*), plotting the real and imaginary impedance components along the horizontal and vertical axes, respectively. Typical Cole–Cole plots obtained by an AC impedance technique for polymer electrolyte samples are shown in Figure 9.9, in which the composition of the sample is PDE600 : PME4000 : B-PEG ($n = 12$) in a weight ratio of 1 : 1 : 5, containing $LiN(CF_3SO_2)_2$ with a molar ratio of Li/EO = 1/8. The plots measured under 40 °C showed a semicircle, whose diameter corresponded to the value of the resistance, probably due to the bulk resistance of the sample. Although such a semicircle was not observed in the plots over 60 °C because of the possible working range for the frequency of the equipment, an intercept on the x-axis indicates the value of the bulk resistance. The impedance that appeared at the lower frequency range (showing spike shape) arises from the electrode|SPE interfaces, and this is usually disregarded, when one discusses the bulk ionic conductivity of SPEs. This interfacial impedance is discussed in the latter part of this chapter.

The measured bulk ionic conductivities of the samples are usually summarized using Arrheniusplots, in which the logarithm of conductivity is plotted as a function of the inverse of temperature (log σ vs $1/T$). Figure 9.10 shows the temperature dependence in Arrhenius-type plots for ionic conductivity of the SPE used in Figure 9.9. An increase in ionic conductivity was observed on increasing the amount

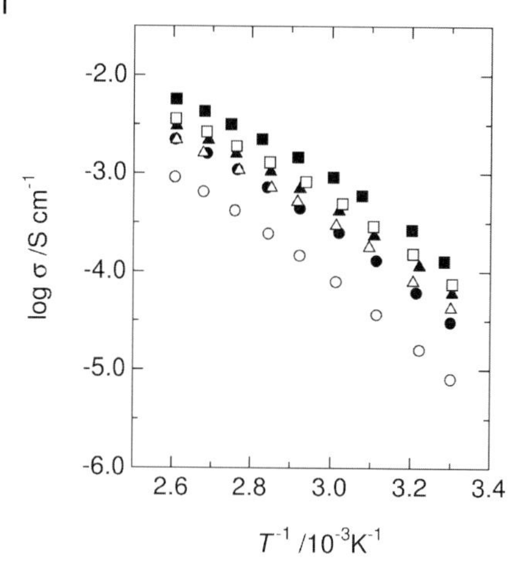

Fig. 9.10 Arrhenius plots for ionic conductivity of polymer electrolytes, PDE600 : PME4000 : PEG-borate ester ($n = 12$) in weight ratio of $1:1:x$ containing $LiN(CF_3SO_2)_2$ (Li/EO $= 1/8$). Weight ratio of PEG-borate ester ($n = 12$) to matrix polymer: ● $x = 1$, △ $x = 2$, ▲ $x = 3$, □ $x = 4$, ■ $x = 5$; ○ without PEG-borate ester.

of the B-PEG ($n = 12$) added to the matrix polymer over the whole temperature range. One can see the curved nature of the plots in the figure. It is generally known that the Arrhenius plots show a straight line (i.e., $\log\sigma$ is proportional to the inverse of temperature), when there is ion-hopping; this is frequently seen in inorganic solid electrolytes [12], solution electrolytes, or gel polymer electrolytes [70]. Rather than applying the hopping mechanism, the free-volume theory of polymers can explain the observed temperature dependence of ionic conductivity for SPEs [71–74]. Free-volume theory of polymers is often indicative of temperature dependence for segmental motion of the polymer chains expressed by the Williams–Landel–Ferry (WLF) relationship [75] The WLF equation is represented as follows:

$$\log\frac{\sigma(T)}{\sigma(T_g)} = \frac{C_1(T - T_g)}{C_2 + (T - T_g)} \tag{9.1}$$

where $\sigma(T)$ and $\sigma(T_g)$ are the conductivity at a temperature T and a glass transition temperature T_g, respectively, and C_1 and C_2 are the WLF parameters for the temperature dependence of ionic conductivity. However, the ionic conductivity at T_g, $\sigma(T_g)$, is difficult to measure in the present experiments because $\sigma(T_g)$ is too low in value to be measured by the complex impedance measurement; therefore, a possible temperature, T_0, is often selected as a reference temperature, for this measurement. Then, Equation 9.2 is rewritten as follows:

$$\log\frac{\sigma(T)}{\sigma(T_0)} = \frac{C_1'(T - T_0)}{C_2' + (T - T_0)} \tag{9.2}$$

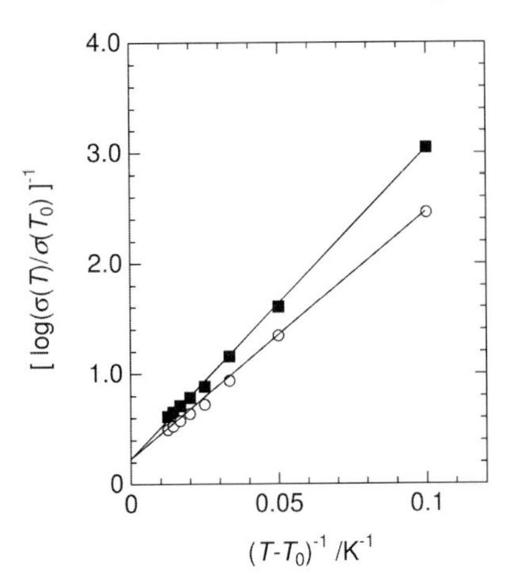

Fig. 9.11 WLF for ionic conductivity of polymer electrolytes, PDE600 : PME4000 : PEG-borate ester ($n = 12$) (1 : 1 : x) containing $LiN(CF_3SO_2)_2$ (Li/EO = 1/8). Weight ratio of PEG-borate ester ($n = 12$): ■$x = 5$; ○ without PEG-borate ester ($x = 0$).

The parameters in Equation 9.3 are calculated as follows:

$$C_1 = \frac{C_1{}'C_2{}'}{[C_2{}' - (T_0 - T_g)]} \tag{9.3}$$

$$C_2 = C_2{}' - (T_0 - T_g) \tag{9.4}$$

Figure 9.11 shows WLF plots for the ionic conductivity of the polymer electrolytes PDE600 : PME4000 added with or without the PEG-borate ester ($n = 12$) (1 : 1 : 5) and containing $LiN(CF_3SO_2)_2$ (Li/EO = 1/8). From the inverse of equation 9.3, the temperature dependence of the ionic conductivity was plotted as $[\log(\sigma(T)/\sigma(T_0))]^{-1}$ versus $1/(T - T_0)$. It was observed, as shown in Figure 9.11, that $[\log(\sigma(T)/\sigma(T_0))]^{-1}$ varies linearly with $1/(T - T_0)$, which indicates that the temperature dependence for the ionic conductivity of both the samples follow the WLF-type equation. The WLF parameters, C_1, C_2, and $\sigma(T_g)$, were estimated from Figure 9.11 and the equations (2)–(4), and are summarized in Table 9.3. The estimated parameters were found to be close to the universal values of WLF parameters, $C_1 = 17.4$ and $C_2 = 51.6$ K, which implies that the temperature dependence of the ionic conductivity for the polymer electrolytes was dominated by that of the segmental motion of polymer chains. Therefore, the ionic conduction in the polymer electrolyte samples is probably caused by segmental motion of polymer chains. In fact, the measured ionic conductivity and the previous DSC results concerning T_g of the polymer electrolytes (Tables 9.1 and 9.3), which is an "indicator" for the mobility of the polymer chains, bring us to the conclusion that the increase in ionic conductivity corresponded to the decrease of the T_g, i.e., the enhanced ionic conductivity by addition of B-PEG may stem from the increase in ionic mobility, or accelerated

Table 9.3 WLF parameters and σ at T_g of PDE600 : PME4000 : PEG-borate ester whose EO chain length is $n = 3$, 6, 12 (1 : 1 : 5 in weight ratio) + $LiN(CF_3SO_2)_2$ (Li/EO = 1/8)

EO chain length (*n*)	T_g (°C)	C_1 ( )	C_2 (K)	$\sigma(T_g)$ (S cm^{-1})
None additive	−36.3	11.8	52.5	3.07×10^{-12}
3	−58.4	14.7	28.6	1.64×10^{-15}
6	−49.1	11.1	40.8	7.72×10^{-12}
12	−47.6	12.0	43.8	1.92×10^{-12}

segmental motion by the plasticization. The general condition for an increase in ionic conductivity was summarized as follows in the case of adding B-PEG plasticizer: (i) increasing the amount of B-PEG and (ii) decreasing the EO chain length of plasticizer (but this does not always apply). One can thus design SPE film with the best ionic conductivity by considering the above conditions. However, the above conditions simultaneously cause the problem of weak mechanical strength of SPE film, since there is an increase in the amount of softer components in SPEs. An SPE film optimization is required to make the ionic conductivity and the mechanical strength compatible with each other for practical LPB use.

The other factors affecting ionic conductivity are the choice of the lithium salt and its concentration. Here, some remarks on the effect of lithium salt concentration are presented. Since the conductivity increases in proportion to the amount of lithium salt, an increase in the salt concentration should show a positive effect in the ionic conductivity. However, as mentioned in the TG results, the lithium salt enhances the polymer chain cross-linking, so that the increase in the salt concentration simultaneously suppress the segmental motion of polymer chain. Thus, both a positive and a negative effect are conceived on adding the lithium salt. These details are described in Ref. 62.

Further, the conductivity at glass transition temperature $\sigma(T_g)$ can be often a useful indicator that qualitatively represents the degree of lithium salt dissociation, i.e., the concentration of carrier ions in SPEs [72]. Since the segmental motion of polymer chains is frozen at the glass transition temperature, it is considered that mobility of the chains at that temperature can be standardized to be the same. Accordingly, one can assign the difference in $\sigma(T_g)$ that arises as due to the dissociation constant of lithium salt, which leads to the difference in concentration of carrier ions. For example, $\sigma(T_g)$ of the SPE with B-PEG listed in Table 9.3 showed that B-PEG-added SPE with EO chain length $n = 6$ and 12 were of the same order of magnitude compared with that of the sample without the ester. Therefore, the B-PEG ($n = 6$ or 12) may be not so effective in the dissociation of the lithium salt, $LiN(CF_3SO_2)_2$. Meanwhile, in the case of the B-PEG ($n = 3$), $\sigma(T_g)$ was found to be lower than that of the sample without the ester by 3 orders of magnitude. The exact reasons are still uncertain, but the too short EO chains in B-PEG probably made the interaction between polymer and salt difficult. In other words, in PEO-based polymer electrolytes, it is generally considered that lithium ions are solvated in the helix of the EO chains. Therefore, the EO chain length of the PEG-borate ester $n = 3$ may be not enough for solvation, leading to a decrease in the dissociation of the lithium salts. Note that it has been suggested that spectroscopic techniques (such as Raman spectroscopy) as well as impedance measurement represent a more direct way of measuring the dissociation. The details are available elsewhere in the literature [76–78].

As mentioned above, the AC impedance technique gives us insightful knowledge of the polymer motion and dynamics as well as the ionic conductivity. Here, we discuss the factors affecting the ionic conductivity and some of the dilemmas regarding conductivity and other factors (such as, safety and mechanical strength) in the design of SPEs at a molecular level.

9.1.6 Transport Number of Lithium Ions

Ionic conductivity of polymer electrolytes containing lithium salts measured by an AC impedance technique includes the conductivity of anions as well as that of lithium ions. However, the conductivity of lithium ions alone is important, especially for application in LPBs. In addition, anion conduction would cause the concentration polarization of SPEs, resulting in a decrease in the power density for LPB usage. Therefore, the estimation of the transport number is important, and molecular design to achieve large transport number of Li ions is demanded. This section describes several attempts or concepts of polymer design to increase the transport numbers of Li ions, and then briefly discusses the measurement techniques for these.

Various techniques for estimating the transport number of lithium ions have been reported by many research groups [79–82], and the principle behind most widespread conventional techniques is based on the combination of the complex impedance and potentiostatic polarization measurements [1, 5]. Watanabe *et al.* [83] suggested one method for the estimation from potentiostatic currents, a bulk resistance of a polymer electrolyte sample and interfacial resistances between the sample and lithium electrodes by the following equation:

$$\begin{aligned} t_{\mathrm{Li^+}} &= \frac{\sigma_{\mathrm{Li^+}}}{\sigma_{\mathrm{Li^+}} + \sigma_{\mathrm{X^-}}} = \frac{\sigma_{\mathrm{Li^+}}}{\sigma_{\mathrm{bulk}}} = \frac{k/R_{\mathrm{Li^+}}}{k/R_{\mathrm{bulk}}} = \frac{R_{\mathrm{bulk}}}{R_{\mathrm{Li^+}}} \\ &= \frac{R_{\mathrm{bulk}}}{R_{\mathrm{total}(\infty)} - R_{\mathrm{interface}}} = \frac{R_{\mathrm{bulk}}}{\Delta V / I_{(\infty)} - R_{\mathrm{interface}}} \end{aligned} \tag{9.5}$$

where X^- is a counteranion, k is a cell constant (can be eliminated), I is the measured direct current by the polarization, ΔV is the applied direct voltage, R_{bulk} is the bulk resistance of an electrolyte sample, and $R_{\mathrm{interface}}$ is the interface resistance between the electrolyte and lithium electrodes (the latter two resistances are measured by the AC impedance technique). In their method, the interfacial resistance is treated as constant during a measurement. Later, Abraham *et al.* [84] estimated the transport number considering the possibility for changes of the bulk

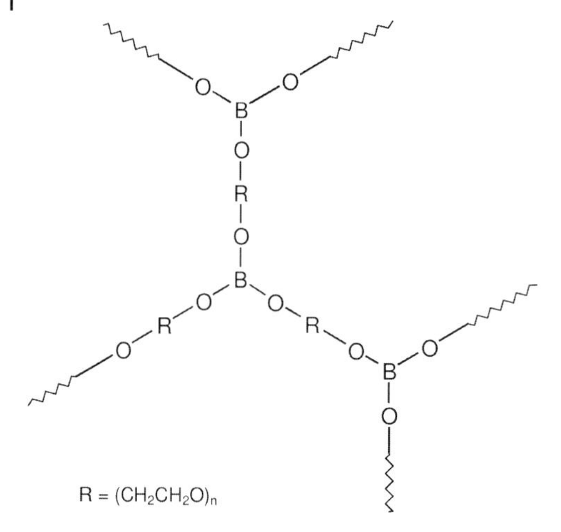

Fig. 9.12 Schematic image of matrix polymer cross-linked with borate ester groups.

resistance and the interfacial resistances during a potentiostatic polarization by the following equation:

$$t_{Li^+} = \frac{I_{(\infty)} R_{b(\infty)} (\Delta V - I_{(0)} R_{e(0)})}{I_{(0)} R_{b(0)} (\Delta V - I_{(\infty)} R_{e(\infty)})} \tag{9.6}$$

where 0 and ∞ refer to the initial state and the steady state, respectively.

The obtained transport number of Li ions t_{Li^+} for typical SPEs added with B-PEG are listed in Table 9.4. The estimated transport number for each sample was rather low, around 0.1, and did not depend on the EO chain length of the B-PEG. A similar tendency was also observed for the Al-PEG added SPE samples. Although these plasticizers were expected to enhance the transport number of Li ions due to their Lewis acidity around B and Al atoms, no significant enhancement was indicated in this study. The exact reasons are uncertain so far, but two ideas that have been suggested are as follows: (i) the acid–base interactions not being strong enough to "trap" the anions and (ii) the large diffusivity of the plasticizer itself despite capturing the anions. The other study regarding SPE cross-linked with B-PEG groups (see Figure 9.12) showed remarkable increase of transport number of Li ions, which reached $t_{Li^+} \sim 0.64$ measured by means of Abraham's technique as above [85]. Thus, the anion-capturing function was indicated by Lewis acidity around boron in the polymer matrix. Compared to the B-PEG plasticizer, the boron, which composed part of the polymer matrix, cannot migrate in the SPE. We infer that the relatively low t_{Li^+} in B-PEG-added SPE arises from the large diffusivity of the plasticizer.

Other attempts to measure transport numbers in SPEs were made by pulsed field gradient nuclear magnetic resonance (NMR) technique and its derivatives. For example, Gorecki *et al.* [86] reported the temperature and salt anion size dependence of transport numbers, and no significant differences in both were observed. Theoretical and technical details on pulsed field gradient NMR are given elsewhere [86, 87].

Table 9.4 Transport numbers of lithium ions t_{Li^+} in PDE600 : PME4000 : PEG-borate ester whose EO chain length is $n = 3, 6, 12$ (1 : 1 : 5 in weight ratio) + $LiN(CF_3SO_2)_2$ (Li/EO = 1/8) measured at 60 °C

EO chain length (*n*)	t_{Li^+}
3	0.11
6	0.14
12	0.12

9.1.7 Electrochemical Stability

The last property of SPEs introduced in this section is the electrochemical stability. In lithium-ion batteries, SPE films place (are placed under) the higher (oxidizing) and lower (reducing) potential conditions around positive and negative electrodes, respectively. The higher potential conditions are, for example, approximately 3.5 V (such as for olivine-type $LiFePO_4$) to about 4.4 V (as for spinel-type $LiMn_2O_4$) versus Li^+/Li, while lower potential ones are around 0 V. Once the decomposition (i.e., electrolysis) occurs beyond the potential range, undesirable side reactions may result in electronic conduction, increase in internal resistance, degradation of cycle performance, and so on. Thus, an SPE that possesses a large potential window is in high demand.

A typical method for evaluating the electrochemical stability is cyclic voltammetry (CV). Li metal sheets were used as both counter and reference electrodes. On the other hand, stainless-steel or Cu foil was used as a working electrode for the measurements of higher or lower potential windows, respectively. Note that the stainless steel and Cu are not stable (or they show large interfacial resistance) in lower and higher potential regions, respectively.)

Figure 9.13 is a typical example that shows the cyclic voltammograms of the SPEs added with B-PEG and $LiN(CF_3SO_2)_2$ (Li/EO = 1/8), at room temperature. In the cathodic region (Figure 9.13(a)), lithium metal deposition occurred below 0 V versus Li^+/Li, and lithium stripping followed on the anodic sweep. The reaction was reversible as a result of integration of the cathodic and anodic current flow in this voltammogram (cycle). Thus, no irreversible electrolysis occurred at in the region with lower potential. On the other hand, Figure 9.13(b) shows that current flow started at approximately 4.5 V versus Li^+/Li, indicating that the SPE would be stable electrochemically up to 4.5 V in the higher potential region. Therefore, the electrochemical stability of the B-PEG-added SPE seems to be enough for LPB applications. It is worth noting that the stainless-steel (SUS) electrode itself is an unrelated electrode for the Li and electron-exchange reactions, so that the potential windows obtained (especially in the higher voltage region) merely displayed approximate decomposition voltage of electrolyte. Knowledge on the redox reaction and its compatibility between positive-electrode materials and

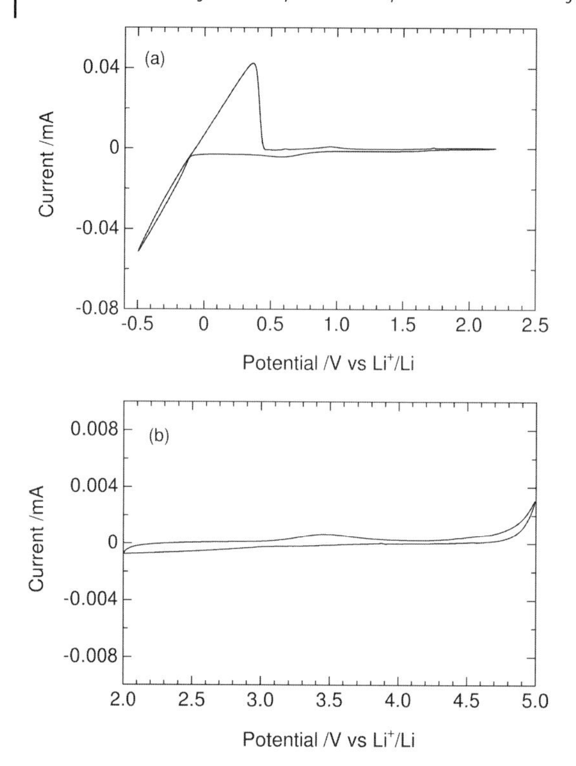

Fig. 9.13 Cyclic voltammograms of polymer electrolytes, PDE600 : PME4000 : PEG-borate ester ($n = 12$) (1 : 1 : 5) containing $LiN(CF_3SO_2)_2$ (Li/EO = 1/8) at room temperature: (a) on a Cu-working electrode swept from −0.5 to 2.2 V versus Li^+/Li, (b) on a SUS-working electrode swept from 2.0 to 5.0 V versus Li^+/Li.

SPE films is needed to discuss details of the electrochemical stability. Recent investigations related to interfacial electrochemical stability between SPE and positive electrodes have reported using impedance techniques for LPB construction. These are described in the following section.

9.1.8 Summary

This section introduced the required physicochemical properties of SPEs and typical measurement techniques for them. SPEs with thermal and electrochemical stability, larger transport numbers of lithium ions, and so on, as well as with large ionic conductivity, are required for the LPB usage. To date, PEO-based SPE is considered as the only option as far as use in future applications is concerned and so various types of PEO-based SPEs have been reported. Among them, SPE added with B-PEG was discussed here as among the promising candidates, along with the methodologies to characterize them.

9.2
Fabrication of All-Solid-State Lithium Polymer Battery

9.2.1
Introduction

The all-solid-state LPB is recognized as the most attractive technology for rechargeable electronic sources, because of higher safety and energy density with formability than that observed in conventional battery using organic solvents. Thus, LPBs meet the demands for large-scale use, such as in electric vehicles and energy-storage systems for day-to-night power shift, as well as for portable devices. As mentioned in the previous section, considerable efforts have been devoted to develop SPEs with high lithium ionic conductivity by many researchers. However, limited attempts have been made to assemble all-solid-state LPBs by using these SPEs [88–90]. This may be owing to technical difficulties in obtaining the desirable physical and chemical contact between the SPE and the electrode materials.

Thus, this section deals with the known and conceivable problems in the fabrication of all-solid LPBs, and the various attempts that have been made to improve the battery performance from the viewpoint of designing solid–solid (polymer–ceramic) interfaces.

9.2.2
Required Ionic Conductivity of SPE

As mentioned above, numerous efforts have been devoted toward the improvement of ionic conductivity of SPEs. Thus, the question arises as to how high an ionic conductivity is required, if one considers only the issue of ionic conduction for fabricating LPBs. Of course, the required conductivity level depends on the purpose of the battery. Let us consider a simple example that is often studied in laboratory-scale charge/discharge tests.

The low ionic conductivity of SPEs causes high internal resistance of the LPB, leading to polarization during charge/discharge. If the requested polarization level were less than 0.01 V under a current density of 1C ($\sim$0.3 mA cm^{-2}, and if one uses the typical cathode sheet for conventional liquid electrolyte battery), then SPE films with 0.1-mm thickness and lithium ionic conductivity of 3×10^{-5} S cm^{-1} would be required, for example. Considering the low transport number of lithium ions observed in PEO-based polymers, the total conductivity of $\sim 10^{-4}$ S cm^{-1} class SPEs may satisfy the above requirement. In fact, many reported SPEs would be applicable to the above LPB usage even at room temperature environment. For example, the SPEs with B-PEG plasticizer (Figure 9.10) developed by us showed a conductivity of roughly 10^{-4} S cm^{-1} around room temperature. However, not only the bulk resistance of SPEs but also interfacial resistances between electrodes and SPE films prevent the Li exchange reaction. The following sections focus on the solid–solid composite interface between electrolyte (polymer) and electrode (ceramics).

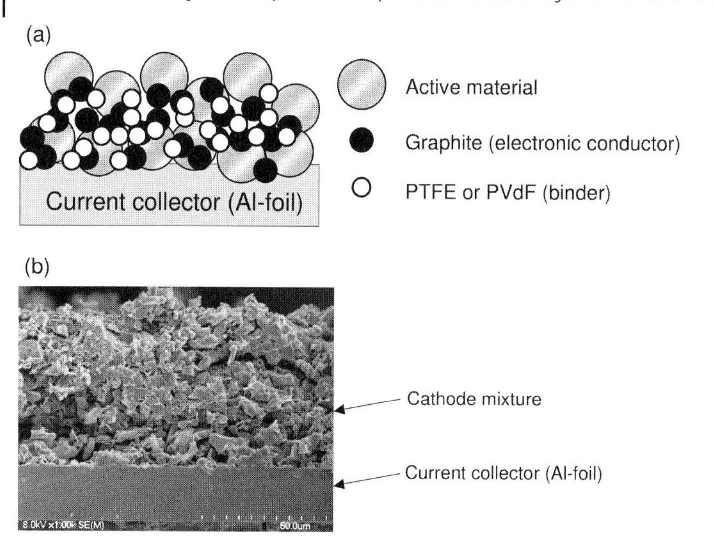

Fig. 9.14 (a) Schematic and (b) SEM images for cross section of cathode sheets for currently used lithium-ion battery.

9.2.3
Difference between Conventional Battery with Liquid Electrolyte and All-Solid-State LPB

Figure 9.14 shows schematic and scanning electron microscope (SEM) images of typical electrode sheets currently used for conventional battery with liquid electrolyte. The electrode was composed of particles of active material, which directly related to the electrochemical reaction of lithium-ion uptake and removal, electron conducting particles (such as carbon), and binder to adhere these particles (such as PVdF = poly(vinyliden fluoride)). As a result, the electrode sheet forms porous structure, even though the volumetric energy density is sacrificed by introducing cavities. These cavities are filled up with liquid electrolyte after a cell is assembled. Thus, the effective diffusion length of lithium ions in the electrode particles decreases, and polarization originating from diffusion is suppressed. Obviously, the cavities are filled up spontaneously with liquid electrolyte when the electrode sheet is immersed in it, whereas this would not occur when one uses the SPE film instead of the liquid electrolyte (see, Figure 9.15). The contact between SPE and ceramic powder of the electrode is limited at the surface of the electrode sheet, leading to increase in diffusion length and a fairly small effective area for the charge-transfer reaction, especially at higher current density. Moreover, the mechanical (or chemical) adhesion between electrode and electrolyte would be weak, so the expansion/shrinkage in volume of active materials due to charge/discharge cycle may cause local exfoliation at the electrode|electrolyte interface. Two methods are described here as examples on how to deal with this issue. The first method used monomer liquid for the SPE. The monomer liquid was poured on the electrode sheet, and then it was polymerized

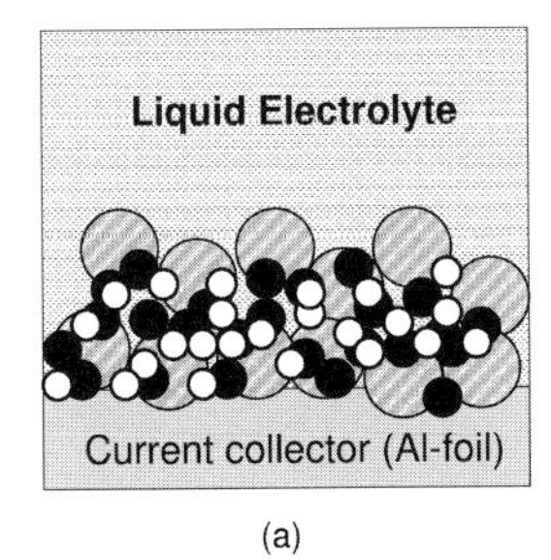

(a)

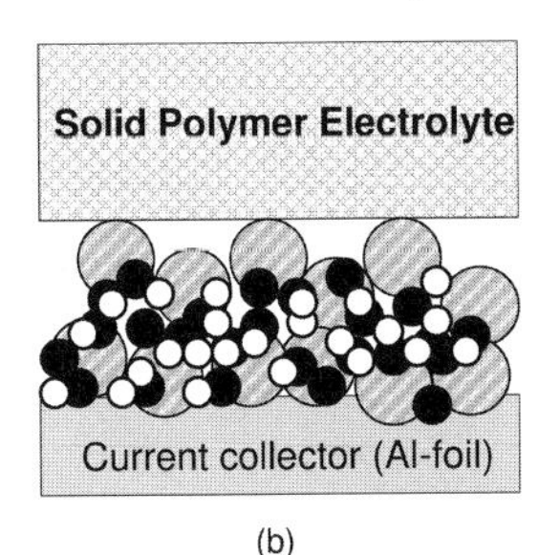

(b)

Fig. 9.15 Schematic images of cathode and electrolyte interface using conventional cathode sheet with (a) liquid electrolyte or (b) solid polymer electrolyte. Since a conventional cathode sheet forms a porous structure, the cavities are filled with liquid electrolytes spontaneously. Since polymer electrolytes do not immerse into the cavity, there will be a reduction in the interfacial area for charge-exchange reaction where the electrolyte and electrode materials are in contact.

by heating, UV irradiation, and so on. The cavities obtained on the electrode sheets would be filled with SPEs, and effective contact may form over the entire surface of active materials with SPEs. By controlling the amount of monomer liquid, the electrode sheet can be self-assembled with appropriate thickness of SPE layer (see Figure 9.16). However, polymerization is sometimes accompanied by an undesirable side reaction with the components of the electrode materials, resulting in the formation of a passivation layer at the electrode|electrolyte interface as can be seen in the magnified image in this figure. The cell showed no capacity for electrochemical charging and discharging. Hence, the charge-transfer reaction might be blocked at the passivation layer. The reaction mechanism of formation of the passivation layer is still ambiguous. However, it is inferred that the cathode materials played the role of catalysts and perhaps enhanced the radical generation. Few studies are available on the phenomena of the passivation layer formation, and one may be able to avoid the formation of passivation layer by choosing a proper combination of cathode material and polymerization reaction. Instead of using monomer liquid, one can prepare polymer solution, since SPEs with normal chain structure can be dissolved in the appropriate solvent. The solution is poured on the electrode sheets, which are heated to evaporate the solvent. Thus the interface between the electrode and the polymer electrolyte maintain much closer contact. The polymer plays the role of a binder, adhering the electrode particles, as well as those of the ionic conductor, which enables effective ion exchange between electrolyte|electrode interface. On the other hand, one of the drawbacks of this method is that the polymer selection is limited, because SPEs with cross-linked structure cannot be dissolved in organic solvents. In addition, the normal chain type SPEs are mechanically weaker than cross-linked polymers, which may have a negative effect on the durability of the LPB. Another method for the preparation of the LPB is a combined method; the electrode sheets consist of active materials and Li^+ conductive polymer with normal chain (to dissolve appropriate organic solvent), and the SPE film with cross-linking polymer matrix is separately made and attached onto the electrode sheet.

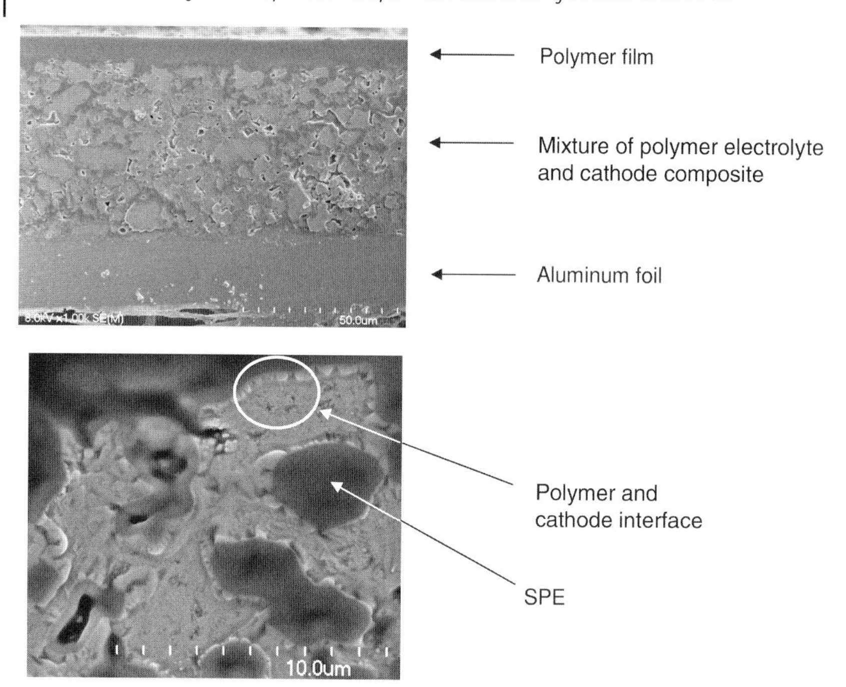

Fig. 9.16 SEM images for a cross section of cathode and SPE composite that was obtained by immersing monomer liquid into cathode sheet (Figure 2.22) and polymerized by UV irradiation. As can be seen, (a) the cavities of the cathode sheet were filled with SPE matrix. (b) A by-product layer was observed at the interface between SPE and the cathode composite (magnified image).

In the above, three methods of preparation of the LPB were introduced as examples. Figure 9.17(a) shows the first charge/discharge profile of LPBs prepared by the conventional method. The LPB with SPE film on the conventional electrode sheet showed relatively large polarization. On the other hand, smaller polarization was seen in the LPB using SPE binder ($LiClO_4$ dissolved PEO). In addition, the capacity retention of the LPB with conventional cathode sheet decayed rapidly (Figure 9.17(b)). Such results could be ascribed to the fact mentioned above, i.e., the elongation of effective diffusion path and limited contact for ion exchange at the surface of cathode sheet. In addition, the weak contact between electrode and electrolyte might cause an exfoliation of the polymer material because of the change in volume of the active material. Meanwhile, one can see the significant improvement in polarization with the use of SPE binder instead of the conventional PVdF binder. Accordingly, it would be effective to fill up the cavities of electrode sheets with the lithium-ion conductive media, although the best method to introduce the ionic conductive media depends on the characteristics and compatibility between the SPEs and the electrode materials.

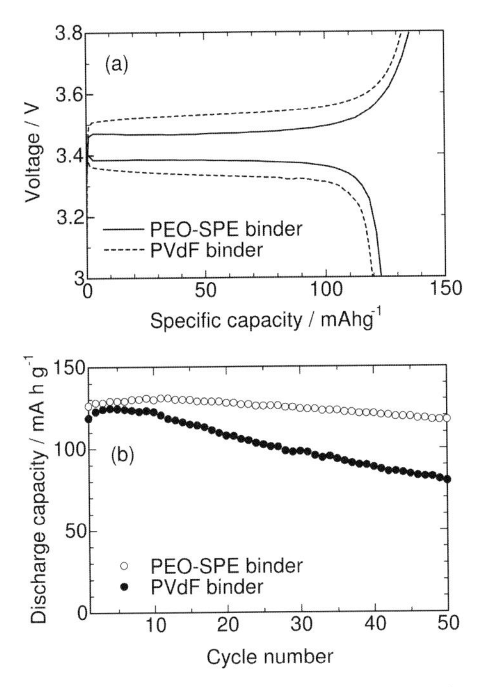

Fig. 9.17 Comparison of (a) first charge/discharge profiles and capacity retention for LPBs ($Li/SPE/LiFePO_4$) using conventional PVdF binder and PEO–SPE binder. Construction of LPBs using PVdF binder corresponds to the Figure 9.14 (b), while that using PEO–SPE binder was prepared as follows: $LiFePO_4$ powder, normal PEO (MW $\sim 10^6$), and $LiClO_4$ salt were dissolved in acetonitrile, and this slurry was cast on Al foil. Current density and temperature were set as 1.0C and 60 °C, respectively, for the electrochemical measurement.

9.2.4 Fabrication and Electrochemical Performance of LPBs Using SPE with B-PEG and/or Al-PEG Plasticizers

In this subsection, the electrochemical properties of LPBs are described, especially those of the LPBs using SPE with B-PEG or Al-PEG plasticizers. Ionic conductivity higher than 10^{-4} S cm^{-1} is required to suppress polarization due to internal resistance. In this respect, thinner SPE film formation and higher temperature operation (such as ~60 °C) are advantageous for increasing ionic conductivity. The addition of B-PEG or Al-PEG plasticizers enhanced the ionic conductivity as mentioned above, so that the improved electrochemical performance could be expected. Figure 9.18 presents the first charge/discharge curves of LPBs using SPEs with or without B-PEG plasticizer. Apparently, polarization was suppressed by using B-PEG plasticizer, which enhance the ionic conductivity of SPE as $\sim 10^{-3}$ S cm^{-1} at 60 °C. Whereas, the SPE without plasticizers ($\sim 10^{-5}$ S cm^{-1} at 60 °C) showed larger polarization, such LPBs also showed poor capacity retention during cycling.

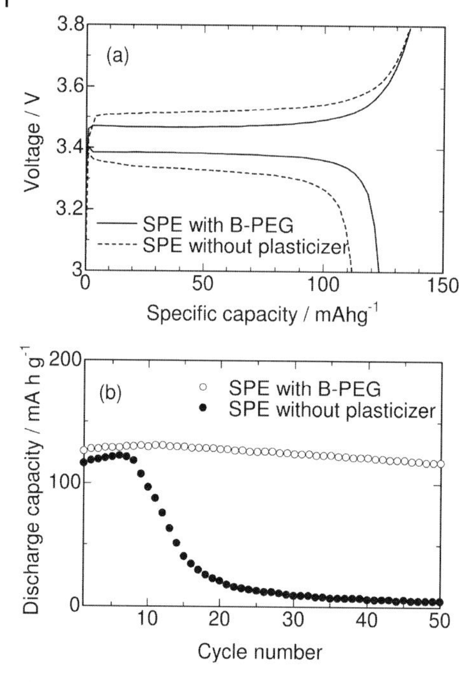

Fig. 9.18 Comparison of (a) first charge/discharge profiles and capacity retention for LPBs (Li/SPE/$LiFePO_4$) using SPEs with and without B-PEG plasticizer. Current density and temperature were set as 1.0 C and 60 °C, respectively, for the electrochemical measurement.

Durability is another important factor for LPB use, so that both electrochemical and mechanical stabilities of SPEs are demanded. As described in the former part of this section, PEO-based SPEs showed sufficient stability for reductive condition or lower voltage region (Figure 9.13), while decomposition may occur above 4.2 V versus Li^+/Li. Figure 9.19 shows the cycle performance of (i) Li/PEO-based SPE/$LiCoO_2$ (cutoff: 3.0–4.1 V) and (ii) Li/PEO-based SPE/$LiFePO_4$ (cutoff: 2.5–3.8 V) [91]. Apparently, poor cycle performance was indicated in the LPB using $LiCoO_2$ cathodes maybe due to the limit of the electrochemical stability of SPE under oxidative conditions. Thus, the selection of cathode materials may be regulated as far as PEO-based polymer is used. However, the potential profiles across the cathode and SPE region can be represented according to the fundamentals of electrochemistry as in Figure 9.20. The sudden potential drop lies in the electric double layer between electrode and electrolytes, indicating that undesirable electrolysis occurs only at the interfacial region. Hence, one can introduce the polymer that is designed with priority for electrochemical stability over oxidative condition, although this partly sacrifices ionic conductivity. In this respect, interesting attempts were made by Kobayashi *et al.* [92]; they coated ceramic materials with lithium conductivity, such as Li_3PO_4 and $Li_{1.5}Al_{0.5}Ge_{1.5}(PO_4)_3$, on the surface of cathode particles. The capacity retention of these LPBs improved significantly,

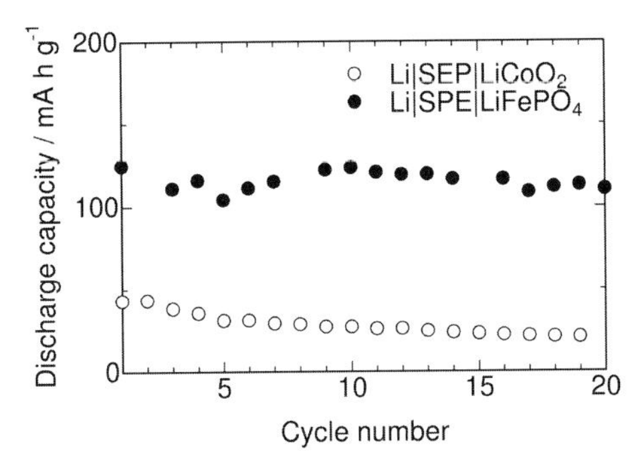

Fig. 9.19 Cycle performance of the LPBs with Li|SPE|$LiCoO_2$ (open circles) and Li|SPE|$LiFePO_4$ (closed circles). Cutoff voltages for these cells were set as 3.0–4.1 V and 2.5–3.8 V for $LiCoO_2$ and $LiFePO_4$ cathodes, respectively. Normal chain type PEO (MW ~ 106) was used as SPE, and the electrochemical charge/discharge was carried out at a temperature of 60 °C and current density of 0.1 C.

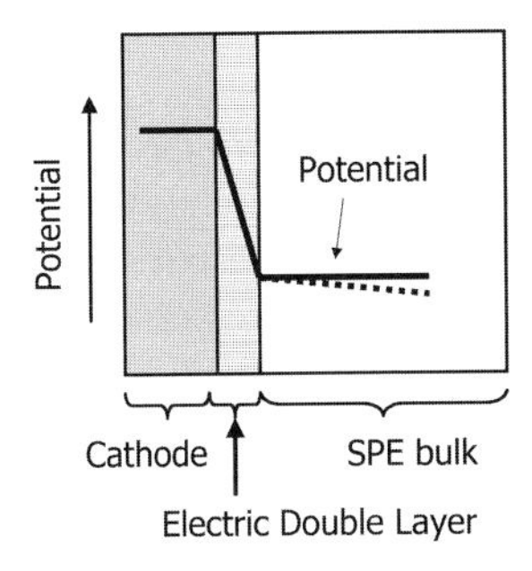

Fig. 9.20 Schematic images of variation of potential at the cathode|SPE interface. The potential drop occurred at the interfacial region solely (electric double layer), so that the SPE is exposed to the oxidative condition only in the vicinity of cathode region. The hatched line indicates the potential drops inside the SPE bulk due to low ionic conductivity. Note that the potentials inside SPE are not detectable, in principle.

and impedance measurements showed suppression of continual increase in the cathode|SPE interfacial resistance. Although the exact reasons are still uncertain (with further discussion in a later section), the ceramics may effectively protect the SPEs from being exposed to oxidative (high) voltages. It is noted that they succeeded in obtaining the charge/discharge cycling using 5-V class cathode of $LiNi_{0.5}Mn_{1.5}O_4$ and PEO-based polymer [93]. Thus, introducing electrochemically stable material at the electrode|electrolyte interface may be beneficial to realize practical LPB with high voltage.

The mechanical stability of SPEs is also an important factor for the durability of LPBs. Figure 9.21 shows the cycle performance of three LPBs using SPEs of (i) chain polymer type PEO (MW ~10^6) of ~0.15-mm thickness, and cross-linked polymer matrix with thickness (b) ~0.15 mm and (c) ~1.0 mm, respectively. All the SPEs were plasticized with B-PEG to enhance the ionic conductivity. The improved cycle performances were observed for the LPBs using cross-linked and thicker

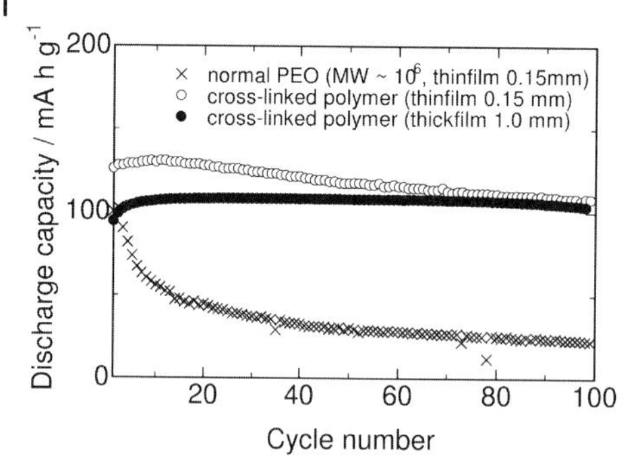

Fig. 9.21 Comparison of the cycle performance among three LPBs using various SPEs. (cross symbol) SPE using normal PEO with 0.15-mm thickness; (open circle) SPE using cross-linked polymer (see figure 9.2) with 0.15-mm thickness; (solid circle) SPE using cross-linked polymer with 1.0-mm thickness. All the SPEs contain B-PEG plasticizers, and the LPBs were cycled at 60 °C at a current density of 1.0 C.

SPEs at the low current density at 60 °C conditions. This result may indicate the importance of mechanical strength and stability of SPEs for the durability of LPBs. The mechanical strength of polymers can be improved by increasing the degree of cross-linking and thickness. However, the improvement of mechanical strength by the above methods caused relatively poor ionic conductivity of SPEs. In fact, slightly smaller reversible capacity was obtained for thicker SPE (Figure 9.21) despite better capacity retention with cycling, whereas no capacity was observed at 30 °C (not shown). Thus, optimization is needed depending on the operation condition.

For further understanding of the decay mechanism of capacity retention, impedance techniques are often used and serve as a diagnostic tool for LPBs. Figure 9.22 shows the variation in capacity retention with cycling in LPBs whose SPE consists of cross-linked polymer and Al-PEG plasticizers. In this case, relatively poor cycle performance was observed, such as rapid decay of capacity retention and large irreversibility in the first 10 cycles. The corresponding voltage profile (not shown) for each cycle showed that the decrease in capacity retention accompanied

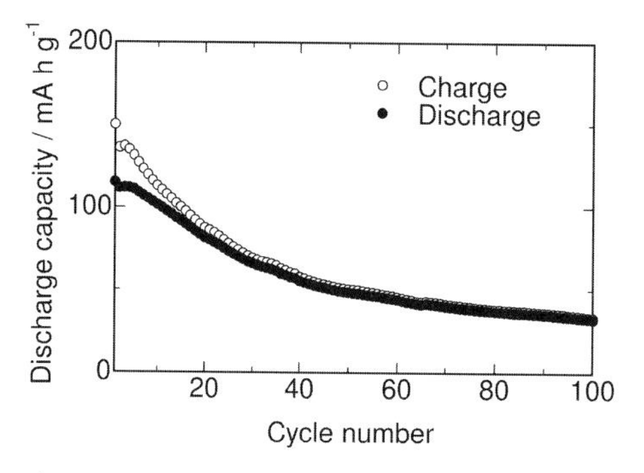

Fig. 9.22 Variation of charge/discharge capacity with cycling for Li|SPE|$LiFePO_4$ cell at 60 °C at 0.5C. The SPE contained Al-PEG plasticizers.

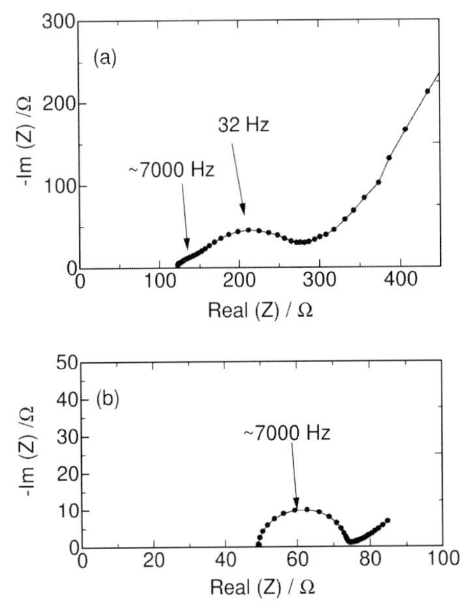

Fig. 9.23 Typical impedance spectra for (a) Li|SPE|$LiFePO_4$ cell after 10 cycles and (b) Li|SPE|Li symmetrical cell at 60 °C at 1.0C. The SPE contained Al-PEG plasticizers.

an increase in polarization. However, the observed polarization in the voltage profiles contains all the components of internal resistance, so that the specifying the crucial factors for decay is usually difficult.

In this respect, AC impedance techniques enable the separation of these resistive components, since each component has its own relaxation time. Figure 9.23(a) shows typical impedance spectra, or Cole–Cole plots, for the above LPB (Figure 9.22) after 10 cycles. The impedance spectra of symmetrical cells, Li|SPE|Li, are also presented in Figure 9.23(b). The impedance of a symmetrical cell consisted of three components. For example, there exists bulk resistance (~50 Ω; the distance from orgin to the end the left-hand side of semicircle on the real axis), following semicircle (~25 Ω; diameter of semicircle on the real axis) at higher frequency region, and Warburg-like slope (the part of straight line) at lower frequency region. The frequency at the top of the corresponding semicircle is ~7000 Hz as shown in the figure. This semicircle could be ascribed to the impedance of charge-exchange reaction at the interface of Li|SPE. The impedance spectra of LPB prior to electrochemical reaction was similar to that of the symmetrical one (not shown), and the frequency at the top of the semicircle agreed with that observed in the symmetrical cell. Therefore, the semicircle appeared in the LPB before cycling could be ascribed to the charge transfer reaction at the interface of Li|SPE. On the other hand, the impedance spectra of LPB after 10 cycles showed an additional larger semicircle at the lower frequency region (32 Hz). This impedance would stem from charge transfer reaction at the SPE|cathode interface (see, for example, Ref. 94). Though the impedance of SPE|cathode interface did not appear in the

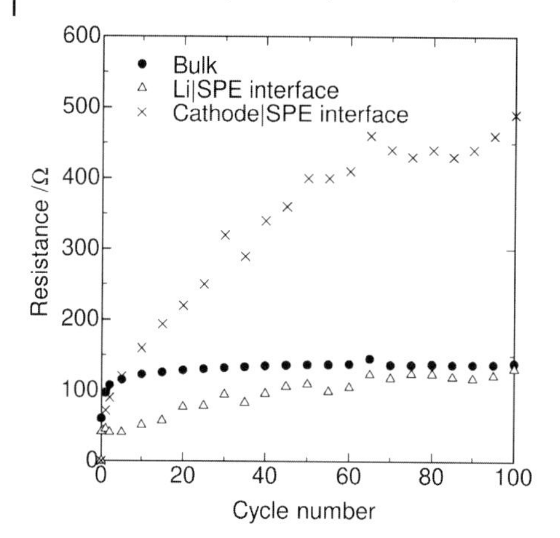

Fig. 9.24 Variation in resistance arising from bulk, Li|SPE, and $LiFePO_4$(Cathode)|SPE in LPB with cycle number. The capacity retention of the corresponding LPB is presented in Figure 9.22.

pristine LPB, it showed a severe increase after charge/discharge cycling. Figure 9.24 summarizes the variation of each impedance component with cycle number. As seen in the figure, a remarkably continuous increase of cathode|SPE interface was observed and this seemed to correspond to the decrease in capacity retention of the LPB. The increase of bulk resistance was also observed as already mentioned. However, this was only indicated during the first 10 cycles, and the bulk resistance kept stable value thereafter. This behavior of bulk resistance corresponds to the irreversibility of charge and discharge capacity, or Coulombic efficiency of each cycle. Figure 9.25 shows the Coulombic efficiency and bulk resistance for the first 30 cycles. One can see clearly see that the irreversible reaction was likely to cause an increase in bulk resistance. Therefore, the bulk of SPEs was decayed with cycling, though severe oxidative or reductive changes are limited to the vicinity of cathode|SPE or anode|SPE interfaces, respectively.

To understand the above electrochemical behavior, SEM/EDS (energy dispersive spectrometry) observations were performed. Figures 9.26 (a) and (b) present the SEM images of the cross section of cathode|SPE interface of the LPB before and after 50 cycles, respectively. As seen in the figure, pulverization of cathode particles was indicated after cycling. Since the cathode|SPE interface was formed as solid–solid contact, the pulverization of cathode may cause a decrease of effective contact area between cathode and SPE. (In case liquid electrolytes are used, the solutions may be immersed in the created cavity by pulverization.) Thus, the pulverization process would be one of the reasons for the relatively poor capacity retention in the LPB. In addition, the irreversible capacity may also come from the pulverization due to the electric connection being cut off for a portion of cathode particles. Since the break in electric connection causes an increase in the internal resistance, the impedance at high-frequency regions (bulk resistance of SPE) may increase as indicated in Figure 9.24. Another aspect

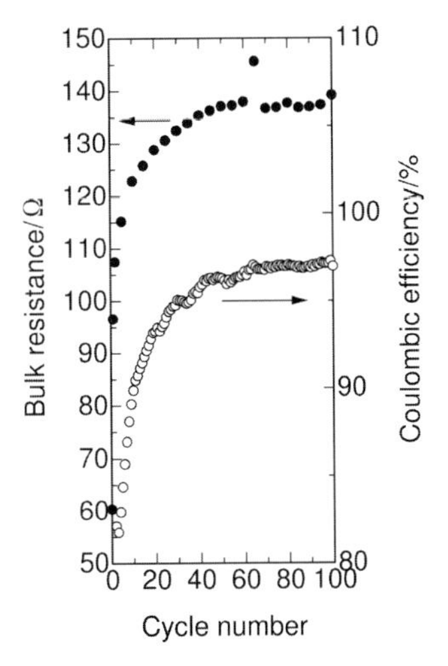

Fig. 9.25 Comparison between bulk resistances and Coulombic efficiency with cycling. The capacity retention of the corresponding LPB is presented in Figure 9.22.

that may be a clue to the degradation of LPB was indicated from EDS analysis. Figures 9.26 (c)–(f) display the elemental distribution of Al and Fe at the vicinity of cathode|SPE interface. The irons resided in the ceramic particles of $LiFePO_4$ before and after cycling. Hence, metal dissolution was unlikely to be occur in the SPE at EDS analysis level, which was reported in the conventional battery system with $LiMn_2O_4$ cathode and liquid electrolyte at elevated temperature [95]. On the other hand, the aluminum was distributed in the electrolyte region due to the plasticizers of Al-PEG. One can see the aluminum distribution was concentrated on the surface of cathode sheet with 10-μm thickness after 50 cycles compared with the LPB before cycling. Such a behavior may indicate that Al-PEG reacted with cathode materials, forming undesirable by-products containing aluminum. Such a side reaction would correspond to irreversible capacity, so that this side reaction might cease after the 10th cycle. In addition, since the growth of the by-products (Al-concentrated region) seems to invade the bulk region of SPEs, the decay of the bulk resistance would correspond to the aluminum concentration or by-product formation on the surface of cathode sheet.

Accordingly, the LPB introduced above showed severe decay of capacity retention due to the pulverization of cathode particle and/or undesirable side reactions at the vicinity of cathode|SPE interface. To improve the above situation, the molar contents of plasticizers were adjusted to attain good mechanical connectivity and suppress the side reaction. By decreasing the plasticizers, the mechanical strength would increase and the side reaction between plasticizers and cathode materials may decrease despite the decrease in ionic conductivity. Figure 9.27 shows the capacity retention during the cycling at 1.0 C (the same as the experimental condition of Figure 9.22 except for a charge/discharge rate that is twice as high). Remarkable improvement in the cycle performance was seen in the LPB with the

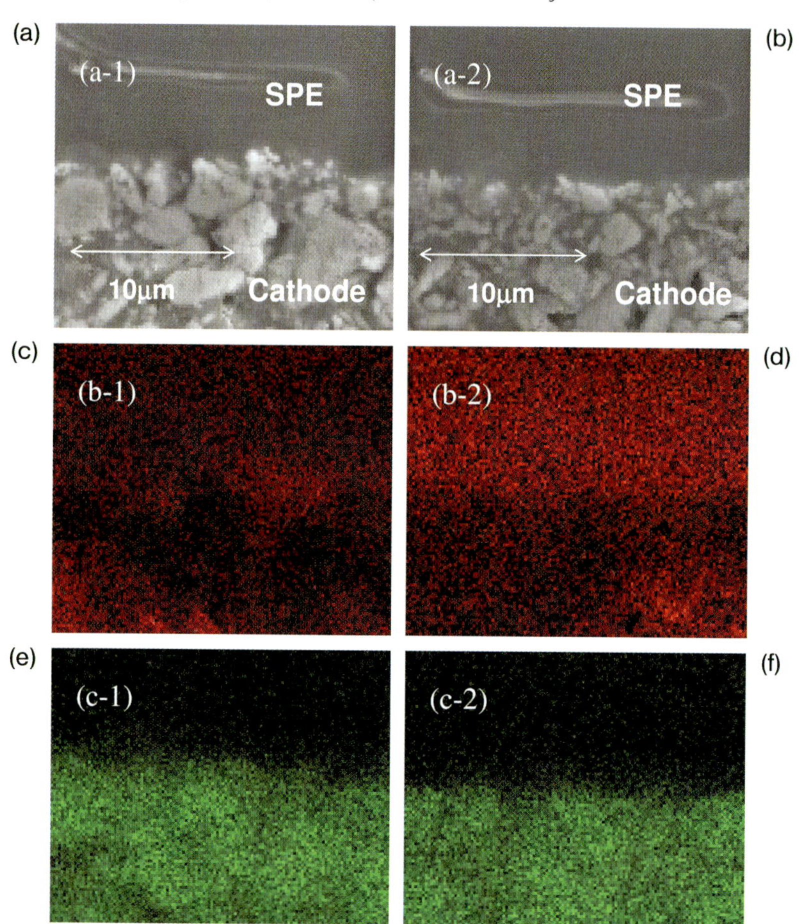

Fig. 9.26 SEM–EDS images for LPB with Li|SPE with Al-PEG plasticizer|$LiFePO_4$ construction. (a) and (b) SEM images for LPB before cycling and after 50 cycles, respectively. (c) and (d) present EDS mapping images for Al, and (e) and (f) show those for Fe in the LPB.

adjusted SPE, with the mixing ratio of plasticizers and polymer matrix becoming important. In addition, molecular design for plasticizers is needed in the future to suppress the side reaction with the surface of cathode materials.

As discussed above, the electrochemical performance of LPBs is deeply related to the interfacial property and reactions between electrode and electrolyte. However, not much is known about the relationship between battery performance and interfacial property. In this respect, the AC impedance technique is a powerful and convenient tool to understand the observed electrochemical reaction. Data related to the interfaces, such as impedance data of various types of LPBs, are needed to elucidate the crucial factors affecting electrochemical performance.

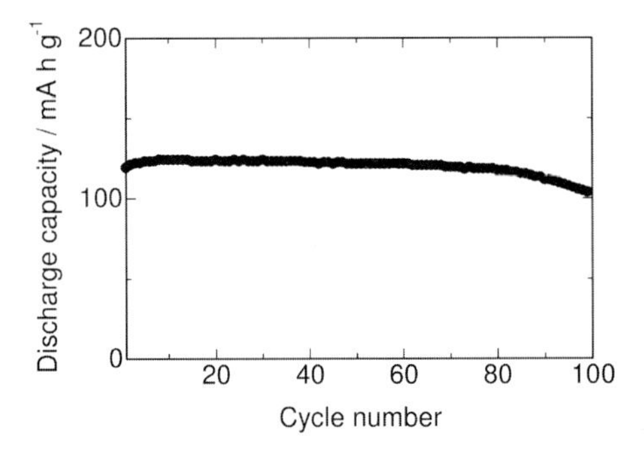

Fig. 9.27 Variation of charge/discharge capacity with cycling for Li|SPE|$LiFePO_4$ cell at 60 °C at 1.0C. SPE contained optimized amount of B-PEG plasticizers.

9.2.5
Fabrication of a Nonflammable Lithium Polymer Battery and its Electrochemical Evaluation

Currently, the development of nonflammable all-solid polymer batteries has become crucial from the point of safety; some battery makers actually recalled all the lithium-ion batteries with particular lot numbers because of recent ignition problems in some of them. In this section, as a good example of practical use with high performance, an all-solid nonflammable lithium polymer battery, Li|polymer electrolyte with B-PEG|$LiFePO_4$, for ambient and elevated temperature applications [96] is proposed from the viewpoint of safety and low cost; further improvement of the battery, with the addition of $AlPO_4$ to the polymer electrolyte leading to surface modification of cathode, is also discussed.

In the previous sections, it has been pointed out that the polymer electrolytes with plasticizers such as B-PEG and Al-PEG show very high thermal stability, but the results of ignition tests were not presented. For purpose of comparison, ignition tests for polymer electrolytes, PEO plasticized by B-PEG and mixed with organic PC, were carried out. The polymer matrix composition before polymerization was PDE-600 : PME-400 = 1: 1. The oligomer mixture was used in two ways.: (i) the mixture was added to the liquid state of B-PEG with a weight ratio of 30 : 70 and then heated at 100 °C for 12 hours to reach complete polymerization; (ii) the above oligomer mixture was polymerized without addition of the plasticizer by heating under the same conditions. The polymer obtained was immersed in PC with a weight ratio of 30 : 70 (matrix polymer : PC). Figure 9.28 illustrates the ignition experiment. The photograph has captured the instant following ignition by a lighter. The sample with PC had a flame at approximately 200 °C, whereas the polymer electrolyte plasticized by B-PEG did not have a flame even during tests over 350 °C; further, it decomposed with color change from a transparent film to a dark gray and black lump [97]. From these tests, it became clear that the polymer electrolyte PEO plasticized by B-PEG shows excellent nonflammable property that leads to fabricating large-scale batteries satisfying safety rquirements.

Fig. 9.28 Photo of "flashing stability" tests of the PC-based (left) and PEG-BE (right) plasticized PDE-600 : PME-400 polymer electrolytes.

Most lithium-ion batteries use layered $LiCoO_2$ as the cathode, which is expensive, toxic, and thermally unstable. From the ecological, safe, and economical point of view, lithium iron phosphate olivine $LiFePO_4$ (space group *Pnma*) [98] seems to be one of the most attractive candidates as the cathode alternatives to Co-containing cathodes, operating within a flat voltage of about 3.4 V. The structure is shown in Figure 9.29. The voltage range exists in the stability window of the present polymer electrolyte plasticized by B-PEG. In the $LiFePO_4$, small tetrahedral PO_4 is isolated and has strong covalency, while octahedrally coordinated lithium and iron ions show strong ionic character. This leads to a favorable high redox potential (~3.4 V) for Fe^{3+}/Fe^{2+} and unfavorable property of insulator ($\sim 10^{-9}$ S cm^{-1} at room temperature.). However, recent efforts to increase its conductivity through

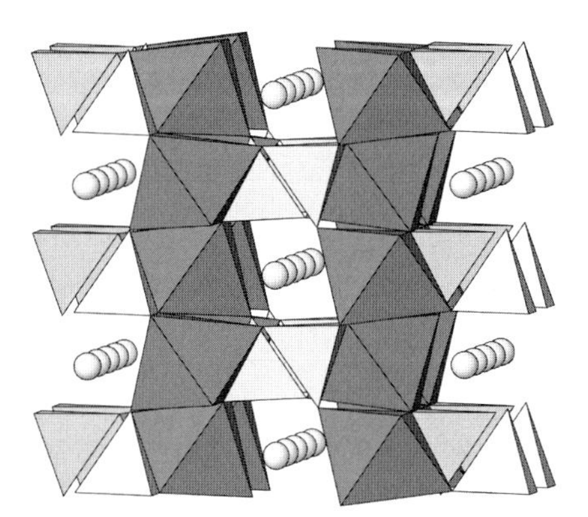

Fig. 9.29 Crystal structure of olivine-type $LiFePO_4$. White spheres indicates Li ions; FeO_6 and PO_4 polyhedra are represented by the darker octahedron and the brighter tetrahedron, respectively.

intimate carbon coating [99] has improved the early rate problem in the Fe material. Usually lithium ion in $LiFePO_4$ diffuses one-dimensionally in the [010] direction channel with a curved trajectory between adjacent lithium sites [100, 101]. Discharge/charge reaction (intercalation/deintercalation of Li in the $LiFePO_4$) also accompanies the phase change [102]; however, it has very small heat flow during the reaction compared with other cathode materials [103] and also maintains high theoretical capacity of 170 mAh g^{-1}. These are the main reasons that we selected $LiFePO_4$ as the cathode.

The electrical performance was evaluated using coin-type cell (CR2032). Preliminarily, the cathode film was prepared by the deposition of slurry of 70% active material ($LiFePO_4$; carbon-coated, Sigma Aldrich, 99.5+% battery grade), 20% of acetylene black conductor, and 10% PVdF binder dissolved in *n*-methyl pyrrolidone (NMP) on an Al foil. The cut-out piece from the foil was welded at a few spots inside bottom of stainless-steel-made coin cell. The preliminary polymeric mixture (PPM) was obtained as follows: the liquid mixture consisting of oligomers poly(ethylene glycol) methacylates (PDE-600 and PME-400) and B-PEG in weight ratio PDE-600 : PME-400 : B-PEG = 15 : 15 : 70, 1 mol kg^{-1} $LiClO_4$ (supporting salt, Fluka), 0.1 wt % benzophenone (Sigma Aldrich) as a polymerization initiator was stirred in a glove box for 24 hours. In some cases, 1 wt % of $AlPO_4$ was added to the PPM and stirred for another 12 hours. The latter was used for checking the effect of the surface modification of the olivine cathode. The PPM was dropped on the cathode surface and UV-irradiated (254 nm) for 15 minutes. After adding several drops of the PPM, the lithium metal foil as the anode was placed on it, which was thereafter UV-irradiated for 1 hour to accomplish the polymerization. Usually metallic lithium maintains good contact with this type of polymer electrolyte. Then the cell (LPB) was sealed. In this system, the use of an expensive separator is unnecessary. The schematics of cell construction are shown in Figure 9.30. The cell was galvanostatically charged/discharged on the multichannel battery testers between 2.5 and 3.8 V [96]. Impedance spectra of the cells were measured before cycling and along with galvanostatic cycling in the frequency range from 50 mHz to 500 MHz.

The polymer electrolyte plasticized by B-PEG appeared as a self-standing solid film without noticeable liquid phase on its surface. Accordingly, the B-PEG would intertwine with matrix PEO. Figure 9.31 shows the disappearance of the C=C double bond peak at 1640–1650 cm^{-1} in infrared (IR) irradiation within 1 hour indicates complete polymerization in the system. Figure 9.32(a) presents the charge/discharge profile at initial cycles of the nonflammable LPB at room temperature and at 60 °C. It shows very flat voltage plateaus about 3.4 V versus Li^+/Li and shows very small irreversibility at the first cycle at 60 °C. Note that the cell contains 1 wt % of $AlPO_4$ in the polymer electrolyte. The cell polarization is also very small especially at 60 °C, which suggests a proper formation of interface between the electrolyte and the electrode. The CV data (Figure 9.32(b)) confirms the fact. Furthermore, this improvement is ascribed to both the enhanced ionic conductivity of Li^+ in the B-PEG polymer electrolyte by the Lewis acidity of B atoms, which interact with anion dissociated from lithium salt ($LiClO_4$) and the

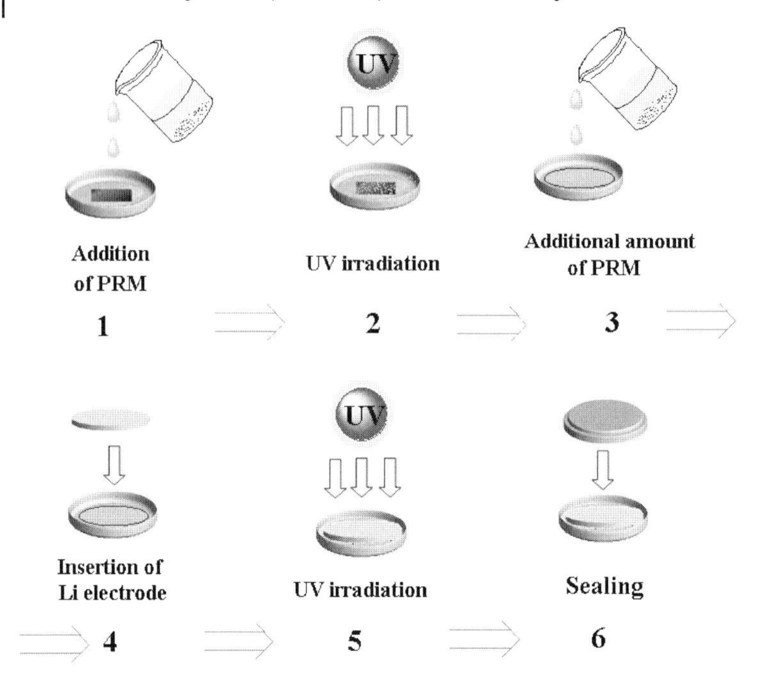

Fig. 9.30 Schematics of the lithium polymer cell construction: **1** addition of several drops of the polymer reacting mixture (PRM), **2** UV irradiation, **3** addition of several "fresh" drops of PRM, **4** – insertion of Li anode, **5** final UV irradiation, and **6** sealing the cell.

decrease of the electrode/electrolyte interfacial resistance, which would positively affect the LPB kinetics.

Figure 9.33(a) presents the cycle performance of the $AlPO_4$-added LPB (AlPO-LPB) and the additive-free LPB. The data show that the addition of $AlPO_4$ leads to increase in the cell-discharge capacity. Figure 9.33(b) shows the comparative data on the rate capacity of the pristine and AlPO-LPB cells. Both cells were cycled at 2C rate till the 50th cycle, then the charge/discharge rate was increased to 5C. After ten cycles at 5C, the both cells were operated again at 2C. The additive-free LPB could not recover the extended capacity and showed poor cyclability after 5C rate, even at 2C rate, while the AlPO-LPB exhibited better stability and acceptable cyclability at the initial stage and successfully operated at 2C rate even after being cycled at 5C (Figure 9.33(b)).

Figure 9.34 shows the influence of the AlPO-LPB impedance development upon cycling at 1C rate. For both the AlPO-LPB and the additive-free cells the impedance plots seem to contain one semicircle arc having a diameter as a combination of the "lithium–polymer electrolyte" (R_a) and "cathode–polymer electrolyte" (R_c) interface impedances, R_{int},

$$R_{int} = R_a + R_c \tag{9.7}$$

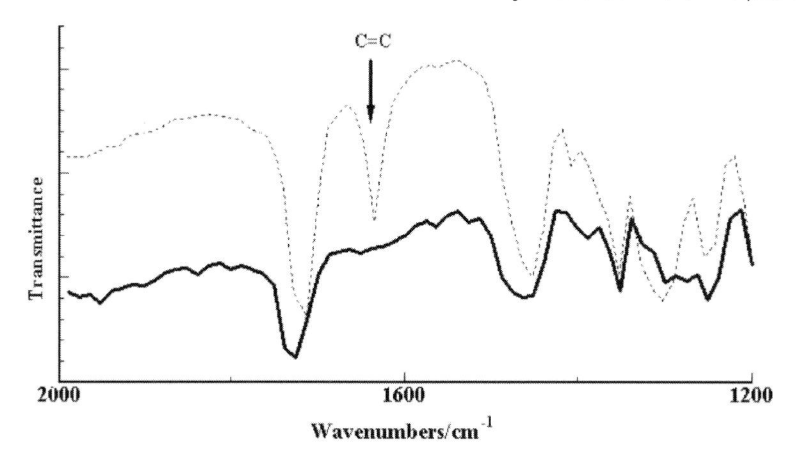

Fig. 9.31 IR spectra of the polymer system at different UV-irradiation time: 15 minutes (hatched line) and 1 hour (solid line).

This behavior is similar to another similar polymer electrolyte in the literature [94, 104]. The high-frequency intercept at the Z′-axis points to the bulk resistance of the electrolyte, R_b. The resistance R_b does not change remarkably upon the polymer cell charge/discharge. However, R_{int} resistance increases during cycling as shown in Figure 9.34. The "fresh" AlPO-LPB has a smaller value than the polymer cell without $AlPO_4$ addition. It can be seen from Figure 9.34(b) that for the additive-free LPB the R_{int} value continuously rises upon cycling. On the other hand, for the ALPO-LPB (Figure 9.34(a)), the difference in the R_{int} from the "fresh" cell and the initially cycled one is relatively small and the semicircle arc does not remarkably grow upon cycling compared with the additive-free LPB. This suggests that the $AlPO_4$ addition to the polymer electrolyte contributes to the formation of a more electrochemically stable electrode|electrolyte interface film with a higher diffusion path of Li^+.

The chemistry of the electrode surface processes in lithium polymer batteries is not well understood so far because the existence of very thin nanolayer is suggested. Sometimes, the nanolayer may not cover all the electrode surface, but locates on some parts of the surface. Here, we discuss a possible model of the processes at the $LiFePO_4$ cathode surface in the presence of $AlPO_4$. Although the solubility of the $AlPO_4$ in the B-PEG and/or the PPM is very low, very small amount of the $AlPO_4$ would dissociate and form Al^{3+} and phosphate anion ${PO_4}^{3-}$ via

$$AlPO_4 \rightarrow Al^{3+} + {PO_4}^{3-} \tag{9.8}$$

on the cathode surface. These ions may form Li_3PO_4 and/or P_4O_{10} (Figure 9.35) and Li_2O oxides during the charge process by the following reaction:

$$4{PO_4}^{3-} + 12Li^+ \rightarrow 6Li_2O + P_4O_{10} \tag{9.9}$$

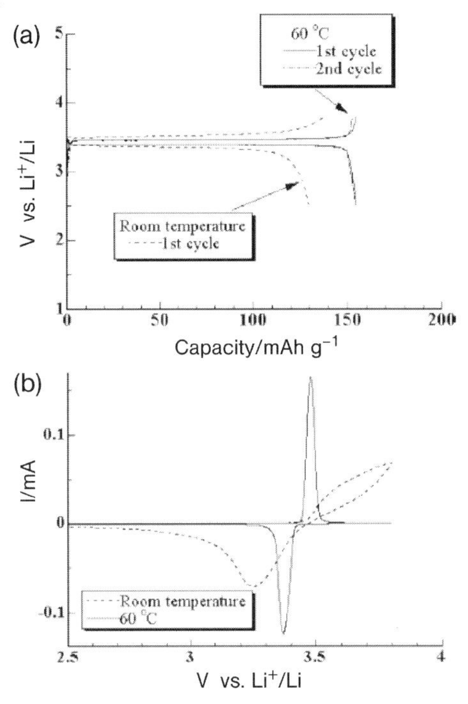

Fig. 9.32 Electrochemical performance of Li–$AlPO_4$ composite PEG–BE–$LiFePO_4$ cell at room temperature (hatched line) and 60 °C (solid line): (a) charge/discharge profiles at 0.1 C, (b) cyclic voltammograms at 0.05 mV s^{-1} scan rate.

$$PO_4{}^{3-} + 3Li^+ \rightarrow Li_3PO_4 \tag{9.10}$$

Nanoparticles (nanolayers) of lithium phosphate Li_3PO_4 and thermodynamically stable phosphorus oxide P_4O_{10} ($\Delta_f G^\circ = -2723$ kJ mol^{-1} at 298 K [105]) may take part in the surface modification of some of the cathode surface, resulting in the enhanced cell performance. As a result, a new diffusion model of lithium ions during the discharge could be proposed on the cathode surface. The model is presented in Figure 9.36. The oxides and phosphate formed, especially P_4O_{10} oxide on the cathode surface, may attract Li^+ ions in the polymer electrolyte due to $-\delta$ charges at the outer oxygen atoms of a P_4O_{10} cluster structure (Figure 9.35). After several cycling, solid electrolyte interface (SEI) would be formed on the surface of the cathode, but the SEI nanolayer that presumably consists of P_4O_{10}, Li_2O, Li_3PO_4 etc., does not cover the entire surface. The easiest lithium ion diffusion will be in bulk cathode, however, the lithium ion diffusion in the nanolayer would be faster than that in the layer of the SEI because of higher concentration of lithium ions on the surface of the nanolayer compared with that on the surface of the SEI. This idea is similar to that of the space charge. In Figure 9.36, l_1, l_2, and l_3 indicate the distance of the bulk cathode, the nanolayer, and the SEI,

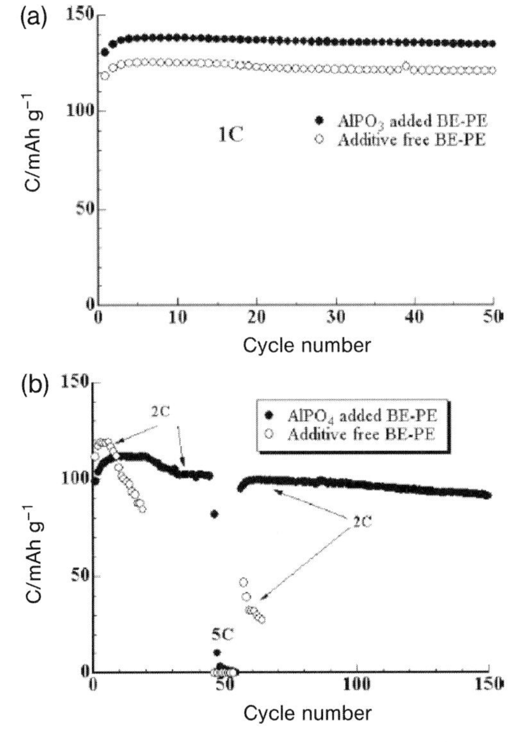

Fig. 9.33 Comparative data on $AlPO_4$ added and additive-free $Li/LiFePO_4$ polymer cell performance at 60 °C, cutoff 2.5–3.8 V, C is the discharge capacity. (a) 1 C rate and (b) cells were initially cycled at 2C for 40 cycles and then at 5C for 10 cycles; thereafter the cells continued to be cycled at 2C.

respectively. ν_1, ν_2, and ν_3 correspond to the velocity of lithium ions in each part. Each arrow in the figure indicates the vector of lithium diffusion in each part. The unit of ν/l corresponds to the frequency $(1/\lambda)$. From the present model, if we can cover the whole surface of the cathode by a suitable nanolayer, the cell performance, especially rate property would improve and much less polarization would be expected.

Summing up, the nonflammable lithium polymer cell Li|polymer electrolyte with B-PEG|$LiFePO_4$ has exhibited a high discharge capacity and stable cycling performance at 60 °C. The olivine $LiFePO_4$ cathode well matched the present all-SPEs and showed a flat curve about 3.4 V. Since the cell does not need an expensive separator, its manufacture would involve low costs and easier fabrication processes. The addition of $AlPO_4$ to the polymer electrolyte could be found to be effective for the electrode|electrolyte interface improvement. A model of the cathode surface processes was proposed, which correlates with the interfacial resistance stabilization in the presence of $AlPO_4$.

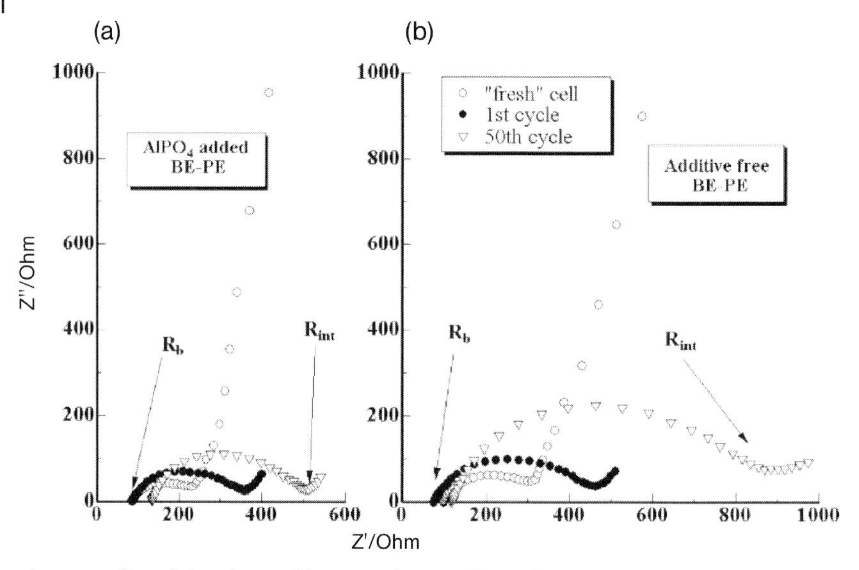

Fig. 9.34 Effect of the $AlPO_4$ addition on the impedance development during the polymeric cell operation: (a) $AlPO_4$ additive, (b) no additive, 60 °C; fresh cell (open circle), first cycle (solid circle), and 50th cycle (reverse open triangle).

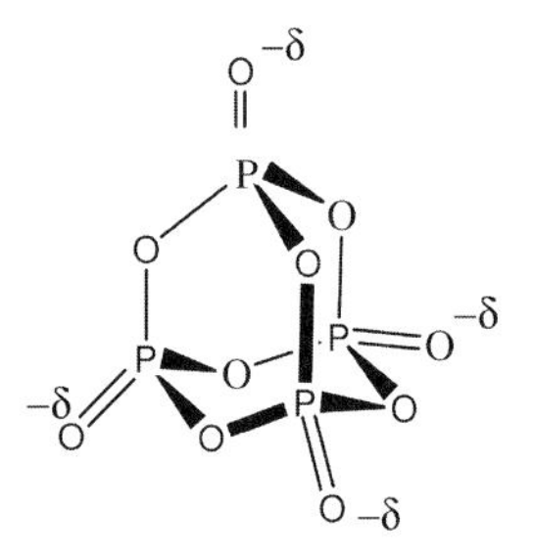

Fig. 9.35 Structure of P_4O_{10} cluster.

9.2.6 Summary

This section overviewed the fabrication and electrochemical performance of LPBs using SPEs with B-PEG or Al-PEG plasticizers as an example. Not only the ionic conductivity of SPEs but also various factors were affected by the electrochemical property of LPBs. In fact, an increase of ionic conductivity is effective to suppress an increase of resistance. However, durability of the LPB requires mechanical and electrochemical stability of SPEs, which sometimes causes a dilemma because of simultaneous decreasing of ionic conductivity. In addition, one can see the clear correspondence between interfacial resistance and capacity retention. Therefore, interfaces play an important role in the electrochemical performance of the LPB,

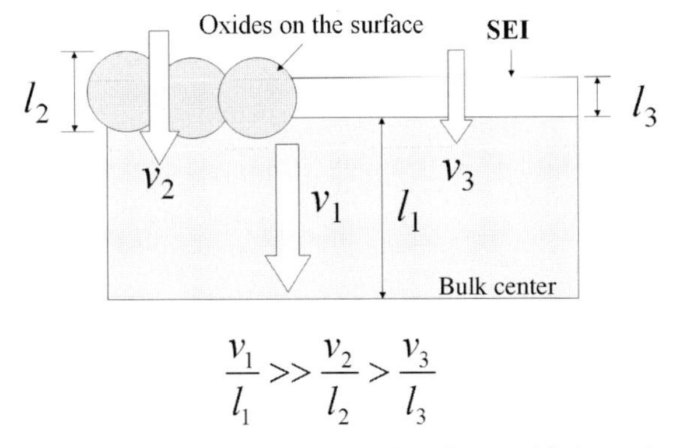

Fig. 9.36 Lithium diffusion model of surface modified samples.

and precise and systematic understanding of the interfacial reaction would be a key to realize practical LPBs.

Up to now, most research work was focused on the bulk properties of individual components of the LPB. On the other hand, not many studies have been conducted on the ceramics|polymer solid–solid electrochemical interfaces. Further research and development of LPBs would require cooperative research on SPEs, electrode materials, and technical design. Such comprehensive research will lead to a new field in materials science and engineering for advanced batteries.

Acknowledgment

The authors wish to thank to Yuki Masuda, Fuminari Kaneko, Shinta Wada, and Dr Zhumaby Bakenov (Tokyo Institute of Technology) for their insightful discussion and help. SEM/EDS images of Figures 9.14, 9.16 and 9.25 were taken by the Center for Advanced Materials Analysis (CAMA), Tokyo Institute of Technology, Japan. The B-PEG plasticizer was kindly supplied by NOF Corporation, Japan. Composite film of cathode and SPE materials for Figure 9.16 and Al-PEG plasticizer were kindly supplied by Nippon-Nyukazai Co., Ltd, Japan. One of the authors, M. N, thanks the New Energy and Industrial Technology Development Organization (NEDO) of Japan for financial aid under the Industrial Technology Research Grand Program.

References

1 Gray, F.M. (1997) *Solid Polymer Electrolytes, Fundamentals and Technological Applications*, Wiley-VCH Verlag GmbH, New York.

2 (a) Fenton, D.E., Parker, J.M., and Wright, P.V. (1973) *Polymer*, **14**, 589; (b) Wright, P.V. (1975) *Br. Polym. J.*, **7**, 319.

3 Armand, M.B., Chabagno, J.M., and Duclot, M. (1979) *Fast Ion Transport in Solids*, (eds P. Vashista, *et al.*), Elsevier, New York.

4 Gray, F.M. (1997) *Polymer Electrolytes, RSC Monographs*, The Royal Society of Chemistry, London.

5 (a) MacCallum, J.R. and Vincent, C.A. (1987) *Polymer Electrolyte Reviews 1*, Elsevier, London; (b) MacCallum, J.R. and Vincent, C.A. (1989) *Polymer Electrolyte Reviews 2*, Elsevier, London.

6 Tadokoro, H., Chatani, Y., Toshihara, T., Tahara, S., and Murahashi, S. (1979) *Macromol. Chem.*, **73**, 109.

7 Wright, P.V. (1998) *Electrochim. Acta*, **43**, 1137.

8 Naoi, K., Mori, M., Inoue, M., Wakabayashi, T., and Yamauchi, K. (2000) *J. Electrochem. Soc.*, **147**, 813.

9 Bruce, P.G. and Vincent, C.A. (1993) *J. Chem. Soc., Faraday Trans.*, **89**, 3187.

10 Meyer, W.H. (1998) *Adv. Mater.*, **10**, 439.

11 Dias, F.B., Plomp, L., and Veldhuis, J.B.J. (2000) *J. Power Sources*, **88**, 169.

12 Kiukkola, K. and Wagner, K. (1957) *J. Electrochem. Soc.*, **104**, 379.

13 Dupon, R., Papke, B.L., Ratner, M.A., Whitemore, D.H., and Shriver, D.F. (1982) *J. Am. Chem. Soc.*, **104**, 6247.

14 (a) Lee, C.C. and Wright, P.V. (1982) *Polymer*, **23**, 681; (b) Payne, D.R. and Wright, P.V. (1982) *Polymer*, **23**, 690.

15 Gorecki, C., Minier, M., Armand, M.B., Chabagno, J.M., and Rigaud, P. (1983) *Solid Sate Ionics*, **11**, 91.

16 Minier, M., Berthier, C., and Gorecki, W. (1984) *J. Physique*, **11**, 307.

17 Papke, B.L., Ratner, M.A., and Shriver, D.F. (1982) *J. Electrochem. Soc.*, **129**, 1694.

18 (a) Watanabe, M., Nagaoka, K., Kanba, K., and Shinohara, I. (1982) *Polym. J.*, **14**, 877; (b) Watanabe, M., Ikeda, J., and Shinohara, I. (1983) *Polym. J.*, **15**, 65; (c) Watanabe, M., Sanui, K., Ogata, N., Inoue, F., Kobayashi, T., and Ohtaki, Z. (1985) *Polym. J.*, **17**, 549.

19 Fontanella, J.J., Wintersgill, M.C., Smith, M.K., Semacik, J., and Andeen, C.G. (1986) *J. Appl. Phys.*, **60**, 2665.

20 Lindsey, S.E., Whitmore, D.H., Halperin, W.P., and Tokelson, J.M. (1989) *Polym. Prepr., Div., Polym. Chem., Am. Chem. Soc.*, **30**, 442.

21 MacCallum, J.R., Smith, M.J., and Vincent, C.A. (1984) *Solid State Ionics*, **11**, 307.

22 Allcock, H.R., Austin, P.E., Neenan, T.X., Sisko, J.T., Blonsky, P.M., and Shriver, D.F. (1986) *Macromolecules*, **19**, 1508.

23 Blannett, J.L., Domback, A.A., Allcock, H.R., Heyen, B.J., and Shriver, D.F. (1989) *Chem. Mater.*, **1**, 14.

24 Cheradame, H., Souquet, J.L., and Latour, J.M. (1980) *Mater. Res. Bull.*, **15**, 1173.

25 Killis, A., LeNest, J.F., Cheradame, H., and Gandini, A. (1982) *Macronol. Chem.*, **183**, 2385.

26 Killis, A., LeNest, J.F., Gandini, A., and Cheradame, H. (1984) *Macromolecules*, **17**, 63.

27 (a) Watanabe, M., Sanui, K., Ogata, N., Inoue, F., Kobayashi, T., and Ohtaki, Z. (1984) *Polym. J.*, **16**, 711; (b) Watanabe, M., Sanui, K., Ogata, N., Inoue, F., Kobayashi, T., and Ohtaki, Z. (1984) *Polym. J.*, **17**, 549; (c) Watanabe, M., Nagano, S., Sanui, K., and Ogata, N. (1984) *Polym. J.*, **18**, 809.

28 Fish, D., Khan, I.M., and Smid, J. (1988) *Br. Polym. J.*, **20**, 281.

29 Ravaine, D., Sominel, A., and Tünker, G. (1986) *J. Non-Cryst. Solids*, **80**, 557.

30 Blonsky, P.M., Shriver, D.F., Austin, P., and Allcock, H.R. (1986) *Solid State Ionics*, **18–19**, 258.

31 Tonge, J.S. and Shriver, D.F. (1987) *J. Electrochem. Soc.*, **134**, 269.

32 Nishimoto, A., Watanabe, M., Ikeda, Y., and Kohjiya, S. (1998) *Electrochim. Acta*, **43**, 1177.

33 Nishimoto, A., Agehara, K., Furuya, N., Watanabe, T., and Watanabe, M. (1999) *Macromolecules*, **32**, 1541.

34 Weston, J.E. and Steele, B.C. (1982) *Solid State Ionics*, **7**, 75.

35 Croce, F., Appetecchi, G.B., Persi, L., and Scrosati, B. (1998) *Nature*, **394**, 456.

36 Capuano, F., Croce, F., and Scrosati, B. (1991) *J. Electrochem. Soc.*, **52**, 1922.

37 Croce, F. and Scrosati, B. (1993) *J. Power Sources*, **43**, 9.

38 Borghini, M.C., Mastragostino, M., Passerini, S., and Scrosati, B. (1995) *J. Electrochem. Soc.*, **142**, 2118.

39 Matsuo, Y. and Kuwano, J. (1995) *Solid State Ionics*, **79**, 295.

40 Kumar, B. and Scanlon, L.G. (1994) *J. Power Sources*, **52**, 261.

41 Rghavan, S.R., Riley, M.W., Fedkiv, P.S., and Khan, S.A. (1998) *Chem. Mater.*, **10**, 244.

42 Kumar, B., Scanlon, L.G., and Spry, R.J. (2001) *J. Power Sources*, **96**, 337.

43 Xiong, H.M., Zhao, X., and Chen, J.S. (2001) *J. Phys. Chem. B*, **105**, 10169.

44 Swierczynski, D., Zalewska, A., and Wieczorek, W. (2001) *Chem. Mater.*, **13**, 1560.

45 Gupta, R.K., Jung, H.Y., and Whang, C.M. (2002) *J. Mater. Chem.*, **12**, 3779.

46 Sun, H.Y., Sohn, H.-J., Yamamoto, O., Takeda, Y., and Imanishi, N. (1999) *J. Electrochem. Soc.*, **146**, 1672.

47 Sun, H.Y., Takeda, Y., Imanishi, N., Yamamoto, O., and Sohn, H.-J. (2000) *J. Electrochem. Soc.*, **147**, 2462.

48 Li, Q., Sun, H.Y., Takeda, Y., Imanishi, N., Yang, J., and Yamamoto, O. (2000) *J. Power Sources*, **94**, 201.

49 Wieczorek, W., Stevens, J.R., and Florjanczyk, Z. (1996) *Solid State Ionics*, **85**, 67.

50 Wieczorek, W., Lipka, P., Zukowska, G., and Wycislik, H. (1998) *J. Phys. Chem. B*, **102**, 6968.

51 Marcinek, M., Bac, A., Lipka, P., Zalewska, A., Zukowska, G., Borkowska, R., and Wieczorek, W. (2000) *J. Phys. Chem. B*, **104**, 11088.

52 Swirczynski, D., Zalewska, A., and Wieczorek, W. (2001) *Chem. Mater.*, **13**, 1560.

53 (a) Armand, M., Gorecki, W., and Andréani, R. (1990) *2nd International Symposium on Polymer Electolytes* (ed. B. Scrosati), Elsevier, London, p. 91; (b) Sylla, S., Sanchez, J.-Y., and Armand, M. (1992) *Electrochim. Acta*, **37**, 1699; (c) Bnrabah, D., Baril, D., Sanchez, J.-Y., Armand, M., and Gard, G.G. (1993) *J. Chem. Soc., Faraday Trans.*, **89**, 355.

54 Appetecchi, G.B., Henderson, W., Villano, P., Berrettoni, M., and Passerini, S. (2001) *J. Electrochem. Soc.*, **148**, A1171.

55 Appetecchi, G.B., Scaccia, S., and Passerini, S. (2000) *J. Electrochem. Soc.*, **147**, 4448.

56 (a) Abraham, K.M., Alamgir, M., and Reynolds, R.K. (1988) *J. Electrochem. Soc.*, **135**, 535; (b) Abraham, K.M., Jiang, Z., and Caroll, B. (1997) *Chem. Mater.*, **9**, 1978; (c) Abraham, K.M., and Jiang, Z. (1997) *J. Electrochem. Soc.*, **144**, L136.

57 Morita, M., Fukumasa, T., Motoda, M., Tsutsumi, H., Matsuda, Y., Takahashi, T., and Ashitaka, H. (1990) *J. Electrochem. Soc.*, **137**, 3401.

58 Ito, Y., Kanehori, K., Miyauchi, K., and Kudo, T. (1987) *J. Mater. Sci.*, **22**, 1845.

59 Morford, R.V., Kelam, E.C., Hofmann, M.A., Baldwin, R., and Allcock, H.R. III (2000) *Solid State Ionics*, **133**, 171.

60 Wang, X., Yasukawa, E., and Kasuya, S. (2001) *J. Electrochem. Soc.*, **148**, A1058.

61 Xu, K., Ding, M.S., Zhang, S., Allen, J.L., and Jow, T.R. (2002) *J. Electrochem. Soc.*, **149**, A622.

62 (a) Kato, Y., Yokoyama, S., Ikuta, H., Uchimoto, Y., Wakihara, M. (2001) *Electrochem. Commun.*, **3**, 128; (b) Kato, Y., Hasumi, K., Yokoyama, S., Yabe, T., Ikuta, H., Uchimoto, Y., and Wakihara, M. (2002) *Solid State Ionics*, **150**, 355; (c) Kato, Y., Hasumi, K., Yokoyama, S., Yabe, T., Ikuta, H., Uchimoto, Y., and Wakihara, M. (2002) *J. Therm. Anal. Cal.*, **69**, 889.

63 Masuda, Y., Seki, M., Nakayama, M., Wakihara, M., and Mita, H. (2006) *Solid State Ionics*, **177**, 843.

64 Yang, L.L., McGhie, A.R., and Farrington, G.C. (1986) *J. Electrochem. Soc.*, **133**, 1380.

65 Zahurak, S.M., Kaplan, M.L., Tietman, E.A., Murphy, D.W., and Cava, R.J. (1988) *Macromolecules*, **21**, 654.

66 Quatarone, E., Mustarelli, P., Tomasi, C., and Magistris, A. (1998) *J. Phys. Chem. B*, **102**, 9610.

67 Wieczorek, W. (1992) *Mater. Sci. Eng.*, **B15**, 108.

68 Lascaud, S., Perrier, M., Vallee, A., Besner, S., Prud'homme, J., and Armand, M., (1994) *Macromolecules*, **27**, 7469.

69 Gorecki, W., Jeannin, M., Belorizki, E., Roux, C., and Armand, M. (1995) *J. Phys.: Condens. Matter*, **17**, 6823.

70 Appetecchi, G.B., Croce, F., and Scrosati, B. (1995) *Electrochim. Acta*, **40**, 991.

71 (a) Killis, A., Nest, J.F.L., Cheradame, H., and Gandini, A. (1982) *Makromol. Chem.*, **183**, 2835; (b) Killis, A., Nest, J.F.L., Gandini, A., and Cheradame, H. (1984) *Macromolecules*, **17**, 63.

72 Watanabe, M., Itoh, M., Sanui, K., and Ogata, N. (1987) *Macromolecules*, **20**, 569.

73 Baril, D., Michot, C., and Armand, M. (1997) *Solid State Ionics*, **94**, 35.

74 Carvalho, L.M., Guégan, P., Cheradame, H., and Gomes, A.S. (2000) *Eur. Polym. J.*, **36**, 401.

75 Williams, M.L., Landel, R.F., and Ferry, J.D. (1955) *J. Am. Chem. Soc.*, **77**, 3701.

76 Schantz, S., Sandahl, J., Borjessn, L., Torell, L.M., and Stevens, J.R. (1988) *Solid State Ionics*, **28–30**, 1047.

77 Kakihana, M., Schantz, S., and Toell, L.M. (1990) *J. Chem. Phys.*, **92**, 6271.

78 Huang, W., Frech, R., and Wheeler, R.A. (1994) *J. Phys. Chem.*, **98**, 100.

79 Gorecki, W., Andeani, R., Berthier, C., Armand, M., Mali, M., Roos, J., and Brinkmann, D. (1986) *Solid State Ionics*, **18–19**, 295.

80 Bhattacharja, S., Smoot, S.W., and Whitmore, D.H. (1986) *Solid State Ionics*, **18–19**, 306.

81 Chadwick, A.V., Strange, J.H., and Worboys, M.R. (1983) *Solid State Ionics*, **9–10**, 1153.

82 Ray, I., Buneel, J., Grondin, J., Servant, L., and Lasségues, J.C. (1998) *J. Electrochem. Soc.*, **145**, 3034.

83 Watanabe, M., Nagano, S., Sanui, K., and Ogata, N. (1988) *Solid State Ionics*, **28–30**, 911.

84 Abraham, K.M., Jiang, Z., and Carroll, B. (1997) *Chem. Mater.*, **9**, 1978.

85 (a) Kato, Y., Suwa, K., Yokoyama, S., Yabe, T., Ikuta, H., Uchimoto, Y., and Wakihara, M. (2002) *Solid State Ionics*, **152–153**, 155; (b) Kato, Y., Suwa, K., Ikuta, H., Uchimoto, Y., Wakihara, M., Yokoyama, S., Yabe, T., and Yamamoto, M. (2003) *J. Mater. Chem.*, **13**, 280.

86 (a) Gorecki, W., Jeannin, M., Belorizky, E., Roux, C., and Armand, M. (1995) *J. Phys.: Condens. Matter*, **7**, 6823; (b) Gorecki, W., Roux, C., Clémancey, M., Armand, M., and Belorizky, E. (2002) *Chem. Phys. Chem.*, **7**, 620.

87 (a) Hayamizu, K., Aihara, Y., and Price, W.S. (2000) *J. Chem. Phys.*, **113**, 4785; (b) Tokuda, H., Tabata, S., Susan, M.A.B.H., Hayamizu, K., and Watanabe, M. (2004) *J. Phys. Chem. B*, **108**, 11995.

88 Wang, C., Xia, Y., Fujieda, T., Sakai, T., and Muranaga, T. (2002) *J. Power Sources*, **103**, 223.

89 Seki, S., Kobayashi, Y., Miyashiro, H., Yamanaka, A., Mita, Y., and Iwahori, T. (2005) *J. Power Sources*, **146**, 741.

90 Miyashiro, H., Kobayashi, Y., Seki, S., Mita, Y., Usami, A., Nakayama, M., and Wakihara, M. (2005) *Chem. Mater.*, **17**, 5603.

91 Masuda, Y., Nakayama, M., and Wakihara, M. (2007) *Solid State Ionics*, **178**, 981.

92 Kobayashi, Y., Seki, S., Tabuchi, M., Miyashiro, H., Mita, Y., and Iwahori, T. (2005) *J. Electrochem. Soc.*, **152**, A1985.

93 Miyashiro, H., Seki, S., Kobayashi, Y., Ohno, Y., Mita, Y., and Usami, A. (2005) *Electrochem. Commun.*, **7**, 1083.

94 Seki, S., Kobayashi, Y., Miyashiro, H., Mita, Y., and Iwahori, T. (2005) *Chem. Mater.*, **17**, 2041.
95 Amatucci, G.G., Schmutz, C.N., Blyr, A., Sigala, C., Gozdz, A.S., Larcher, D., and Tarascon, J.M. (1997) *J. Power Sources*, **69**, 11.
96 Bakenov, Z., Nakayama, M., and Wakihara, M. (2007) *Electrochem. Solid-State Lett.*, **10**, A208.
97 Bakenov, Z. (2007) Doctor Thesis, Tokyo Institute of Technology, Tokyo.
98 Radhi, A.K., Najundaswamy, K.S., and Goodenough, J.B. (1997) *J. Electrochem. Soc.*, **144**, 1189.
99 Huang, H., Yin, S.-C., and Nazar, L.F. (2001) *Electrochem. Solid-State Lett.*, **4**, A170.
100 Morgan, D., Van der Ven, A., and Ceder, G. (2004) *Electrochem. Solid-State Lett.*, **7**, A30.
101 Islam, M.S., Driscoll, D.J., Fisher, C.A.J., and Slater, P.R. (2005) *Chem. Mater.*, **17**, 5085.
102 Yamada, A., Koizumi, H., Nishimura, S., Sonoyama, N., Kanno, R., Yonemura, M., Nakamura, T., and Kobayashi, Y. (2006) *Nat. Mater.*, **5**, 357.
103 Miyashiro, H. (2006) Doctor Thesis, Tokyo Institute of Technology, Tokyo.
104 Seki, S., Kobayashi, Y., Miyashiro, H., Yamanaka, A., Mita, Y., and Iwahori, T. (2005) *J. Power Sources*, **146**, 741.
105 Barin, I. (1989) *Thermochemical data of pure substances*, Verlag Chemie, Weinheim, p. 1116.

Further Reading

Dean, J.A. (1999) *Lange's Handbook of Chemistry*, 15th edn, McGrow-Hill, Inc.

10
Thin-Film Metal-Oxide Electrodes for Lithium Microbatteries

Jean-Pierre Pereira-Ramos and Rita Baddour-Hadjean

10.1
Introduction

The reversible electrochemical lithium insertion–extraction reaction is a process that induces two important fields of fundamental and applied research for thin-film transition metal oxides (TMOs). Indeed electrochromic materials are able to change their optical properties upon insertion/extraction induced by an external voltage but the same process, when a larger amount of Li ions can be reversibly accommodated in the metal oxide host lattice, gives rise to possible electrode materials for secondary Li batteries and/or Lithium-ion batteries (LIBs). Electrochromic materials and devices have been reviewed quite often [1–3] and they are considered to be a subset of "the solar energy materials." Starting 1990, years of assiduous effort and research to improve LIB performance has enabled LIBs to play a leading role in the market of rechargeable batteries especially for portable equipment like cellular phones, video cameras, notebook computers, and portable minidisks etc.

The ongoing improvements in the microelectronics industry and the miniaturization of electronic devices have reduced the current and power requirements of some devices to extremely low levels. Therefore there is an increasing need for new lightweight batteries with long life, and high energy density. In addition, unlike conventional batteries, these batteries have to be deposited directly onto chips or chip package in any shape or size. One approach to fulfilling this need is the development of thin-film batteries (TFBs) in which all the components, i.e., the anode, electrolyte, and cathode materials, as well as suitable current collectors, should be fabricated into a multilayered thin film (Figure 10.1). The lithium microbatteries have potentially many applications such as non-volatile memory backup, smart cards, microelectromechanical systems (MEMSs), sensors, actuators, and implantable medical devices. Their total thickness must not exceed few tens of micrometers. Intensive efforts are being made to obtain high-performance cathode materials for TFBs. About 20 years ago, TiS_2 was the most widely studied material as positive electrode for rechargeable Li batteries and it was proposed as cathode in the first TFBs [4]. Its interest has been confirmed by other works [5, 6], but it

Lithium Ion Rechargeable Batteries. Edited by Kazunori Ozawa

ISBN: 978-3-527-31983-1

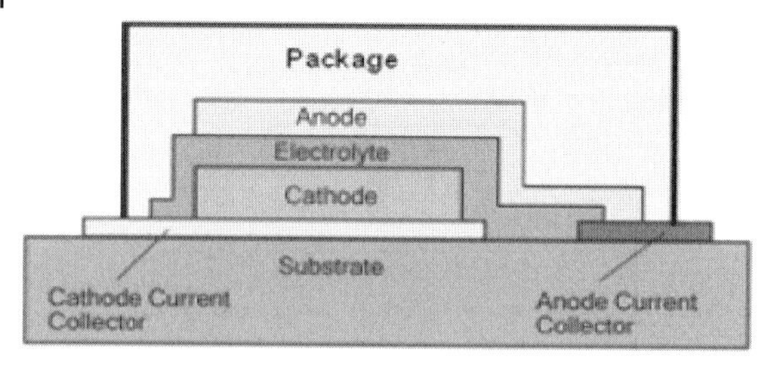

Fig. 10.1 Schematic layout of a TFB (from *www.oakridgemicro.com/tech/appl*).

has been now abandoned to the benefit of Ti, W, and Mo oxysulfide thin films, which exhibit attractive properties [7–9]. Movement into TMOs started from 1990 [10, 11] after a decade of intense research on various Li intercalation compounds. A critical point for the development of TFBs is the use of a high-performance solid electrolyte; some data on various solid electrolytes used in TFBs can be found in [12, 13], as well as a brief overview of the main TFB systems investigated up to 2000.

In spite of numerous research efforts, the real challenge still consists in finding appropriate cathode materials to fit the requirements of the electronic devices in terms of specific capacity, cycle life, reliability, and safety. This chapter is mainly concerned with recent developments on thin-film oxides usable as positive electrodes in lithium microbatteries. In the case of secondary lithium batteries and LIBs, numerous TMOs have been investigated and evaluated as possible positive electrodes with the development of modified or new synthesis routes. However, only a limited number of TMOs have proved to be of interest as thin-film materials because of the film-deposition technology required. To prepare TMO thin films, several techniques such as vacuum thermal evaporation, sputtering, electrostatic spray deposition (ESD), electron-beam deposition, chemical vapor deposition (CVD) and pulsed laser deposition (PLD) can be used. In addition various solution techniques can be applied for obtaining thin-film oxides.

Of course, the appropriate synthesis of high-performance thin-film oxides is of utmost interest to fabricate TFBs. However, at the same time, electrochemical study of pure thin-film materials without any binding and conductive agent as in conventional composite electrode Li batteries allows to reach the intrinsic properties of the oxide. In particular, the data extracted from well-defined geometries are of fundamental interest as they reveal the influence of morphology and orientation. In other respects, the film configuration provides a high quality of structural and electrochemical characterization. The information collected from pure thin-film material will then facilitate also the understanding of the composite electrode because most of thin-film materials are first investigated in liquid electrolytes. This work focuses on the main data that have recently become available on the most promising thin-film oxides such as 4- and 3-V positive electrodes for TFBs i.e., $LiCoO_2$, $LiNiO_2$, and $LiMn_2O_4$ as well as their substitutive forms, V_2O_5 and MoO_3. For each material, the electrochemical properties in liquid electrolyte are described first before their behavior in an all solid-state device is addressed. We are aware that it is difficult to cover all thin-film oxides and each aspect of their properties in an exhaustive manner.

10.2 Lithium Cobalt Oxide Thin Films

$LiCoO_2$ thin films have received considerable attention because $LiCoO_2$ is the dominant electrode material for the positive electrode in commercial LIBs. $LiCoO_2$ thin films can be prepared by sputtering techniques [14–41], PLD [42–53], chemical vapor deposition (CVD) [54, 55], and chemical routes including electrostatic spray pyrolysis [56–67]. The reversible capacity per unit area and thickness is 69 μA h cm^{-2} μm^{-1} for exchange of 0.5 Li per mole of oxide (137 mA h g^{-1}) in a fully dense film. Hence, depending on the effective density of the deposit, the maximum specific capacity can range from 55 to 69 μA h cm^{-2} μm^{-1}.

10.2.1 Sputtered $LiCoO_2$ Films

10.2.1.1 Liquid Electrolyte

Direct-current DC or radio frequency (RF) magnetron sputtering is the most widely used technique for $LiCoO_2$ preparation investigated in liquid or solid electrolytes. Recent studies report a high difference in the preferential orientation of deposits produced by PLD and RF sputtering using identical $LiCoO_2$ targets [14, 15]. The deposition technique largely determines the $LiCoO_2$ unit-cell alignment with the substrate and thus, the preferential orientation of the annealed films: RF-sputtered and PLD films exhibit (110) and (003) lattices planes parallel to the substrate surface respectively (Figure 10.2). However, the RF film can also develop the (003) reflection if a prolonged annealing procedure at high temperature is applied [14]. The preferential (110) orientation of the sputtered film is shown to be favorable to fast Li diffusion in the oxide while the process of relaxation potential in PLD films is slow and the Li diffusivity lower by a few orders of magnitude [15]. In spite of interesting data on the influence of the preferred orientation, any conclusions that are arrived at must be regarded cautiously since other parameters like crystallinity, stoichiometry, and purity of the films could also play a significant role on the electrochemical properties. The electrochemical efficiency of the film is strongly dependent on the preferred orientation; for instance, the theoretical capacity is obtained using cyclic voltammetry at a scan rate of 0.1 mV s^{-1} for the (110) orientation, whereas only a small part of the expected faradaic yield is recorded for the PLD (003)-oriented film. In the latter case, this shows that only a small part of the material is working. The authors explain this discrepancy by the perpendicular (accessible) and parallel (inaccessible) alignment of the lithium diffusion planes toward the electrolyte solution respectively for the (110) and (003) orientations.

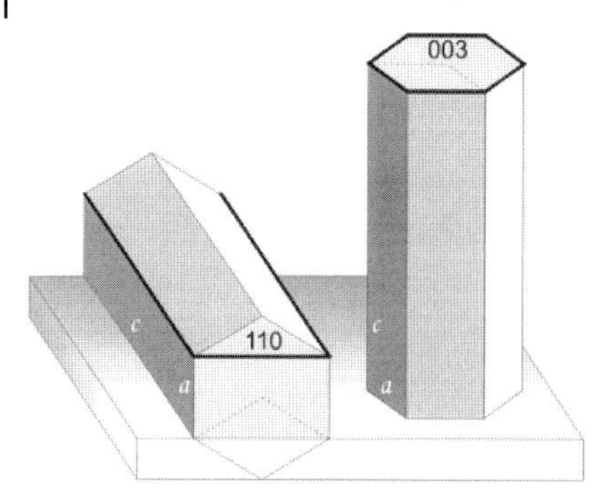

Fig. 10.2 Schematic drawing showing the $LiCoO_2$ unit-cell orientation when deposited with RF sputtering (left) and with PLD (right). From [14].

A major characteristic of the sputtered films is their low crystallinity and low electrochemical efficiency when they are used as-prepared in Li batteries. The synthesis of high-performance thin films, in general, requires a postannealing treatment at high temperature (600–700 °C) to improve the crystallinity of $LiCoO_2$ thin films. However, such a high-temperature process is often not compatible with devices requiring the direct integration of the battery and many desirable substrates such as flexible polymer materials and low-temperature substrate materials, which can limit the application and fabrication of the TFBs. Therefore, significant efforts have been made to optimize the film crystallinity and electrochemical properties. Several ways of improvement have been proposed, such as the use of a moderate annealing treatment (<450 °C) [16, 17], the use of a heated substrate (300–500 °C) during deposition [18], the optimization of the power [19], and the use of a bias substrate as process parameter [20] without any postannealing procedure.

As-deposited cathodes made using a RF power of 150 W exhibit a (101) and (104) orientation with attractive properties since a capacity of about 50 μA h cm^{-2} μm^{-1} is obtained in the 4.2–3 V range (Figure 10.3) [19]. The use of a high RF power decreased the presence of Co_3O_4 impurity phase. In such a case, a high electrochemical efficiency is obtained for as-deposited films of various thickness in the range 0.6–1.6 μm as evidenced by the linear relation in the capacity dependence versus thickness (Figure 10.4). However, cycling experiments performed between 4.2 and 3 V or in the restricted range 4–3 V indicate a capacity loss of 0.2% per cycle but no clear conclusion on this behavior has been proposed. Bias sputtering is proved to be a promising method [20] owing to microstructural changes induced by the ion bombardment delivering energy to the growing film and, in spite of a low crystallinity, interesting discharge capacities are reached when an optimized substrate bias value of −50 V is applied. In that case no heat treatment is required and high capacity values close to the theoretical one are obtained for a film 0.35 μm thick. A compromise must be found between the Li/Co molar ratio and the presence of Co_3O_4. The advantage of this effect has been confirmed by the promising cycle life of as-prepared films in an all solid-state microbattery [20].

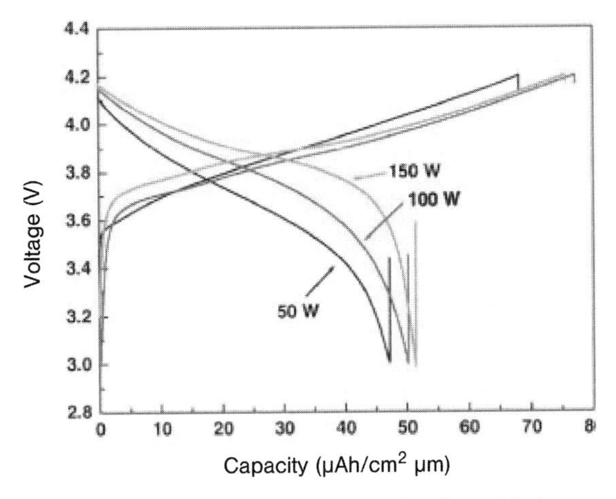

Fig. 10.3 Influence of the RF power on the charge/discharge curves of $LiCoO_2$. From [19].

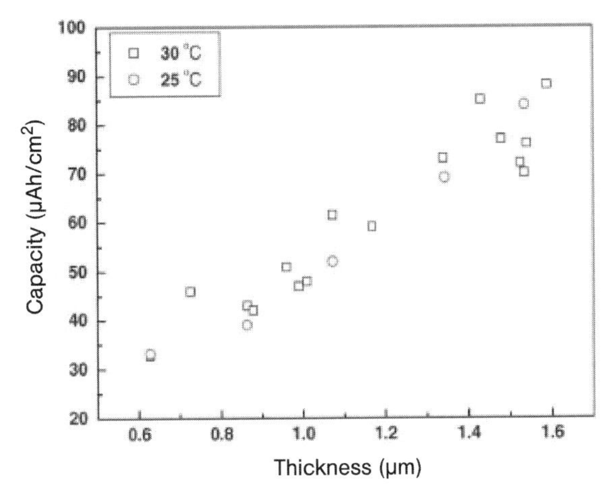

Fig. 10.4 Discharge capacity versus thickness of RF-sputtered $LiCoO_2$ films (RF power = 150 W) in the potential range 4–3 V at 30 μA cm^{-2}. From [19].

Other experimental parameters have to be taken into account for controlling high-quality thin films. For instance the influence of target history and deposition geometry on RF magnetron sputtered $LiCoO_2$ thin films is described in [21]. Li-deficient film oxides can be obtained with heavily used targets. The selection of an appropriate substrate is important in terms of deposit morphology, electrical resistance, and rate capability [22].

Besides the complicated optimization of the numerous experimental parameters involved for the sputtering deposition process and the postannealing treatment often required, some authors proposed to improve the specific capacity by raising the charge cutoff voltage up to 4.4 V [23–25]. The sol–gel coating of $LiCoO_2$ powder by high-fracture-toughness oxides has been proved to decrease or to suppress, in some cases, the changes in the c lattice constant for an Li extraction occurring in the composition range $0 < x \leq 0.7$ for x in $Li_{1-x}CoO_2$ [23]. This finding has been successfully applied to the coating of sputtered films, 600 nm thick, by sputtered Al_2O_3 of few tens of nanometers: which ensured a significant decrease of the capacity fading from 37 μA h cm^{-2} as the initial capacity to only 8 μA h cm^{-2} for the bare $LiCoO_2$ film; but still 27 μA h cm^{-2} for the Al_2O_3-coated film after 100 cycles between 4.4 and 2.75 V at 0.2 mA cm^{-2} in 1 M $LiPF_6$ EC/DEC electrolyte [24]. The coating has been found to suppress cobalt dissolution during electrochemical cycling and allow high Li diffusivity to be maintained upon cycling. $AlPO_4$ coating has also been proved to lead to similar results [25].

A specific structural response of $LiCoO_2$ thin film exists even when the charge cutoff voltage is limited to the conventional value of 4.2 V corresponding to the $Li_{0.5}CoO_2$ material. Indeed, surprisingly, in contrast to the $Li_{1-x}CoO_2$ powders exhibiting ≈3% c-axis expansion at $x \approx 0.5$ (from ≈14.05 to ≈14.45 Å), the XRD patterns of charged films (500 nm thick) at 4.2 V show negligible change in the c lattice parameter [26]. Five or 10 cycles are needed to provoke the appearance of the conventional $Li_{0.5}CoO_2$ expanded phase. This phenomenon seems to be strongly dependent on the preferred orientation. Interesting works have been devoted to the study of local structure of $LiCoO_2$ during electrochemical process. XAS [27], XPS [28], RBS, PIGE [29], IR [30], and Raman [31] studies deal with the charge compensation mechanism during the charge, the chemical composition of oxide films and their spectroscopic characteristics. The oxygen ions would be involved in the oxidation process in parallel to the cobalt system. Investigation of stresses generated during lithium transport through the RF sputter-deposited $Li_{1-x}CoO_2$ film has been performed by a double quartz crystal resonator (DQCR) technique [32].

10.2.1.2 **Solid-State Electrolyte**

Many TFBs with crystalline $LiCoO_2$ cathodes have been prepared and investigated at Oak Ridge National Laboratory and a wide and deep knowledge has been provided by the group of J. B. Bates [33–36]. Figure 10.1 shows a schematic cross section and plan view of a TFB. Lithium cobalt oxide films were deposited by planar RF magnetron sputtering of $LiCoO_2$ targets in an $Ar-O_2$ gas mixture. The films are then annealed in air between 500 and 700 °C. The as-prepared films, with film thickness from 0.05 to 1.8 μm were found to contain a small amount of Li and O excess as Li_2O. The cathodes are covered with a 1–2 μm amorphous lithium phosphorus oxynitride (LiPON) electrolyte deposited by RF magnetron sputtering of a Li_3PO_4 in N_2 ($Li_{2.9}PO_{3.3}N_{0.46}$) with a conductivity of 2×10^{-6} S cm^{-1}. This electrolyte is stable up to 5.5 V versus lithium metal. Lithium anode films about 3 μm thick were then deposited over the electrolyte by thermal evaporation of lithium metal contained in a Ta crucible at a pressure of 0.1 mPa. Ten years ago, this group clearly established the advantage of using well-crystallized material rather than an amorphous phase and indicated good electrochemical properties for films heated at 700 °C after deposition. Figure 10.5 [33] shows that such solid-state batteries can sustain high discharge rate, up to 1 mA cm^{-2} and deliver a large fraction of the maximum capacity expected in the 4.2–3 V potential range. The

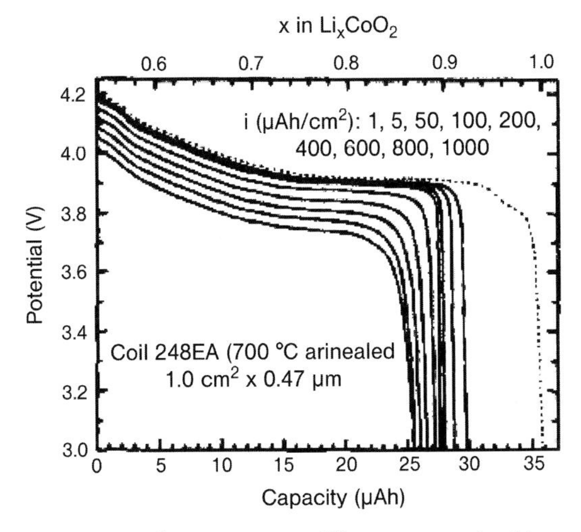

Fig. 10.5 Discharge curves at different current densities for an Li/LiPON/$LiCoO_2$ thin cell. Cathode thickness is 0.47 μm. From [33].

average specific capacity obtained is 64 μA h cm^{-2} μm^{-1}. Long cycle life of more than a few hundred of cycles and a capacity loss between 0. 0001 and 0.002% per cycle is reported [33]. A detailed study of cycling properties of such batteries has been recently been published for a wide range of film thicknesses [34]. For a 4.2-μm-thick film, a high capacity of 300 μA h cm^{-2} can be achieved.

Owing to the large number of experimental parameters required in the deposition procedure, some questions have been raised regarding the control of predominant factors to achieve high-performance electrodes and reproducible synthesis. After annealing at 700 °C, $LiCoO_2$ films exhibit a strong preferred orientation of the (003) type for thickness <1 μm while the (101)–(104) orientation is predominant for films thicker than about 1 μm. Of the variables such as the gas mixture and pressures, deposition rate, substrate temperature, and substrate bias, the substrate temperature (200–300 °C) is thought to induce large changes in texturing the thick films from predominantly (101)–(104) to (003) orientation [35]. However, no clear influence of the preferred orientation on the kinetic response of the cathode is reported. Another key parameter is the cell resistance; the cathode resistance can be minimized by using low-temperature deposition because high-temperature deposition leads to larger grains and increased void fraction, and then byreducing the contact area between the electrolyte and cathode, and contact between grains, resulting in higher cell resistances.

The optimization of the battery capacity can be performed by preparing thicker positive oxide films by using a process that does not involve postannealing allowing then to deposit the electrode material on flexible polymer substrates. The preparation of such a cathode film, 6.2 μm thick has been realized by electron cyclotron resonance (ECR) plasma sputtering method [37] and a high discharge capacity of about 250 μA h cm^{-2} with good cyclability is reported.

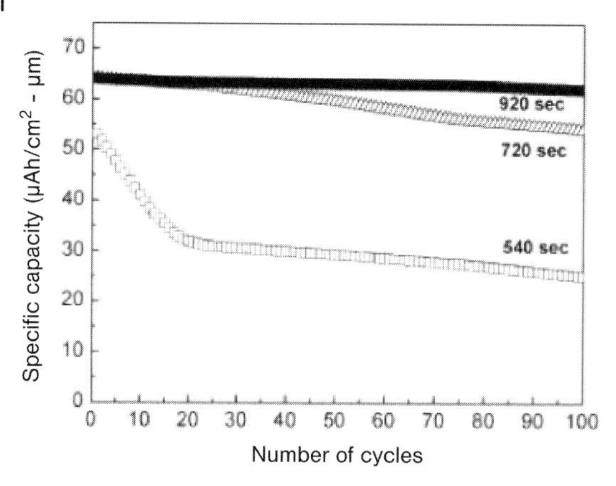

Fig. 10.6 Cycling behavior (about 1 C rate) of Li/LiPON/$LiCoO_2$ cells with films prepared at different RTA times. From [39].

Some integrated circuits (ICs) are fabricated using solder reflow (surface mount) assembly in which the IC is heated to 250–260 °C for a short time, causing all of the components to be soldered at once. Because of the low melting point of lithium, 180 °C, lithium batteries fail when heated to these temperatures. Therefore, some attempts have been made to replace lithium [36, 38] by sputtered anode materials such as silicon–tin oxynitride ($SiSn_{0.87}O_{1.20}N_{1.72}$) and nitrides ($Zn_3N_2$, Sn_3N_4). However, a significant decrease of the energy delivered (50–60%) is observed [36]. The most promising approach seems to be the use of a lithium-free battery with *in situ* plated Li anode [38]. A copper current collector is deposited over the electrolyte and cycling then involves repeated stripping and plating of lithium under the Cu. The electrochemical properties of Li-free TFBs compare very well with those of conventional Li TFBs.

To overcome the drawbacks of conventional heat treatment to get crystalline thin films, a rapid thermal annealing (RTA) treatment system installed in a vacuum chamber, the heat source being a halogen lamp, applied at 650 °C for 500–900 seconds (Figure 10.6), enables c-$LiCoO_2$ films to be achieved [39]. Attractive results are reported for $LiCoO_2$ films 2.3 µm thick prepared with an RTA time of 900 seconds since a stable specific capacity close to the theoretical one is obtained over 100 cycles at C rate and 70% of the maximum capacity is recovered at high rate (20 C). In addition to less-crystallized samples, a too short RTA time probably promotes the formation of the LT spinel phase leading to a capacity fading upon cycling.

Many attempts have been made to decrease the heat treatment temperature or to suppress this step. For instance, sputtered (104) oriented films can be obtained after a heat treatment performed as low as 300 °C [16]. At the same time, what is more interesting is the optimization of the substrate bias, which has been shown to positively influence the electrochemical behavior of 0.35-µm-thick films [20] thus avoiding a postannealing treatment; a satisfactory specific capacity of

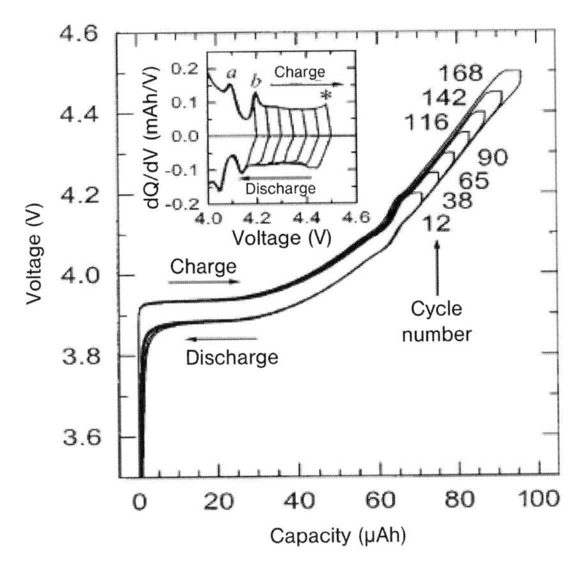

Fig. 10.7 Charge/discharge curves with the various charge cutoff voltages (0.1 mA cm^{-2} at 25 °C). From [40].

50 μA h cm^{-2} μm^{-1} is reached after 100 cycles. As for thin films used with liquid electrolytes, one possible way for enhancing the specific capacity consists in increasing the charge cutoff voltage up to 4.4 V [40, 41]. Using LiPON electrolyte in the microbattery, it has been shown that the capacity fade of a film oxide 1.2 μm thick increases a lot from 4.4 V compared to 4.2 V, but still remains acceptable to deliver 90 μA h cm^{-2} (Figure 10.7), i.e., 98% of the initial capacity after 100 cycles between 3 and 4.4 V at 1.4 C rate [40]. This degree of capacity retention is remarkably high, since, in similar conditions, only 50% of the initial capacity is recovered after 50 cycles with uncoated $LiCoO_2$ powders in liquid electrolyte.

10.2.2 PLD $LiCoO_2$ Films

With RF sputtering, PLD is one of most widely used techniques to prepare thin metal oxides. The PLD technique has many advantages for producing good-quality thin films with a high-deposition rate, and for retaining the original stoichiometry of a target to the deposited films and the capability of making a multilayer film by exchanging the targets inside a chamber. Moreover, it is possible using PLD to synthesize solid electrolytes [68, 69]. In the case of $LiCoO_2$, solid-state thin-film LIBs consisting of crystalline $LiCoO_2$ and amorphous SnO as positive and negative electrodes respectively with an amorphous $Li_2O-V_2O_5-SiO_2$ (LVSO) solid electrolyte were fabricated [42, 43].

The first interesting results obtained in liquid electrolyte indicated that amorphous $LiCoO_2$ could be deposited on unheated stainless steel substrates using the

$R\bar{3}m$ layered structure as target [44]. Therefore, a postannealing treatment at temperature >500 °C in air allows the samples to crystallize but the authors mention the presence of a mixture of the high- and the low-temperature phases characterized by a lower capacity and cycle life. Striebel *et al.* reported the synthesis of $LiCoO_2$ thin films on heated (600 °C) stainless steel substrates [45]. These films have thicknesses from 0.2 to 1.5 μm and are crystalline without postdeposition annealing. A 200-nm-thick film exhibits a maximum capacity density of 62 μA h cm^{-2} μm^{-1} in the 4.2–3.7 V voltage range, which is very close to the theoretical value, but it has a poor cycling life with a capacity loss of 50% after 50 cycles. When the target is a pure sintered $LiCoO_2$ powder, the films are often lithium deficient. An optimization, i.e., an increase of both the substrate temperature and the oxygen partial pressure, is necessary to decrease the amount of the Co_3O_4 impurity in this phase [46]. Figure 10.8 illustrates the strong influence of the substrate temperature and oxygen partial pressure on the film morphology and the grain size of active material.

To compensate the lithium loss in the film, the use of a sintered target made of $LiCoO_2$+ 15 wt% Li_2O has been successfully proposed with a heated substrate at 300 °C [47]. This method has been recently applied with promising results with films 300–500 nm thick deposited on stainless steel substrates heated at 600 °C [48, 49]. The oxygen pressure has also to be optimized and its influence on the orientation and morphology is well investigated in [48]. It is shown that a randomly oriented film induces a higher capacity and coulombic efficiency, and better cycling stability than the (003)-oriented film. Indeed Figures 10.9 and 10.10 clearly indicate a capacity of 68 μA h cm^{-2} μm^{-1} for the cathode deposited at 50 mTorr of oxygen (520 nm thick) compared with that of 45 μA h cm^{-2} μm^{-1} achieved at 300 mTorr (320 nm thick) characterized by the stacking of oxide sheets parallel to the substrate.

Actually, two preferred orientations are commonly found [48–51]: the (003) and the (104) orientations, which correspond to a smooth and a rough surface, respectively. The different textures can be reached by changing the substrate (stainless steel or Si, for instance) [49–51] or the partial pressure of oxygen [48]. Surprisingly, in spite of a lower utilization and a lower Li diffusivity in the (003)-textured thin films by one order of magnitude than in the (104)-textured films, the (003)-textured thin films exhibit a slightly better capacity retention upon cycling (Figure 10.11) [51]. This is attributed by the authors to their better structural stability and smooth surface. The electrochemical performance of the (003) films is thought to indicate that fast Li diffusion through the grain boundaries is comparable to or even faster than that through the grains [51]. One way of possible improvement in terms of capacity would be to enlarge the potential window for cycling up to 4.5, 4.7, or 4.9 V. Recent studies address the phase transitions, the kinetics of Li transport of PLD $LiCoO_2$ films in this high-voltage region [52].

Cycling properties of PLD $LiCoO_2$ thin films are relatively rare. A significant improvement has been made when considering the fast capacity fading reported in [46] and the recent data in [48, 51]. High capacity values of 60 μA h cm^{-2} μm^{-1} (Figure 10.9) or 50 μA h cm^{-2} μm^{-1} (Figure 10.11) can be reached in the 4.2–3 V potential range with a satisfactory stability over 20 cycles at 50 and 20 μA cm^{-2} for films in the range 400–500 nm thick. Nevertheless, the effective capacity available

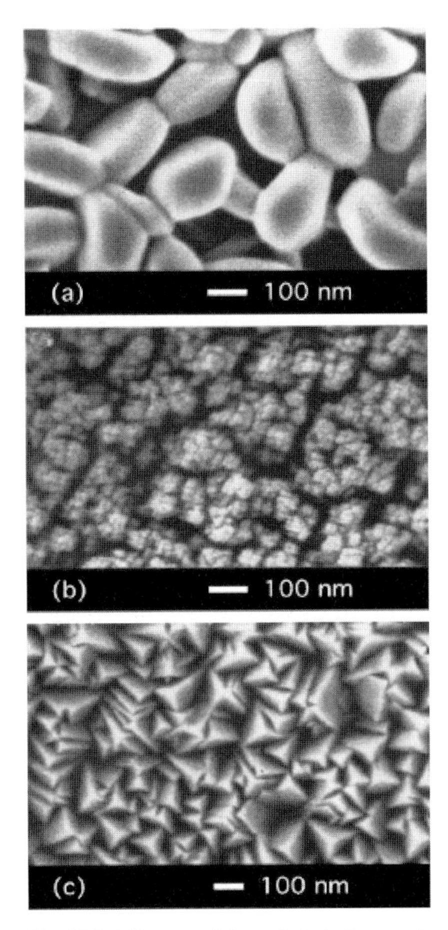

Fig. 10.8 Influence of the substrate temperature and oxygen partial pressure on the morphology of PLD $LiCoO_2$ films grown on (200)-textured SnO_2-coated glass; (a) $T_s = 700\,^\circ C$, $pO_2 = 2000$ mTorr; (b) $T = 300\,^\circ C$, $pO_2 = 2000$ mTorr; and (c) $T = 400\,^\circ C$, $pO_2 = 300$ mTorr. From [46].

is then only of 30–20 μA h cm^{-2} due to relatively thin films with thickness ≤500 nm. There is a lack of data on extended cycling properties and on the rate capability of PLD thin films.

It comes out that the influence of the preferred orientation on the electrochemical behavior in terms of utilization, capacity retention, and rate capability must be clarified, since many literature data actually seem contradictory, especially in regard to Li diffusivity.

Two types of solid-state TFBs, one using an amorphous LVSO solid electrolyte [42, 43] and the other, an LiPON electrolyte, have been investigated [53]. The battery consisting of 150 nm SnO/1.1 μm LVSO electrolyte/400 nm $LiCoO_2$ operates at 2.5 V for a capacity of 9 μA h cm^{-2} that decreases continuously with cycling to reach

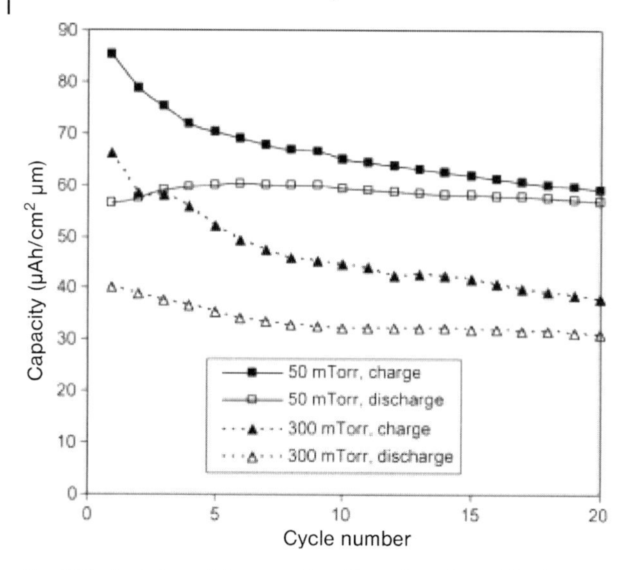

Fig. 10.9 Evolution of the capacity with cycles of two $LiCoO_2$ thin films deposited at 50 and 300 mTorr with current density of 50 μA cm^{-2}. From [48].

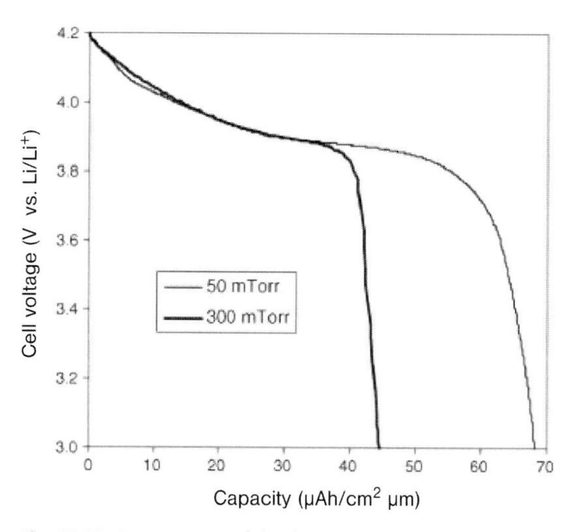

Fig. 10.10 Comparison of discharge curves of two $LiCoO_2$ thin films deposited at 50 and 300 mTorr with current density of 10 μA cm^{-2}. From [48].

4 μA h cm^{-2} after 100 cycles at 44 μA cm^{-2}. A thinner $LiCoO_2$ film, 100 nm thick $\times 0.25$ cm^2 has been used with LiPON deposited by RF sputtering and a lithium negative electrode. A discharge capacity of 2 μA h is obtained which shows a high electrochemical efficiency; however, no cycling data have been given for such a TFB.

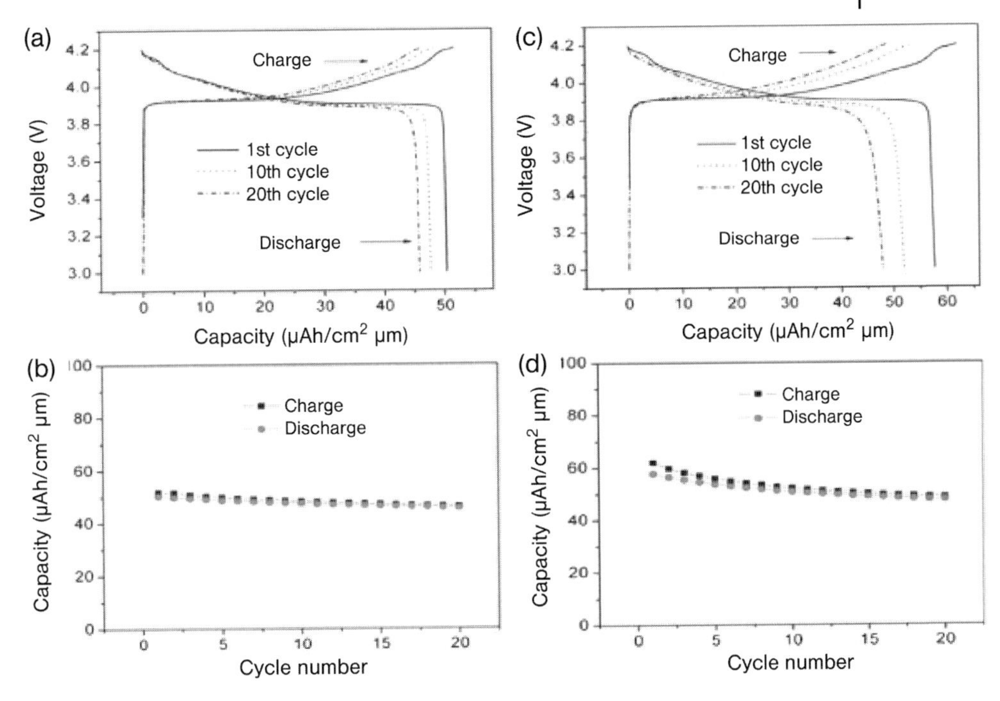

Fig. 10.11 Charge/discharge curves and capacity versus cycle number of the (003)-textured $LiCoO_2$ film (a), (b) and the (104)-textured $LiCoO_2$ film (c), (d). From [51].

10.2.3
CVD $LiCoO_2$ Films

The CVD technique has allowed to synthesize crystalline $LiCoO_2$ thin films in the range 100–200 nm by controlling deposition parameters such as deposition temperature (270–450 °C) and input Li/Co mole ratio [54, 55]. This does not avoid the use of a postannealing treatment at 700 °C. Nevertheless, the specific capacity achieved for these thin films does not exceed about only 50% of the theoretical one with a limited rechargeability leading to a poor cycle life. The CVD process could be of interest to improve the discharge capacity of cathode films using a trench structure in place of planar–cathode films.

10.2.4
$LiCoO_2$ Films Prepared by Chemical Routes

Even when the physical deposition methods such as the sputtering and the PLD techniques are widely used, various chemical methods for the synthesis of thin-film cobalt oxides have been proposed. The available data are rather recent and the as-prepared $LiCoO_2$ films have been investigated in liquid electrolyte, but not in a solid-state microbattery. Ten years ago, some authors demonstrated that thin films

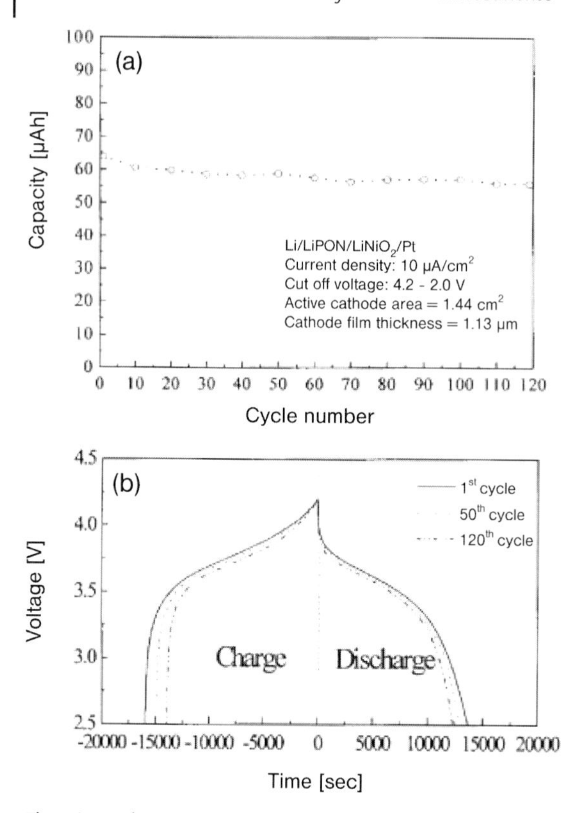

Fig. 10.12 The galvanostatic cycling properties of ESD SiO_2-modified $LiCoO_2$ films in the 4.3–2.7 V voltage range at 0.1 mA cm^{-2}; the SiO_2 content is indicated in the graphs. From [58].

0.1–0.3 µm thick could be achieved using a mixture of appropriate sols sprayed on to heated substrates from 200 to 600 °C with, sometimes, a postannealing treatment [56, 57]. This spray pyrolysis method did not give satisfactory results in terms of cycling properties. More recently, promising electrochemical properties of nano-SiO_2 modified $LiCoO_2$ films fabricated by the ESD technique have been reported [58]. A mixture of an SiO_2 sol and a solution of lithium and cobalt salts were used for the deposition on a heated Pt substrate kept at 350 °C, and a postannealing process at 700 °C was applied. The promoting influence of the SiO_2 content in the as-prepared films is demonstrated and a stable specific capacity of 130 mA h g^{-1} is recovered over 60 cycles between 4.3 and 2.7 V at 0.1 mA cm^{-2} current density when the film contains 15 wt% SiO_2 (Figure 10.12).

The sol–gel spin-coating method combined with a postannealing process in the range 650–800 °C has been applied by a few groups [59–61]. One of the major drawbacks of this method is that the above process, the "deposition-heat treatment", must be repeated, for instance, from three up to 10 times to get a thickness of 1 µm [59, 61] with the annealing temperature [61] and substrate [62] greatly influencing the cycling properties. One of the best results is reported for a 0.5-μm-thick film, prepared with poly(vinylpyrrolidone) as an additive to sol, which exhibits a stable capacity of 60 μA h cm^{-2} μm for 10 cycles between 4.3 and 3.5 V at about 1 C in 1 M $LiClO_4$/EC + DEC [59]. Actually, the true long-term cycling behavior of sol–gel films is not known.

Under galvanostatic conditions, $LiCoO_2$ thin films can be also produced on Co substrate from an appropriate aqueous solution of cobalt and lithium ions at 100 °C, but only preliminary but interesting voltammetric curves are available [63]. It must be recalled that a $LiCoO_2$ thin film, 0.3 μm thick can be prepared by the chemical oxidation of metallic Co in molten $LiCO_3/KCO_3$ at 650 °C [64].

The screen-printing technique could afford a possible alternative to the use of thick films to significantly increase the specific capacity of the Li-ion microbattery. Indeed, the feasibility of preparing doped M-$LiCoO_2$ (M = Zr, C) thick films (6–20 μm) by the screen-printing technique is demonstrated in [65, 66]. A stable capacity as large as 270 μA h cm^{-2} is obtained over 40 cycles at 100 μA cm^{-2} current density with a 5-μm-thick Ag-added Zr-$LiCoO_2$ film printed on to a platinum-coated alumina substrate [65]. This value corresponds to about 87% of the theoretical capacity, but the degree of utilization of the electrode material decreases with increasing thickness; it drops for instance to 43 and 36% for the same film 10 and 20 μm thick, respectively.

Preliminary interesting results have been obtained for an all solid-state LIB using a thick electrolyte layer of $Li_{0.33}La_{0.56}TiO_3$ and $Li_4Ti_5O_{12}$ and $LiCoO_2$ thin films deposited by spray pyrolysis as anode and cathode materials [67] but no cycling data are available for this device.

10.2.5 Conclusion

Most of the fundamental studies on the electrochemical properties of $LiCoO_2$ thin films have been performed in liquid electrolyte and aim at getting thermodynamic [15], kinetic [15, 45, 49] and structural [14, 26–28, 52] data. A demonstration of the influence of the diffusion plane orientation for sputtered films illustrated by detailed kinetic data is reported in [14, 15]. As far as the cycling performance is considered, it comes out that the cycling behavior is subject to deterioration. Some works suggest the capacity loss is attributed not only to the growth of the SEI layer, which decreases the electrode surface area and poses an additional barrier for lithium intercalation, but also to a loss of active material or to a peeling off process even when this phenomenon is not often taken into account in the literature.

The $Li_{1-x}CoO_2$ system with $x_{max} = 0.5$ is usually cycled between 4.2 and 3 V. In spite of the optimization of deposition conditions, the vacuum techniques like RF sputtering and PLD were not successful to systematically produce high-performance thin films since most of the studies reveal a continuous more or less pronounced capacity decrease upon cycling in the 4.2–3 V range. A rigorous optimization of the deposition and crystallization processes must be found to get interesting film properties. Among the best results, the capacity of sputtered films deposited on stainless steel without any postannealing treatment exhibits a decline of 0.2% per cycle leading to a value of 35 μA h cm^{-2} $μm^{-1}$ after 100 cycles between 4 and 3 V [19]. In the 4.2–3 V region, 4-μm sputtered thick films only deliver, at 50 μA cm^{-2}, 27 μA h cm^{-2} $μm^{-1}$ (≈ 50% of the theoretical value) as stable value of capacity over 150 cycles, i.e., in fact 108 μA h cm^{-2} [22]. With PLD films, 0.4–0.5 μm thick, stable capacities in the range 50–60 μA h cm^{-2} $μm^{-1}$ can be reached [48, 51] in very close cycling conditions showing that only 25–30 μA h cm^{-2} are available. Though the solution techniques cannot be easily applied to the multilayer step process to fabricate TFBs, they offer new possibilities for deposition of attractive $LiCoO_2$ films. For instance, the ESD SiO_2 modified-$LiCoO_2$ films with 15 wt% SiO_2 show a remarkable and stable specific capacity of 130 mA h g^{-1} over 60 cycles at 100 μA cm^{-2} between 4.3 and 2.7 V [58]. The sol–gel process also allows to reach a capacity of 60 μAh cm^{-2} $μm^{-1}$, i.e., 30 μA h cm^{-2} for a 0.5-μm film, over 10 cycles [59], whereas the screen-printing technique enables a high value of 270 μA h cm^{-2} to be stabilized over 40 cycles with an Ag-added Zr-$LiCoO_2$ film, 5 μm thick [65]. One interesting way to improve the cycle life of the $LiCoO_2$ system in the 4.2–3 V range consists in protecting the film by a thin coating layer of few tens of nanometers. Such trends must also be investigated more deeply to enable higher capacities to be reached with a cutoff voltage above 4.2 V.

In all solid-state TFBs, the usual cycling behavior of a $LiCoO_2$ thin film is more stable than in liquid electrolyte showing the particular high reactivity of the active material toward the liquid electrolyte mainly based on $LiClO_4$ or $LiPF_6$ in PC or DEC–EC solvents. The available capacities are already high, reaching values between 200 and 300 μA h cm^{-2} depending on the current density (20–1000 μA cm^{-2}) for a cathode 4.2 μm thick. This finding is combined with long cycle life of over a few hundred cycles in the 4.2–3 V voltage range. The main objective now concerns the crystallization process of the cathode whose temperature and time must be lowered, for instance, by using an RTA system, or by optimizing the substrate bias or the substrate temperature. Of course, the next step for improved TFBs is to suppress the evaporated Li anode. Lithium-free batteries based on *in situ* Li plating have been proposed, but intercalated thin-film compounds constitute a safer solution and hence an exciting field of research.

10.3 $LiNiO_2$ and Its Derivatives Compounds $LiNi_{1-x}MO_2$

In regard to LIBs, the limited capacity of $LiCoO_2$, its high cost, and toxicity have been regarded as drawbacks. Moreover, $LiNiO_2$ is usually deposited as nonstoichiometric $Li_{1-x}Ni_{1+x}O_2$ because of the difficult oxidation of Ni^{2+} to Ni^{3+}. Therefore, not only $LiNiO_2$ but also $LiNi_{1-x}Co_xO_2$ have been recently investigated for all solid-state TFBs. This class of materials must allow to reach maximum capacities in the range 70–100 μAh cm^{-2} $μm^{-1}$.

10.3.1 Solid-State Electrolyte

Electron-beam evaporated $LiNi_{0.5}Co_{0.5}O_2$ cathode films have been shown to be suitable in Li microbatteries without any successful application [70].

The best results have been reported for RF-sputtered $LiNiO_2$ [71] and $LiNi_{0.5}Co_{0.5}O_2$ [72], 1.1 and 0.3 μm thick, respectively, and deposited on Pt substrates. From the RBS and ICP results, it follows that the as-prepared films are lithium deficient with a chemical composition of $Li_{0.84-0.86}MO_2$. Figure 10.13 shows the voltage profiles of a thin-film microbattery corresponding to the stacking of the layers Li/LiPON/$LiNiO_2$ or $LiNi_{0.5}Co_{0.5}O_2$/Pt/MgO/Si. The RTA treatment applied (700 °C for 3 minutes in flowing oxygen) allows to avoid the surface oxide layer (Li-O) to be formed, and a high specific capacity of about 40 μA h cm^{-2} μm is stabilized after 100 cycles in the 4.2–2.20 V potential range for $LiNiO_2$ (Figure 10.13), the effective capacity being only in the order of 20 μA h cm^{-2} for the Li–Ni–Co mixed oxide, 300 nm thick [72]. The RF sputtering deposition process can be modified to decrease the temperature of the postannealing treatment by controlling the substrate temperature. For instance well-crystallized thin films, 500 nm thick were prepared on ITO substrates

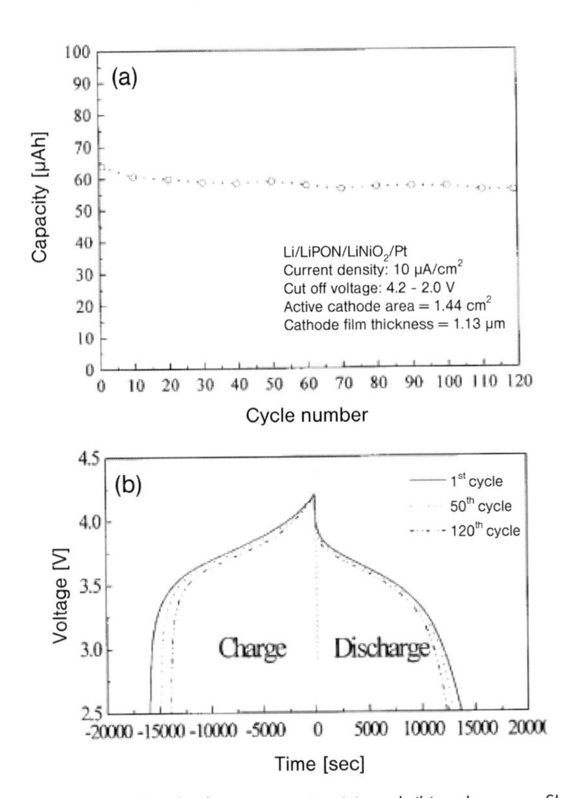

Fig. 10.13 The discharge capacity (a) and (b) voltage profiles of a Li/LiPON/$LiNiO_2$ (1.44 cm^2/Pt) TFB. From [71].

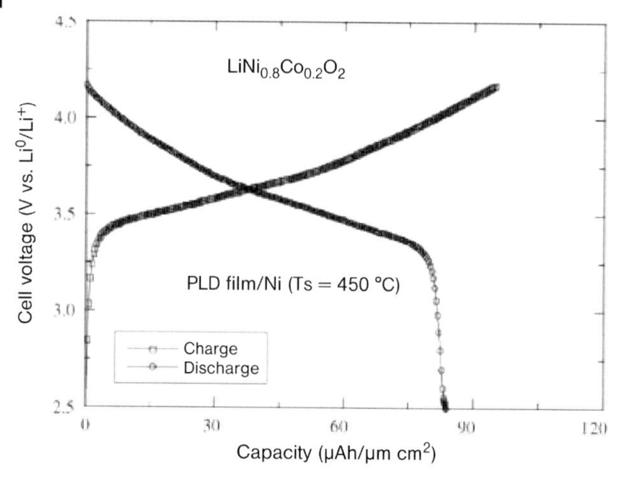

Fig. 10.14 Initial charge/discharge curves for a PLD $LiNi_{0.8}Co_{0.2}O_2$ film; $p(O_2) = 150$ mTorr and $T_s = 450\,^\circ C$. From [75].

heated as low as 350–450 °C which is approximately half the usual postannealing temperature for Li–Ni–O and Li–Ni–Co–O thin films [17].

10.3.2 Liquid Electrolyte

Other compositions in the $LiNi_{1-x}Co_xO_2$ system have been proposed and tested in liquid electrolytes. In particular, pulsed laser deposited $LiNi_{0.8}Co_{0.2}O_2$ and $LiNi_{0.8}Co_{0.15}Al_{0.05}O_2$ are of interest [73–75]. These cathode materials deposited on heated Ni substrates (450 °C) exhibit a (00 l) preferred orientation and attractive capacities of 83 and 92 μA h cm^{-2} μm^{-1}, respectively (Figure 10.14). The Al-containing film is reported to deliver high capacities with a good rate capability since 95, 92, and 80 μA h cm^{-2} μm^{-1} are recovered in liquid electrolyte after 36 cycles at C/20, 2 C, and 10 C discharge rate (C/2 charge rate). This behavior is reported to be related with the decreasing particle size provided by the Al-doping effect [74]. These works are in good accord with previous data established on PLD $LiNi_{0.8}Co_{0.2}O_2$ films, 0.6 μm thick [76]. Finally, some preliminary results outline the possible interest of ESD to prepare $LiNiO_2$, $LiNi_{1-x}M_xO_2$ (M = Co, Al) films which are relatively thick from at least 1 μm to a few micrometers [77]. Soft chemistry routes like the sol–gel synthesis and the electrochemical–hydrothermal method have been considered for deposition of thin films in the range 10–115 nm [78, 79].

10.3.3 Li – Ni – Mn Films

RF magnetron sputtering has been successfully applied to synthesize Li–Ni–Mn oxide films 1.5 μm thick [80, 81]. The cathode with the $Li_{1.12}Ni_{0.44}Mn_{0.44}O_2$

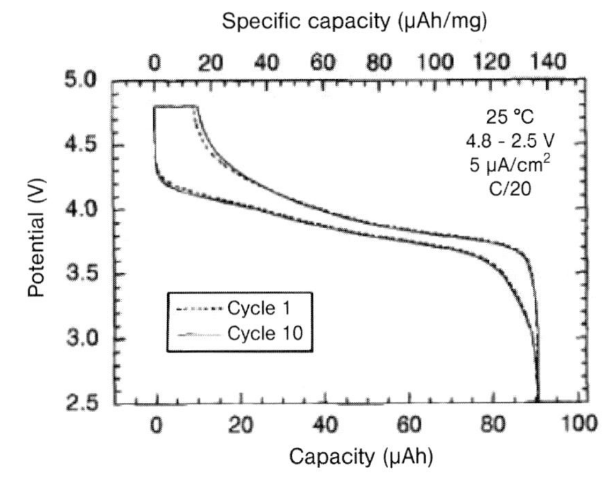

Fig. 10.15 Cycle 1 and 10 of the $Li/Li_xNi_{0.44}Mn_{0.44}O_2$ battery between 4.8 and 2.5 V at C/20 rate. From [81].

composition was deposited on an Au/Ni/alumina substrate from a target with the composition $Li_{1.23}Mn_{0.37}Ni_{0.40}O_2$ and needed a further heat treatment at 750 °C under O_2 for 3 hours. After charging up to 4.2 V in an all solid-state battery using LiPON as electrolyte, the discharge capacity was only 63% of the charge capacity. The authors proposed to enhance the capacity with the enlarged 4.8–2.5 V potential window (Figure 10.15) and a specific capacity of 90 μA h, i.e., 60 μA h cm^{-2} μm^{-1} has been achieved with a good stability on 10 cycles at C/20 rate.

10.3.4 Conclusion

From the use of the layered host lattices, $LiNiO_2$, $LiNi_{1-x}Co_xO_2$, and $LiNi_{1-x}Mn_xO_2$, large specific capacities in the range 40–60 μA h cm^{-2} μm^{-1} in solid electrolyte and 60–90 μA h cm^{-2} μm^{-1} in liquid electrolyte can be expected with a good compromise on the working voltage, which indicates a possible improvement compared to $LiCoO_2$ thin-film oxides. However, basic data related with the electrochemical and structural behavior of the film in the $LiNiO_2$, $LiNi_{1-x}M_xO_2$ systems are not available as yet in spite of the promising results described above.

10.4 $LiMn_2O_4$ Films

Assuming a fully dense film, the maximum specific capacity is 63 and 50 μA h cm^{-2} μm^{-1} for the reversible extraction/insertion of 1 Li and 0.8 Li per mole of oxide, respectively. First attempts for $LiMn_2O_4$ film deposition have involved the electron-beam evaporation. The reactive electron-beam evaporation of the preheated bulk oxide in a low oxygen pressure followed by an annealing procedure as low as

400 °C leads to the formation of well-crystallized films, 1 μm thick, deposited on stainless steel substrate [36, 82]. Promising results were obtained between 4.3 and 3 V with capacity close to the theoretical one and 0.8 Li reversibly intercalated in liquid electrolyte. At 100 μA cm^{-2}, such films produce specific capacities of about 130 mA h g^{-1} (≈ 88% of the theoretical capacity) and 100 mA h g^{-1} even after 220 cycles at 55 °C. Surprisingly, in spite of its attractive properties, no further work using this technique has been published. $LiMn_2O_4$ thin films, 0.5 μm thick, have been evaluated in all solid-state batteries using LiPON as solid electrolyte. Over the very large potential window 5.3–1.5 V, 1 Li per mole of oxide is extracted at high potential but no clear cycling data is available [83]. The cell delivered a low capacity of 8–10 μA h cm^{-2} between 4.5 and 3.8 V.

10.4.1 Sputtered $LiMn_2O_4$ Films

Contrary to the $LiCoO_2$ and V_2O_5 thin-film cathodes, $LiMn_2O_4$ has been little investigated and prepared using RF magnetron sputtering. Indeed, only a few papers have dealt with the electrochemical properties of sputtered $LiMn_2O_4$ thin films in liquid [84–87] and solid-state electrolytes [88–91]. A promising behavior has been reported for 200-nm thick $LiMn_2O_4$ films working in the 4.3–4.2 to 3.7–3.8 V potential range in liquid electrolyte [84, 85]. A rapid postannealing treatment at 750 °C for 60 seconds (RTA) in an oxygen atmosphere is required to optimize the cycling behavior of the film oxide. The rechargeability of these films has been demonstrated with a stable capacity of 50 μA h cm^{-2} μm^{-1} recovered after 1000 cycles (Figure 10.16) at 200 μA cm^{-2} [84]. This capacity value shows that 0.8 Li could be deintercalated reversibly at a nearly constant voltage of 4 V. An electric bias applied on the substrate during deposition with or without a postannealing treatment can be used to modify the morphology, the crystallinity, and the discharge/harge profile of the $LiMn_2O_4$ film [86]. Surprisingly, the possible use of the spinel Li–Mn–O thin film in the enlarged potential window 4.3–2.0 V has been reported for 400-nm-thick films treated for 1 hour in O_2 at 600 °C [87]. Indeed, a capacity stable over at least 20 cycles and estimated at 200 mA h g^{-1} is obtained, whereas the cycling behavior of the oxide powder is poor due to the large volume change of about 5.6% during lithium insertion in the 3.0-V plateau region.

All solid-state rechargeable TFBs were fabricated with the cell structure of $LiMn_2O_4$/LiPON/Li using sequential thin-film deposition techniques [88–90]. For amorphous films obtained as-deposited, 660 nm × 1.2 cm^2, the cell delivers a capacity of 40 μA h at low current density (2 μA cm^{-2}) between 4.5 and 2.5 V and 18 μA h at 40 μA cm^{-2} [88]. Some works indicate that crystalline sputtered films are Mn deficient and that the lithium content can widely vary with the experimental deposition and annealing conditions [89]. A solid-state cell ($LiMn_2O_4$/LiPON/Li) working at a nearly constant potential of 4 V and using crystallized $LiMn_2O_4$ films, 300 nm thick, deposited on Pt current collectors and heat-treated at 750 °C in O_2 can sustain current densities from 50 to 800 μA cm^{-2} for a specific capacity of 50–45 μA h cm^{-2} μm, respectively [90], with a small ohmic drop across the solid

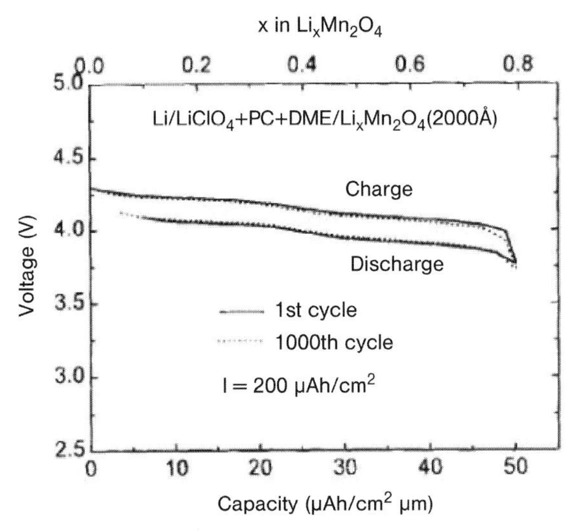

Fig. 10.16 Typical charge/discharge curves of a 0.2-μm $LiMn_2O_4$ spinel film prepared by RF magnetron sputtering followed by RTA at 750 °C in O_2. From [84].

electrolyte. This indicates fast lithium-intercalation kinetics within the $LiMn_2O_4$ spinel structure. A remarkable capacity retention of 96% is reached after 100 cycles at 100 μA cm^{-2}. Moreover, by interconnection of eight unit cells in series, higher voltages were successfully obtained with two voltage plateaus at 33 and 31 V. A new approach has been recently proposed by Park *et al.* [91] with a cell using a thin-film polymer electrolyte PEO–$LiClO_4$ directly deposited on the $LiMn_2O_4$ cathode film ≈0.3 μm thick by spin coating; 0.83 Li ions can be removed from the film oxide during the charge process and the capacity decreases slowly to 85% of the initial value (53 μA h cm^{-2} $μm^{-1}$) after 100 cycles at 6 C rate. Hence the spin-coated solid polymer electrolyte PEO–$LiClO_4$ was shown to be as stable as the liquid electrolyte.

10.4.2
PLD $LiMn_2O_4$ Films

The first work on $LiMn_2O_4$ thin film as positive electrode produced by PLD reports on a thin (0.3-μm) and a thick (1.5-μm) crystallized film obtained without any postannealing treatment but using a heated substrate at 600 °C [92]. While the cycling behavior in liquid electrolyte was rather poor for the thicker film since only 28 μA h cm^{-2} $μm^{-1}$, i.e., 50% of the maximum capacity is recovered after 100 cycles between 4.25 and 3.8 V, promising results were found for the 0.3-μm-thick film with a stable capacity near 50 μA h cm^{-2} $μm^{-1}$ over 300 cycles. At the same time, very thin films, 100 nm thick, were examined as positive electrode and lithium-ion sensor [93]. The synthesis of stoichiometric $LiMn_2O_4$ thin films requires to carefully adjust the temperature substrate, the oxygen partial pressure, and the composition of the target [94–96]. For instance, the oxidation state for Mn determined by RBS and XAS is 3.5 ± 0.1 for an oxygen pressure of 0.2 mbar on a Pt substrate heated at 500 °C [94], but the well-crystalline and stoichiometric form of the spinel oxide can also be prepared using a target with Li in excess, $LiMn_2O_4$ + 15% Li_2O, on a silicon substrate heated as low as 300 °C in an oxygen pressure of 0.13 mbar [95]. In the latter case, a 300-nm film exhibits only a capacity of 33 μA h cm^{-2} μm, which is about half the maximum capacity expected for the extraction of one Li ion from the spinel structure. This capacity slowly declines at very low current density to reach 30 μA h cm^{-2} μm after 35 cycles between 4.2 and 3 V.

Structure, morphology, and electrochemical properties are strongly dependent on the temperature of the heated substrate during film growth [95–97]. It follows that a moderate substrate temperature of 400 °C at 200 mTorr of oxygen leads to a very dense film made of nanocrystals of about 100 nm with relatively rough surfaces (Figures 10.17 and 10.18) and constitutes a good compromise between the amorphous oxide and the highly crystallized compound with the best cycling behavior (▲) as shown in Figure 10.19 [96, 97]. Indeed, a nearly stable capacity of 60 μA h cm^{-2} $μm^{-1}$ is observed at 20 μA cm^{-2} for a 360-nm-thick film deposited on a stainless steel substrate and a capacity of nearly 45 μA h cm^{-2} $μm^{-1}$ is still available at 100 μA cm^{-2} over 200 cycles. An interesting point raised by results of [98] is the possible use of such films at 55 °C in liquid electrolyte. Figure 10.20 shows an attractive capacity retention from cycle 1 to cycle 100 at 20 μA cm^{-2} and from cycle 101 to cycle 500 with an average capacity loss rate of 0.07% per cycle.

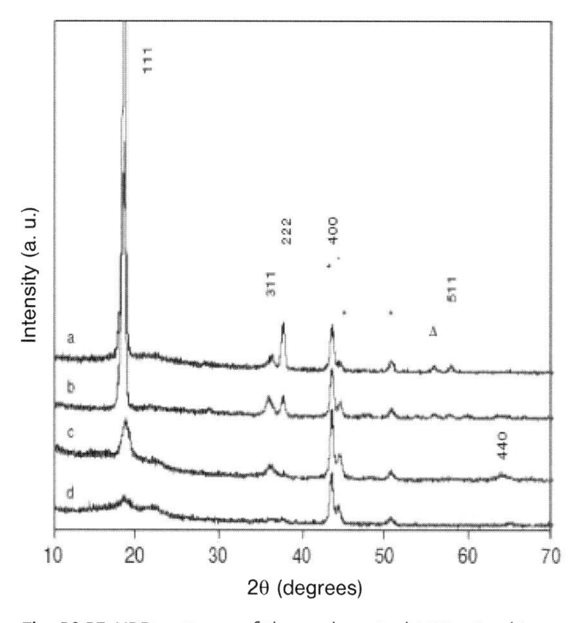

Fig. 10.17 XRD patterns of the as-deposited $LiMn_2O_4$ thin films on stainless steel substrates at different temperatures and oxygen pressures; (a) 625 °C and 200 mTorr, (b) 600 °C and 100 mTorr, (c) 400 °C and 200 mTorr, and (d) 200 °C and 200 mTorr; *, substrate. From [96].

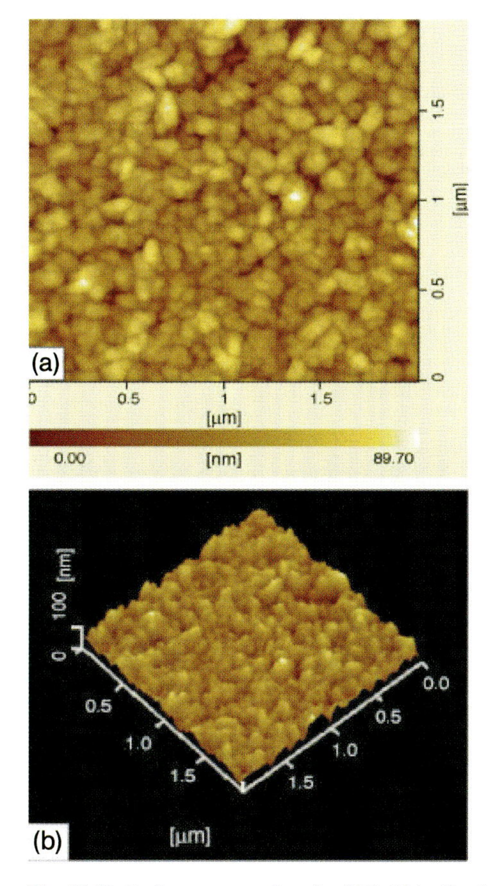

Fig. 10.18 Surface topography of a PLD $LiMn_2O_4$ thin film deposited on stainless substrate grown at 400 °C and 200 mTorr of oxygen. From [97].

No structural damage is found from XRD and Raman examination after 500 cycles between 4.5 and 3 V. The nanocrystalline film is thought to inhibit the Jahn–Teller effect and to limit the structural stress induced by the Li extraction/insertion reaction, and is not very sensitive to overdischarge cycles. The crucial aspect of the interface between the liquid electrolyte and the film oxide has been investigated by AC impedance spectroscopy [99, 100]. The charge-transfer resistance first decreases with increasing voltage before reaching a minimum value near 4 V, then increases up to a potential of 4.2 V; however, the meaning of such R_{ct} variation versus the potential has not been established [99]. Another impedance analysis as a function of time, state of charge, and cycling history suggests the disproportionation reaction of part of $LiMn_2O_4$ into $Li_2Mn_2O_4$ at the surface and a lithium deficient oxide $Li_{1-\delta}$ Mn_2O_4 in the rest of the film that explains the degradation of $LiMn_2O_4$ electrodes [100].

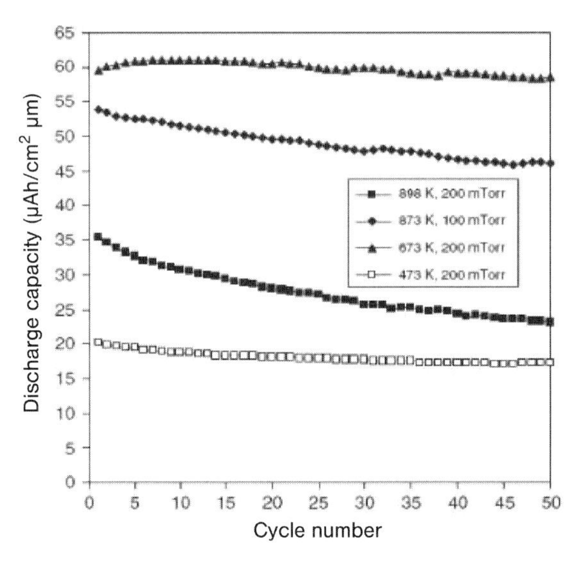

Fig. 10.19 Evolution of the discharge capacity versus cycle number of PLD $LiMn_2O_4$ thin films, 360 nm thick, as a function of the substrate temperature and oxygen pressure; $j = 20\ \mu A\ cm^{-2}$. From [96].

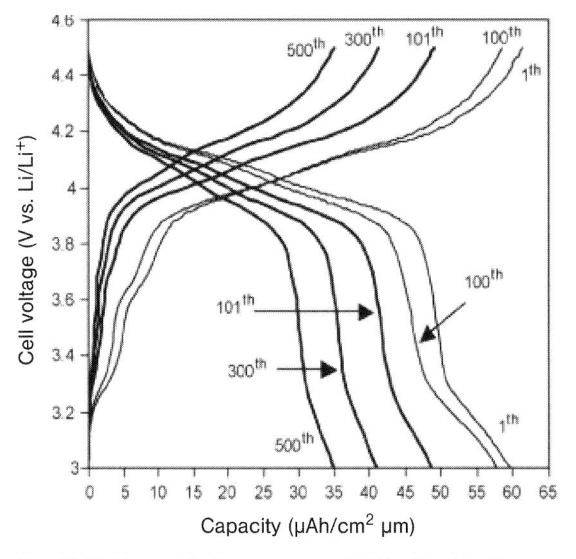

Fig. 10.20 Charge/discharge curves of $LiMn_2O_4$ thin films during cycling experiments at 55 °C between 4.5 and 3 V; $j = 20\ \mu A\ cm^{-2}$ for cycles 1–100 and $100\ \mu A\ cm^{-2}$ for cycles 101, 300, and 500. From [98].

Only one paper indicates that LiPON electrolyte can be successfully used with a PLD $LiMn_2O_4$ thin film, only 60 nm thick in an all solid-state battery [101]. A stable capacity of 2.5 μA h cm^{-2} (41 μA h cm^{-2} $μm^{-1}$) is achieved over 500 cycles at 1 μA cm^{-2} in the 4.25–3.25 V potential range.

An LIB prepared only by PLD has been fabricated using an $LiMn_2O_4$ thin film with a postannealing treatment (600 °C), an amorphous LVSO electrolyte ($Li_{2.2}V_{0.54}Si_{0.46}O_{3.2}$), and an amorphous SnO anode respectively 0.1, 1, and 0.15 μm thick [102]. The rechargeability of the battery was shown only for three cycles with a low capacity of about 1.5 μA h cm^{-2} (15 μA h cm^{-2} $μm^{-1}$) obtained between 3 and 1 V at a current density of 4.4 μA cm^{-2}.

10.4.3 ESD $LiMn_2O_4$ Films

It is clear that ESD did not induce electrochemical performance as high as that exhibited by PLD or sputtered films but this technique strongly contributes toward obtaining well-characterized films used for fundamental studies in terms of kinetics and mechanism.

In spite of a great effort to develop the ESD synthesis of $LiMn_2O_4$ films, only a few works deal with the cycling properties of such films and, once again, no structured discussion or comparison of experimental data with the data from other techniques can be found. Moreover, the number of papers devoted to the preparation of $LiMn_2O_4$ thin films is significantly higher than that observed for other 4- and 3-V metal oxide films. The electrochemical quartz crystal microbalance (EQCM) has been successfully applied to demonstrate [103–105] (i) the reversibility of the lithium insertion/extraction reaction in $LiMn_2O_4$ films, 200 nm thick, deposited on Pt or Au; (ii) the formation of a passivating layer during charge/discharge; and (iii) manganese dissolution during storage at 50 °C in liquid electrolyte. By combining the results of the EQCM technique and the *in situ* bending beam method (BBM) to measure the strain of the oxide, it has been possible to show that after removing of the initial surface film, the formation of a new surface film takes place in the potential range 4.03–4.1 V and during cycling [105].

Well-crystallized dense films can be obtained when deposition takes place on substrate heated at 400 °C or by using a heating treatment in the range 300–800 °C when the substrate temperature is as low as 300 °C. The main kinetic parameters of the Li-insertion reaction into ESD $LiMn_2O_4$ thin films have been widely investigated at room temperature and at 55 °C using EIS and cyclic voltammetry [106–111]. The exchange current intensity does not significantly change with the Li content and the chemical diffusion coefficient for lithium decreases when charging up to a minimum value for $x \approx 0.5$ before increasing up to the end of the charge. The variation of D_{Li} versus Li content, in particular, the higher value found for Li-rich compositions, is not explained or discussed in terms of structural response. The ESD technique allows to prepare crystalline $LiMn_2O_4$ films with a thickness in the range 0.1–1 μm with a rapid growing rate of 2 μm h^{-1}. Figure 10.21 displays the cyclic voltammograms at 0.5 mV s^{-1} for various film thicknesses and a linear

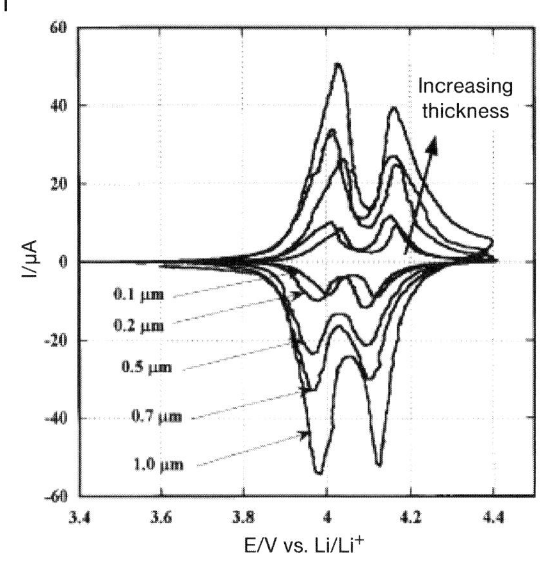

Fig. 10.21 Cyclic voltammograms at several thicknesses of $LiMn_2O_4$ films cycled between 3.4 and 4.4 V at 0.5 mV s^{-1} in 1 M $LiClO_4$/PC solution. From [110].

relation that almost passes through zero is observed for the 4-V anodic and cathodic peaks with a 98% efficiency in terms of rechargeability for all thicknesses [110].

The suppression of structural fatigue in films, which consists in the onset of the Jahn–Teller effect at the surface particles, can be performed by using substituted spinel oxides $LiMn_{2-y}M_yO_4$ where $y = 0.05$; 0.10 and M = Ni, Co [112]. Another solution consists in the deposition of a thin layer of $LiCoO_2$, a few tenths of nanometers thick, on the film to preserve the surface of the spinel oxide from the electrolyte and then to impede Mn dissolution during cycling at 50 °C, allowing a capacity gain of about 40–50% [113].

10.4.4
$LiMn_2O_4$ Films Prepared Through Chemical Routes

The solution-based route is an alternative procedure fort the deposition of thin-film $LiMn_2O_4$ cathodes owing to its simple, easy, and low-cost process. The solution containing the Li and Mn precursors is spin coated on Pt substrate and several layers are successively deposited and dried in the range 250–400 °C before a final annealing treatment between 600 and 750 °C takes place. Well-crystallized lithium manganese spinel oxide films are then obtained in the ranges 200–300 nm and 1–2 μm thick and a stable specific capacity corresponding to only ≈ 60% of that expected for the extraction of 0.8 Li is obtained over 100 cycles between 4.3 and 2.8 V [114–117].

A significant improvement in terms of electrochemical efficiency has been recently proposed by spin coating a chitosan- [118] or a polyvinylpyrolidone- (PVP-)

[119] containing precursor solution on a Pt or Au substrate followed by a two-stage heat treatment procedure. In both cases, attractive capacities close to the theoretical one are achieved; in particular 50 μA h cm^{-2} $μm^{-1}$ can be recovered after 200 cycles at 50 μA h cm^{-2} with a 1-μm-thick film [119]. This represents a significant capacity loss of 16% after 200 cycles. Close values are obtained with very thin film of Al-doped material [120]. From a mechanistic point of view, an *in situ* spectroscopic ellipsometry study has revealed the SEI layer thickness increased in proportion to a linear function of the number of cycles [121].

10.4.5 Substituted $LiMn_{2-x}M_xO_4$ Spinel Films

As shown for bulk materials, one excellent method for improving the cycle performance of $LiMn_2O_4$ is the substitution of other transition metals for Mn at the expense of the initial capacity. In other respects, it allows to significantly increase the working potential of the spinel material. Therefore, some attempts to improve the capacity and the cycle life of films have been made with the preparation of substituted spinel thin films with the composition $LiMn_{1.90}Ni_{0.10}O_4$ and $LiMn_{1.75}Co_{0.25}O_4$ using the PLD [122]. More interesting results can be achieved with 0.3–0.5 μm thick PLD $LiMn_{1.5}Ni_{0.5}O_4$ deposited on stainless steel substrate heated at 600 °C [123, 124]. Nonstoichiometric $LiMn_{1.5}Ni_{0.5}O_{4-\delta}$ exhibit a stepwise voltage profile near 4.7 V and a small plateau in the 4-V region (Figure 10.22), and then an attractive stable specific capacity of 55 μA h cm^{-2} μm available between 5 and 3 V. The good rate capability of these films is illustrated by Figure 10.23, indicating a high kinetics for Li diffusion, which is comparable to that of layered $LiCoO_2$. Thinner films (0.2 μm) can be synthesized by the ESD technique on gold substrate with a further annealing treatment at 700 °C [125]. In that case, a higher utilization is found, but the capacity retention seems slightly lower than with PLD films. A thin-film lithium-ion microbattery using a Ag thin-film anode coupled with a $Li_{1.2}Mn_{1.5}Ni_{0.5}O_4$ thin film in liquid electrolyte has been reported to deliver a stable capacity of ≈ 25 μA h cm^{-2} at 4.8 V over 2500 cycles [126].

10.4.6 Conclusion

$LiMn_2O_4$ thin films are characterized by relatively low practical capacity in the order of 30 μA h cm^{-2} during cycling, whatever the deposition technique. This is mainly due to the thickness of the deposited films, which rarely exceeds 0.3 μm. ESD and spin-coating methods have been used to investigate the electrochemical Li-insertion mechanism, but the cycling behavior of as-deposited films is rather poor and calls for deeper investigation. The best cycling performances in liquid electrolyte are provided with RF-sputtered and PLD films, 200–360 nm thick, with 60 μA h cm^{-2} $μm^{-1}$ stable over 1000 cycles and 100 cycles for the RF-sputtered and PLD films, respectively. Hence the practical capacity is found to be limited to only 10–20 μA hcm^{-2}. Most of the thin films prepared by RF sputtering or PLD

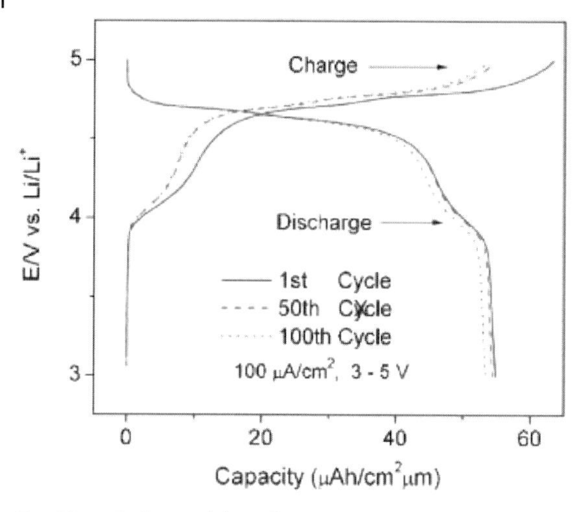

Fig. 10.22 Cycling stability of a PLD $LiMn_{1.5}Ni_{0.5}O_4$ between 3 and 5 V in 1 M $LiPF_6$/EC/DEC electrolyte. From [124].

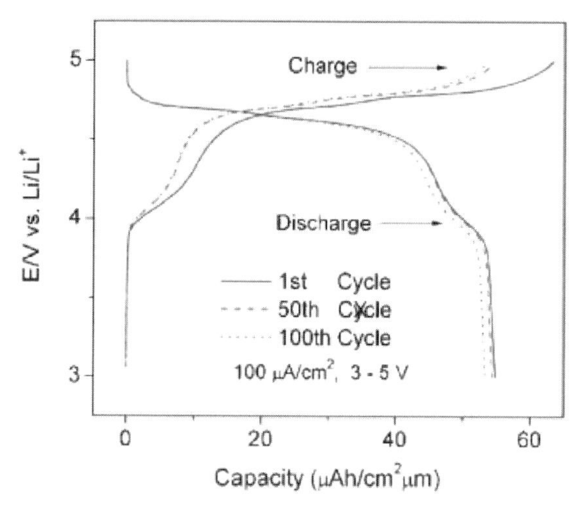

Fig. 10.23 Rate capability of a $LiMn_{1.5}Ni_{0.5}O_4$ thin-film electrode. From [124].

techniques exhibit low electrochemical efficiency, the reason for which needs to be investigated. Moreover, the poor cycling properties of films deposited by ESD and solution techniques are not yet understood even if the theoretical value of capacity is reached with these materials.

In terms of solid-state TFB, once again, very thin films (300 nm) have been integrated into a device using LiPON as electrolyte with a capacity of 50 μA h cm^{-2} μm^{-1} between 4.3 and 3.7 V and only 4% of capacity loss after 100 cycles at a current density of 100 μA cm^{-2}; such a cell can sustain high discharge rate up to 800 μA cm^{-2} without significant capacity decrease. One of the key factors that enhance the capacity of TFBs based on films is to make thicker films at least in the range 1–2 μm to compete with other thin-film electrode materials and to deeper study the fundamental aspects of the Li-insertion reaction and the cathode reactivity versus the electrolyte.

One attractive stimulating trend stems from the use of PLD $LiMn_{1.5}Ni_{0.5}O_4$ films characterized by a larger thickness (0.5 μm), a higher working voltage (4.5 V), and a good rate capability allowing to get 92% of the maximum capacity at 3 C rate. A stable capacity of 55 μA h cm^{-2} μm^{-1} over 100 cycles is thus obtained at 100 μA cm^{-2} between 5 and 3 V in liquid electrolyte.

10.5
V_2O_5 Thin Films

Vanadium pentoxide is probably the most widely investigated cathode material for TFB. Its layered and open structure (Figure 10.24) makes this oxide well suited for electrochemical insertion reactions with Li ions owing to the equation: $V_2O_5 + x\bar{e} + xLi^+ \leftrightarrow Li_xV_2O_5$ with $0 < x \leq 3$. Indeed vanadium pentoxide is attractive because of its large specific capacity (theoretical value: 440 mA h g^{-1}) in the potential range 3.8–1.5 V and a postannealing treatment, which, when it is required, takes place at moderate temperatures (<300 °C). For one lithium ion reversibly inserted and extracted, the maximum specific capacity of a V_2O_5 thin film is in the range 49–37 μA h cm^{-2} μm^{-1} whether the film is very dense or porous. Another reason is that this oxide can be prepared using various chemical and physical synthesis routes including RF and DC sputtering, PLD, thermal evaporation, CVD, atomic layer deposition, electrodeposition, thermal oxidation, and the sol–gel method. Finally, both the amorphous and crystalline phases are of interest with different potential ranges and specific electrochemical properties.

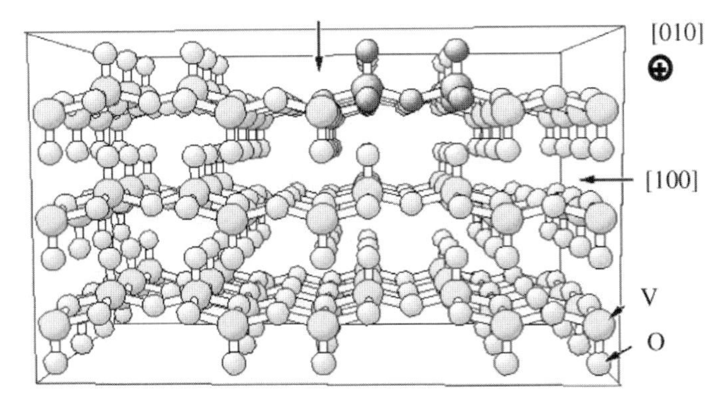

Fig. 10.24 Crystal structure of orthorhombic V_2O_5 in the *ac* plane. The stacking of V_2O_5 layers can be viewed in the 00l direction; the Li diffusion takes place between the oxide layers along the *b*-axis, perpendicular to the plane of the sheet.

10.5.1
Sputtered V_2O_5 Thin Films

Vacuum processes such as sputtering are known to allow low-temperature deposition of vanadium oxides, which is considered to be beneficial to on-chip application. Sputtered V_2O_5 thin films are of great interest since the most attractive solid electrolyte, LiPON is also deposited by RF sputtering for fabricating all solid-state devices.

10.5.1.1 Liquid Electrolyte

In spite of scarce data, several works dealing with electrochemical properties of crystallized V_2O_5 (c-V_2O_5) thin films prepared by RF or DC sputtering and investigated in liquid electrolytes clearly demonstrate the practical interest of crystalline films [127–135]. Some discrepancies in the results outline the need to rigorously control the experimental deposition conditions owing to their influence on morphology, structure, and electrochemical performances of c-V_2O_5 films. Some data about the influence of the preferred orientation and morphology can be found in [129, 132, 133, 135].

Here we describe our recent results, which illustrate the influence of the preferred orientation for vanadium oxide films deposited on a nickel substrate by DC or RF magnetron reactive sputtering using a vanadium metal target [132–134].

The experimental DC sputtering conditions for the deposition of four vanadium oxide films are summarized in Table 10.1. The thickness of all samples was 800 nm except for sample A, 550 nm thick. The V_2O_5 mass was determined by weighing experiments and by analysis for vanadium content (atomic absorption spectrometry) after dissolution of the deposit in 2 M H_2SO_4. Figure 10.25 shows the XRD diffraction patterns of vanadium oxide films as obtained. Depending on the sample, seven diffraction lines corresponding to the orthorhombic V_2O_5 phase (space group *Pmmn*) appear [JCPDS 41–1426], the 200, 001, 101, 110, 400, 002 and 600 diffraction peaks. In Table 10.2, the values of the lattice parameters of the orthorhombic cell for the different samples are reported. These values are consistent with that found for the bulk oxide. However, a comparison with the bulk data shows that the films have an enlarged c parameter (interlayer separation) of $\approx$4.42 Å which could indicate a mechanical stress built into the film during its cooling to room temperature and/or a small deviation from the ideal V_2O_5 stoichiometry [136].

Table 10.1 Preparation conditions for DC sputtering of V_2O_5 thin films deposited on Ni foils.

Sample	DC power (W)	O_2 flow rate (sccm)	O_2 partial pressure (Pa)
A	1000	40	0.16
B	1000	80	0.45
C	500	40	0.28
D	500	80	0.50

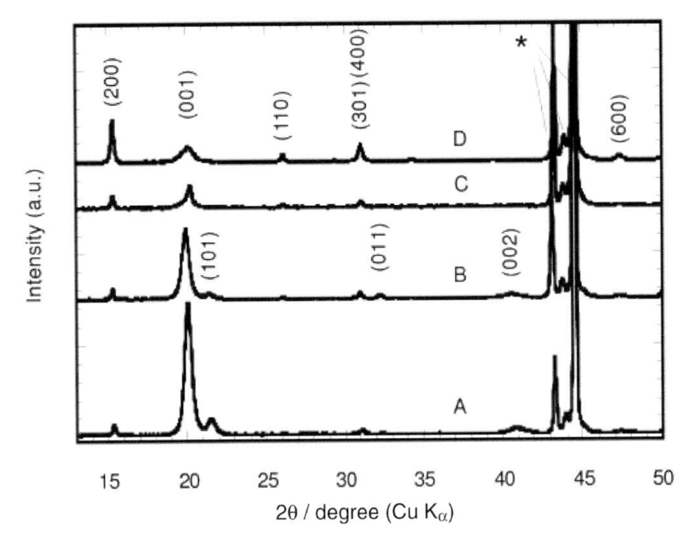

Fig. 10.25 XRD patterns of 800-nmV_2O_5 thin films deposited on Ni foil by DC sputtering (DC power = 1000 W for samples A, B and 500 W for samples C, D; * (asterisk), Ni substrate). From [133].

Table 10.2 Lattice parameters of the orthorhombic unit cell of V_2O_5 thin films deposited as described in Table 10.1.

Sample	*a* (Å)	*b* (Å)	*c* (Å)	*I*(200)/*I*(001)
A	11.495	–	4.425	0.07
B	11.520	3.558	4.443	0.14
C	11.514	3.567	4.396	0.53
D	11.513	3.569	4.414	2.92
Polycrystalline powder	11.516	3.565	4.372	0.33

Whatever the oxygen flow, the use of a DC power of 1000 W leads to the formation of well-crystallized V_2O_5 films strongly oriented with the *ab* planes parallel to the substrate (samples A and B). Conversely, in regard to samples C and D, the unusually high intensity of the 200 and 400 lines compared with that of the 001, as well as the presence of the 110 line, clearly indicate another kind of preferred orientation of the deposited thin film. The ratio in intensities $I(200)/I(001)$ for samples C and D increases from 0.33 for the bulk material to 0.53 and 2.9, respectively. In that case, the crystallites of V_2O_5 films preferentially grow with the *a* direction and the *ab* planes perpendicular to the substrate.

The specific morphology of each thin film elaborated at a DC power of 1000 and 500 W and corresponding to the two different preferred orientation, (00 l) for samples A, B and (*h*00) for C and D, is seen in Figure 10.26. At high power

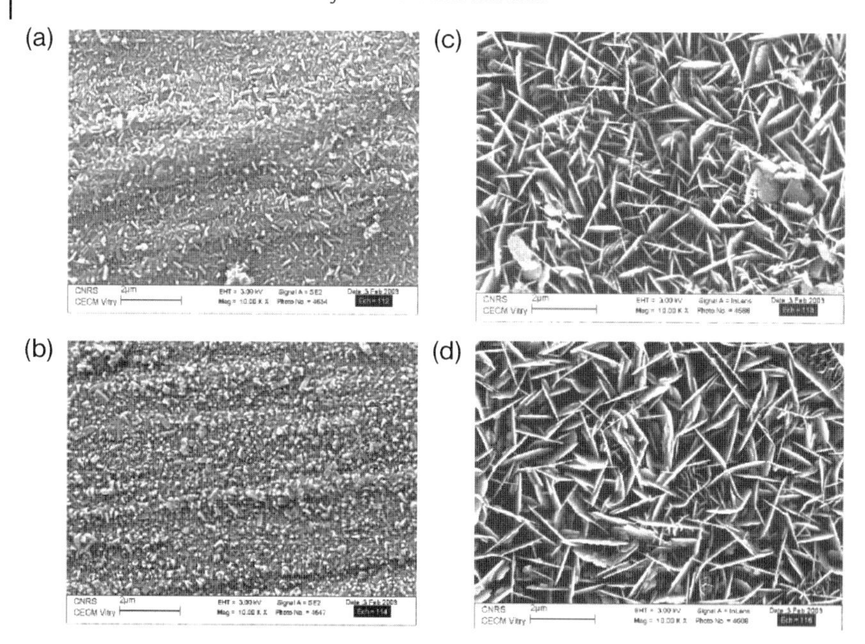

Fig. 10.26 SEM micrographs of 800 nm V_2O_5 thin films (samples A, B, C, and D). From [133].

(samples A, B), the deposit consists in a stacking of more or less long platelets 200 nm × 50 nm all arranged parallel to the substrate; the surface is made of platelets homogeneous in size. In that case, the deposit appears as relatively dense.

Samples C and D are viewed as platelets standing perpendicular to the surface of the substrate and from which the thin cross section, in the order of 20 nm, can be observed (Figure 10.26). This specific morphology is in good accord with XRD data presented above (the (h00) orientation) and induces a high porosity. The films preferentially have grown along the a direction, perpendicular to the substrate, and the stacking of the ab planes is probably performed on a very short distance that never exceeds 10–20 nm in the c direction.

The strong influence of the morphology and the preferred orientation is clearly demonstrated when the chronopotentiometric behavior of the h00 and 00 l deposits with the same thickness are compared at different current densities (Figure 10.27). Taking the deposit mass available into account, the faradaic yield achieved at slow rate corresponds to about 0.84–0.87 F mol^{-1}. The capacity exhibited by the h00 deposit is not significantly sensitive to the current density and is larger than that of the 00 l deposit when the C rate increases. Of practical interest is the capacity of 24 μA h cm^{-2} still available at 20 μA cm^{-2} without any significant polarization. Conversely, in the case of the 00 l deposit, both the working potential and the specific capacity strongly depend on the C rate with a drastic decrease in the capacity from 28 to 19 μA h cm^{-2} when the current density increases from 2 to 10 and 20 μA cm^{-2}. The promoting role of the (h00) preferred orientation was also pointed out by H. Miyazaki *et al.* [129] for 250-nm thin films.

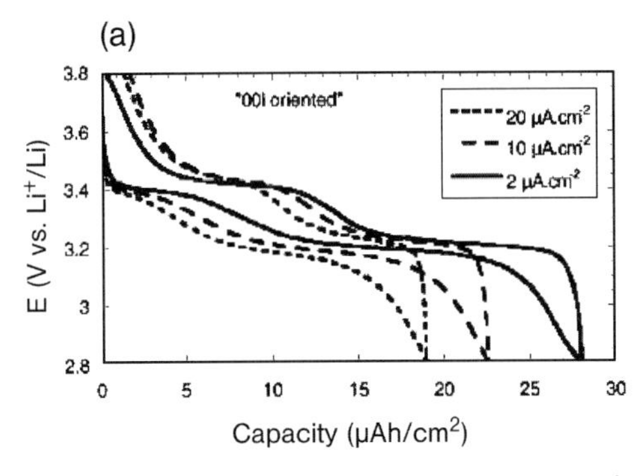

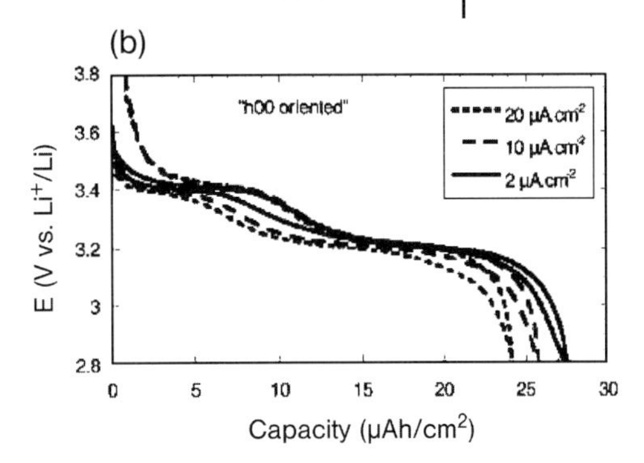

Fig. 10.27 Influence of the current density on the first discharge/charge cycle of 800 nm V_2O_5 thin films characterized by a (h00) and a (00 l) preferred orientation. (3.8–2.8 V). From [133].

Another important point of the specific properties of the V_2O_5 thin films is their structural response summarized in Figure 10.28. Indeed, while the electrochemical Li-insertion reaction in a V_2O_5 bulk sample is known to induce successive phase transitions corresponding to the $\alpha, \varepsilon, \delta$ phases with the composition ranges $0 \leq x \leq 0.15$, $0.3 \leq x \leq 0.7$, $0.9 \leq x \leq 1$ respectively, separated by two-phase regions [137], the emergence of a new phase when the electrochemical lithiation proceeds in the V_2O_5 thin film is not observed. Li insertion in (h00) oriented V_2O_5 thin film induces a linear expansion of the interlayer spacing from 4.396 to 4.68 Å as x increases from 0 to 0.95. At the same time, the a parameter decreases from 11.51 to 11.37 Å. Finally the δ phase characterized by a and $c/2$ parameters of 11.25 and 4.95 Å is not observed. All these results clearly show the structural response of the (h00) thin film, a solid solution behavior, differs from that exhibited by bulk materials. These results are consistent with scarce data reported on sputtered and sol–gel thin films characterized by a (00 l) preferred orientation [136, 138–140]. This finding is consistent with results obtained from a Raman study recently performed in our group (see the chapter 6 by R. Baddour-Hadjean *et al.*).

Raman microspectrometry has been used to investigate the local structural changes induced by the electrochemical lithium-intercalation reaction in crystalline sputtered V_2O_5 thin films in liquid electrolyte. The use of pure V_2O_5 thin film without any conductive and binding agent allows a homogeneous Li insertion in the material and a high quality of Raman signatures to be obtained. The Raman spectra of $Li_xV_2O_5$ compounds for $0 < x < 1$ are examined as a function of the lithium content and discussed in relation with the XRD data pertinent to these (h00)-oriented thin films and literature data. An assignment of all Raman bands is proposed and the Raman fingerprint of the ε-type phase whose interlayer distance continuously increases with x, is clearly evidenced all along the Li-insertion process: lithium ions rapidly produce an orthorhombic ε phase characterized by a vanadyl stretching mode at 984 cm^{-1} for $0 < x < 0.5$ and further Li accommodation induces a splitting into two stretching modes, the first one shifting from 984 to 975 cm^{-1}, the second from $x = 0.7$ being located at a fixed wavenumber of 957 cm^{-1}. Both

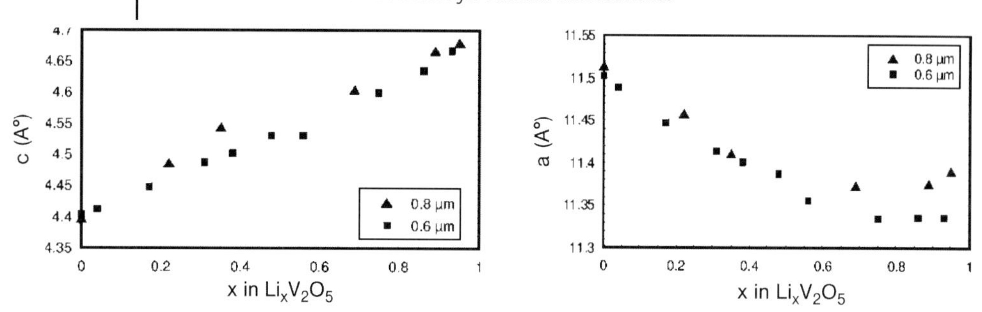

Fig. 10.28 Variation of the unit-cell parameters c and a of (h00) oriented $Li_xV_2O_5$ thin film as x increases for 800- and 600-nm-thick films. From [133].

modes are consistent with the local structure of the ε lithium-rich phase called ε' and reflect the existence of two different lithium sites. This work illustrates that the structural changes, in terms of long- range order and local structure, are strongly dependent on the microstructure and morphology of the material.

This structural response induces a high-rechargeable behavior for 800-nm-thick deposits investigated in liquid electrolyte. Such deposits exhibit specific capacities of 25 and 40 μA h cm^{-2} after 50 cycles at C/5 rate in the potential windows 3.8–2.8 and 3.8–2.15 V, respectively. One common method for optimizing the capacity available with V_2O_5 films consists in the use of the largest depth of discharge, allowing samples to reach a capacity between 2 and 3 Li per mole of oxide, i.e., by cycling in the potential windows 4–3.8 to 2–1.5 V. Nevertheless, this approach has never been successful [127, 128, 130, 141] and has resulted in a strong capacity decrease, especially for thickness higher than 1 μm [127, 128]. For instance, some authors report that for thick films (1.6 and 3.2 μm) the capacity dramatically decreases by 60% after only 30 cycles in the potential range 3.8–1.5 V at a moderate discharge/charge rate [128]. Furthermore, the sputtered films with thickness of over 800 nm are easily taken off from the substrate after a few cycles, which impedes the development of high-capacity thin-film electrodes. In short, no satisfactory electrochemical behavior has been reported as yet for crystalline V_2O_5 thin film with thickness higher than 800 nm [127, 128].

The possibility of using high-capacity thin films has been detailed in [134] for deposited films in the range 0.6–3.6 μm thick working in the 3.8–2.8 V voltage range. Whatever the thickness, the efficiency of the charge process is 100% and the specific capacity obtained linearly increases with the thickness, changing from 22 μA h cm^{-2} for 0.6 μm to 124 μA h cm^{-2} for 3.6 μm without any polarization.

The influence of the thickness and the discharge/charge rate on the capacity recovered is reported in Figure 10.29. The capacity value of $\approx$ 35 μA h cm^{-2} μm obtained at C/15 rate is very close to the theoretical value expected for these oriented and porous films. The specific capacities achieved are only slightly dependent on the C rate. This indicates that the kinetics of lithium transport into these oriented films is very high. At C/5 and C/15 rates, the maximum uptake of 0.94 Li^+ per

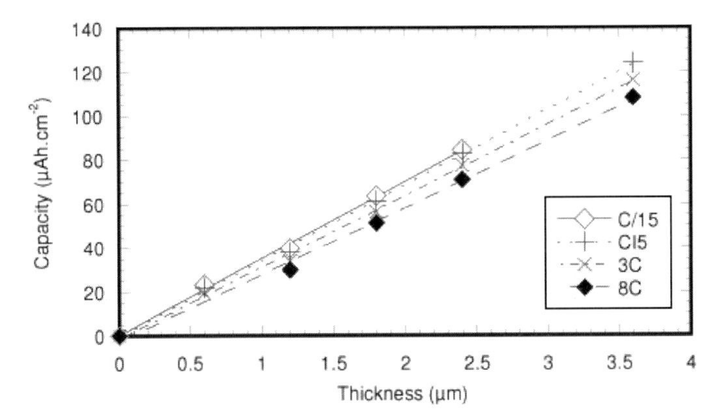

Fig. 10.29 Influence of the thickness and the discharge/charge rate on the capacity recovered in the voltage window 3.8–2.8 V. From [134].

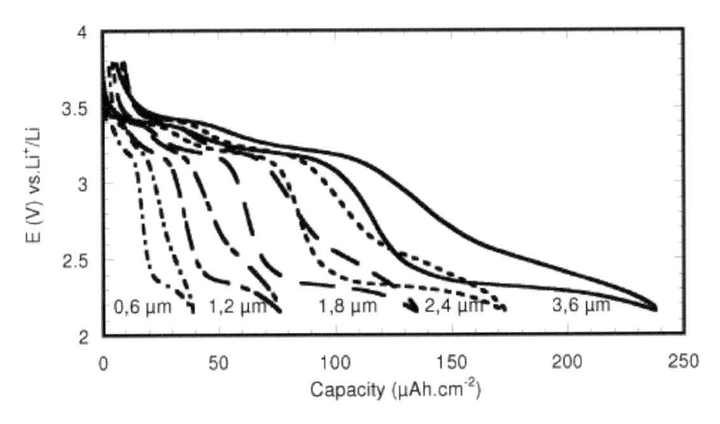

Fig. 10.30 Comparison of the discharge/charge profile of the sputtered V_2O_5 films in the extended potential range 3.8–2.15 V at C/2 rate. From [134].

mole of oxide is reached. The evolution of the specific capacity as a function of the thickness for various discharge/charge rates can be represented by a straight line and this linear relationship demonstrates that the high electrochemical efficiency of the films is due to a good homogeneity and a high porosity, which ensures a good contact between electrolyte and particles of active material. The same effect is practically observed in the enlarged 3.8–2.15 V potential window in an attempt to increase the specific capacity (Figure 10.30) since it regularly increases with the thickness from 40 µA h cm^{-2} for 0.6 µm up to 237 µA h cm^{-2} for 3.6 µm at C/5 rate.

Cycling tests on the thickest film (3.6 µm) between cycling limits 3.8–2.8 V at 100 µA cm^{-2} and between 3.8 and 2.15 V at 1 mA cm^{-2} have been reported (Figure 10.31). After a slow capacity decline, a high and stable capacity of 100 µA h cm^{-2} is achieved after 90 cycles at high voltage. The capacity of 150 µA h cm^{-2}

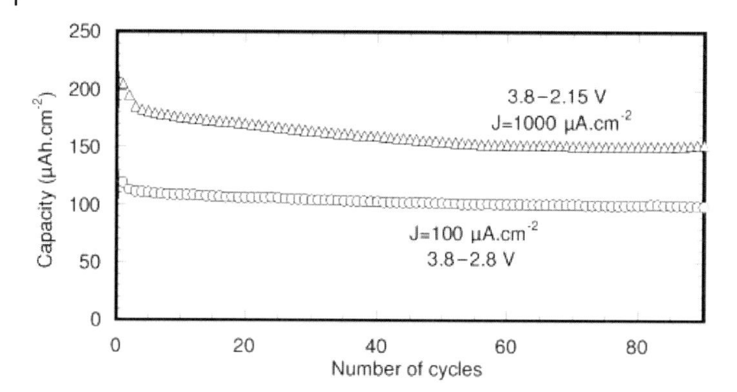

Fig. 10.31 Evolution of the specific capacity as a function of cycle number for a 3.6-µm-thick V_2O_5 film in two voltage ranges; (o) 100 μAcm^{-2}; 3.8–2.8 V (D) 1 mA cm^{-2}; 3.8–2.15 V. From [134].

found to be stable from the 50th cycle in the extended voltage range illustrates the performance of such a material at high current density (8 C), as expected from the good rate capability demonstrated in Figure 10.28. These results overcome those reported for thin or thick crystalline V_2O_5 films investigated in the same or wider potential window in liquid electrolytes [128, 142] and indicate an excellent adherence to the substrate. Between 3.8 and 2.15 V, at least half the capacity loss can be explained by Li trapping in the material [131, 133].

Figure 10.32 reports two typical Nyquist plots for the 2.4-µm thick and the thinnest film (0.6 µm) with the same composition $Li_{0.4}V_2O_5$. Two major points must be outlined: (i) the charge-transfer resistance decreases by approximately a

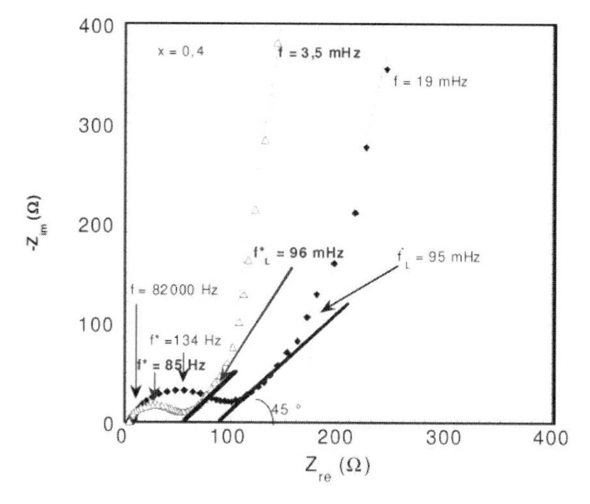

Fig. 10.32 AC impedance diagrams for $Li_xV_2O_5$ films 600 nm (♦) and 2.4 (△) µm thick; $x = 0.40$. (Our work.).

factor two but the characteristic frequency does not greatly change. (ii) For each Li content, the limiting frequency has practically the same value for the thin (0.6 μm) and the thick (2.4 μm) film. Therefore, it can be thought that the diffusion pathway does not correspond to the height of the platelets but to the length of the edges, i.e., as expected along the *b* direction ($L \approx 1$ μm) (see Figure 10.26). Assuming the Li diffusion takes place along the thickness would imply that for the same Li content in the film, the limiting frequency ($f^*_L = D_{Li}/L^2$) for the 0.6- and 2.4-μm thick films should be different by a factor 16, whereas, in fact, practically the same value is found for f^*_L. The little change in the capacity and the polarization reported for these films can be then understood to be due to a fast mass transport phenomenon that always takes place along the *b* direction, i.e., along the direction parallel to the edges of V_2O_5 platelets which practically have the same size whatever the thickness. In that case, the time required for Li to homogeneously diffuse in the host material is the same for the sputtered films investigated here in the range 0.6–3.6 μm thick. This could explain the excellent rate capability and good cycling properties of these (*h*00) and (110) oriented films. The activation energy for Li diffusion has been calculated from the slope of the straight line – Ln $D_{Li} = f(1/T)$ and a value of 0.98 eV is found. This value is in good agreement with previous studies on the diffusivity of lithium in V_2O_5 crystal using molecular dynamics simulation [143]. Indeed, the activation energy for the Li diffusion along the *b*-axis was found to be equal to 0.87 against 2.47 eV along the *c*-axis.

Depending on the deposition parameters, a heat treatment can be required from 200 °C [130] to get crystalline V_2O_5 films. The effect of total gas and oxygen partial pressure can allow to get amorphous or crystallized films from a V_2O_5 target under various Ar–O_2 atmospheres [144–146] without heat treatment. Amorphous films exhibit a smooth surface and are quite dense in comparison with the crystalline films. The typical chronopotentiometric behavior for amorphous films consists in a featureless shape and the cycling limits are 3.8–1.5 V for a maximum uptake of 2 Li per mole of oxide. A better stability of the capacity with cycles is reported for amorphous films with about 200 mA h g^{-1} over 50 cycles (Figure 10.33).

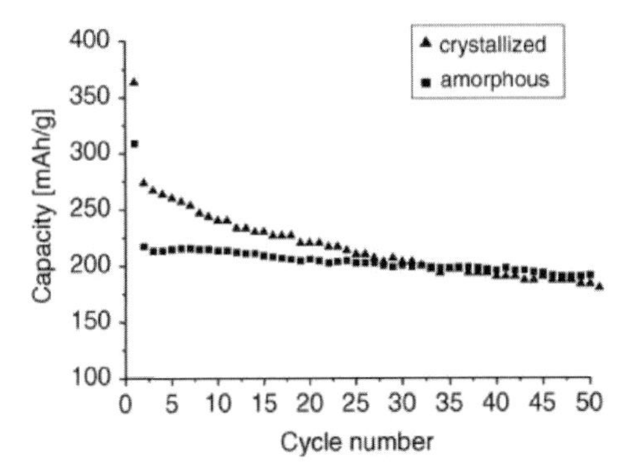

Fig. 10.33 Evolution of the discharge capacity versus cycle number between 3.7 and 1.5 V for sputtered crystallized and amorphous thin films; $j = 15$ μA cm^{-2}. From [145].

XPS analyses of crystallized and amorphous V_2O_5 (a-V_2O_5) films indicate that an interface is growing upon the discharge and partially dissolves during the charge, the layer being composed by Li_2CO_3 and alkyl carbonate. Amorphous $MoO_3-V_2O_5$ and $In_2O_3-V_2O_5$ composite films have been prepared by the cosputtering method using two different corresponding targets [147, 148]. Some results are of interest with $MoO_3-V_2O_5$ composite films, 200–250 nm thick [147]; for instance, a stable capacity of 80 μA h cm^{-2} μm^{-1} is obtained after 70 cycles between 3.9 and 1.5 V at 20 μA cm^{-2}.

10.5.1.2 **Solid-State Electrolyte**

Rechargeable thin-film microbatteries were first fabricated with sputtered V_2O_5 thin films 10 years ago [88, 149, 150]. As far as the c-V_2O_5 thin films are concerned, evaluation of such cathodes has been directly performed in all solid-state devices using LiPON as electrolyte at room temperature or PEO-based electrolytes at 100 °C [88, 149–151]. In both cases, the results probably included transport problems across the electrolyte–cathode interface. In the case of the c-V_2O_5–LiPON system, an excessive polarization was observed in comparison with that achieved with amorphous V_2O_5 (a-V_2O_5). Therefore, most efforts were focused on a-V_2O_5 thin films. In the case of the c-V_2O_5/PEO–$LiCF_3SO_3$ system at 100 °C, the discharge capacity reported in the potential range 4–1.5 V strongly decreases by 50% and more after 25 cycles at C/15 rate at a current density of 50 μA cm^{-2}. The authors reported crystallized solid films do not possess the necessary compliance to accommodate changes in volume of the electrode material during deep cycling.

In the typical TFB constructed in Oak Ridge National Laboratory, LiPON is used as solid electrolyte. The solid electrolyte LiPON corresponds to the chemical compositions $Li_{3.1}PO_{3.8}N_{0.16}$, $Li_{3.3}PO_{3.8}N_{0.22}$, and $Li_{2.9}PO_{3.3}N_{0.46}$ characterized by a conductivity of about 2×10^{-6} S cm^{-1}. The amorphous cathode material and solid electrolyte are 1 μm thick and the thickness of the evaporated lithium anode is about 5 μm. A discharge capacity of ca 30 μA h cm^{-2} is obtained between 3.75 and 2.75 V but it can be extended to 130–140 μA h cm^{-2} μm^{-1} when the potential range is enlarged to 3.6–1.5 V [150]. Typical discharge curves for a 0.13-μm-thick film are reported in Figure 10.34 as a function of current density. Long-term cycling experiments with a thinner film, 0.13 μm thick $\times$ 1.21 cm^2, give capacities that strongly depend on the depth of discharge: 12 and 4 μA h cm^{-2} in the potential windows 3.4–1.5 V and 3.1–2 V with a capacity loss decreasing from 0.15 to only 0.034% per cycle [88]. This continuous capacity loss was reported to be associated with the entrapment of Li into the film oxide. More recently XRD and TEM examinations have demonstrated that the gradual degradation of the capacity of TFBs was induced by the occurrence of V_2O_5 crystallites embedded in the LiPON and V_2O_5 materials [152].

The doping effect of a-V_2O_5 thin film by silver using the simultaneous cosputtering of V_2O_5 and silver from a pure vanadium metal and pure silver metal targets has been evaluated [153, 154]. The as-prepared films are amorphous and a gain in the specific capacity by 30% per V_2O_5 is claimed for the compound containing 0.8 Ag per mole of oxide leading to 80 μA h cm^{-2} μm^{-1}, i.e., about 13 μA h cm^{-2} for

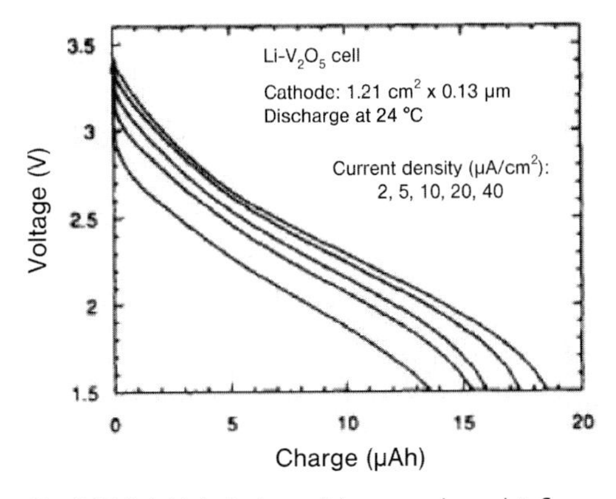

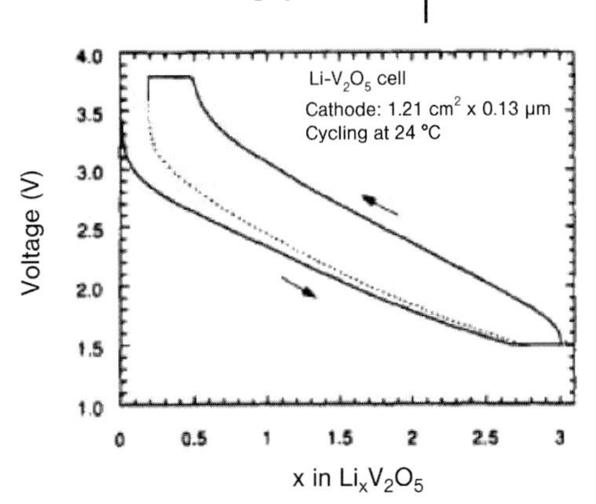

Fig. 10.34 Initial discharge/charge cycle and influence of the current density on the discharge curves of a Li/LiPON/a-V_2O_5 (0.13 μm thick). From [88].

Table 10.3 Electrochemical characteristics of some solid-state Li/LiPON/a-V_2O_5 TFBs as a function of the V_2O_5 film thickness.

Film thickness (μm)	Potential range (V)	Current density (μA cm^{-2})	Capacity (μA h cm^{-2})[a]	Ref.
1	3.75–2.75	20	30(n20)	[150]
1	3.60–1.50	15	130(n8)	[150]
0.13	3.40–1.5	10	10(n100)	[88]
0.13	3.1–2	10	3(n300)	[88]
0.3	3.6–1.5	20	15(n450)	[152]
0.27	3.6–2.7	20	3.5(n250)	[155]
0.16	3.6–1.5	20	13(n200)	[154]

[a] n is the cycle number for which the capacity is given.

a 0.16-μm-thick film [154]. In fact, the active material is made of a mixture of AgO and Ag_2O embedded in the V_2O_5 framework. Table 10.3 summarizes the main data related with the electrochemical performance of amorphous V_2O_5 thin films used with LiPON as solid-state electrolyte.

Rocking chair TFBs have been proposed with the use of two amorphous V_2O_5 and $Li_xV_2O_5$ thin films used as cathode and anode materials, the anode being electrochemically lithiated before operating the cell [156]. Such a cell with positive and negative electrodes 0.3 μm thick exhibits a stable but limited capacity of 6 μA h cm^{-2} over a few hundreds of cycles between 3.5 and 1 V. Another kind of lithium-free metal TFB has been successfully fabricated using a V_2O_5 film 0.3 μm thick as a negative electrode and a $Li_{2-x}Mn_2O_4$ film 0.8 μm thick as a positive electrode [157]. Large-sized TFBs batteries 100 mm × 100 mm can be then

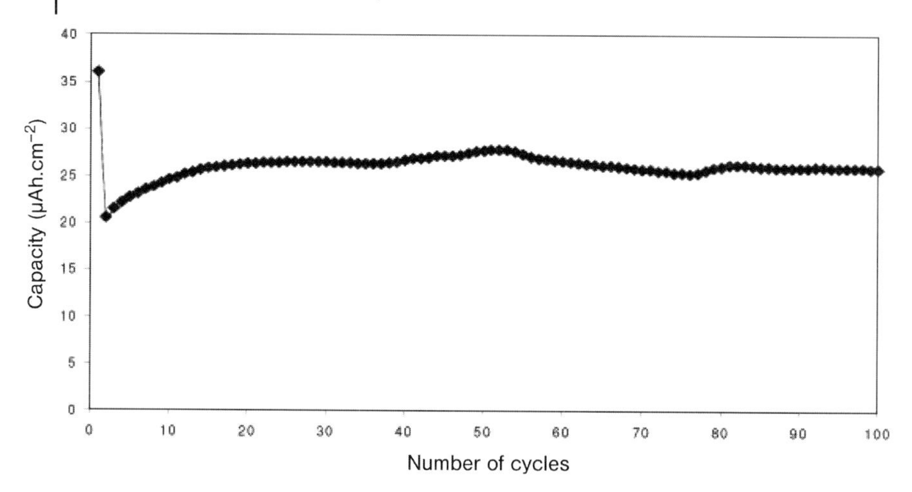

Fig. 10.35 Cycling behavior of a c-V_2O_5 (1 μm)/LiPON (1.4 μm)/Li (3.5 μm) TFB between 3.8 and 2.15 V at 100 μA cm^{-2}. (Our work).

fabricated with a typical charge/discharge capacity of about 0.9 mA h with a good rechargeability between 3.5 and 0.3 V.

Figure 10.35 illustrates the cycling performance of a V_2O_5 (1 μm)/LiPON (1.4 μm)/Li (3.5 μm) battery at 100 μA cm^{-2} in the potential range 3.8–2.15 V, with a crystalline V_2O_5 dense deposit obtained by RF magnetron sputtering using a vanadium target. The stable specific capacity of 26 μA h cm^{-2} compares very well with the results quoted for the best amorphous phases. In other respects, at a lower current density of 10 μA cm^{-2}, a large capacity of 50 μAh cm^{-2} can be obtained upon cycling in a higher voltage range than the amorphous phase, which makes c-V_2O_5 also attractive for TFBs.

10.5.2
PLD V_2O_5 Thin Films

Crystalline (*h*00) or (00 l) oriented V_2O_5 thin films can be prepared at temperature as low as 200 °C in an oxygen environment when a V_6O_{13} target [141, 158] or a V_2O_5 target [159–161] is used while an amorphous phase is obtained at room temperature. However the efficiency of the Li-insertion process in (*h*00) textured films tested in liquid electrolyte in the range 0.15–0.8 μm thick does not seem as high as in *h*00 sputtered films since the maximum capacity is between 50 and 80% of the theoretical value [141, 158] depending on the potential range. However, amorphous films exhibit an attractive and stable capacity corresponding to 1.2 F mol^{-1} when cycled between 4.1 and 1.5 V.

Surprisingly, no attempt to apply the PLD V_2O_5 films in a TFB has been undertaken except with a thin composite AgV_2O_5 film cathode, 100 nm thick, using LiPON electrolyte prepared by electron-beam reaction evaporation and evaporated Li as anode [162]. A low capacity of 40 μA h cm^{-2} μm^{-1}, i.e., only 4 μA h cm^{-2} μm^{-1} is recovered after 20 cycles at low current density.

10.5.3 CVD V_2O_5 Films

CVD has been rarely used to deposit V_2O_5 thin films [163–165]. Of interest is the fabrication of LiV_2O_5 thin films with a high stable capacity of about 200 mA h g^{-1} over a lot of cycles. V_2O_5 films are synthesized by CVD from pure or diluted $VO(OC_3H_7)_3$ precursors and in most cases a heat treatment of the deposits is needed, for instance at 500 °C, to get well-crystallized oxides. Nevertheless, the films suffer from a low electrochemical efficiency, of about 50% of the expected value, in spite of a good reversibility in the 3.8–2.8 V range.

Atomic layer epitaxy which is a technique derived from CVD, consists in a sequential introduction of the reaction precursors. This procedure allows to control the growth on the substrate at the atomic layer level, hence leading to an excellent coverage and uniformity of the deposit. As-grown films are amorphous. This method has been applied to prepare c-V_2O_5 films 700 nm thick, obtained after an appropriate heating at 400 °C in air and they can incorporate Li reversibly (Figure 10.36) between 3.8 and 3 V [166, 167].

10.5.4 V_2O_5 Films Prepared by Evaporation Techniques

Of the evaporation techniques, *thermal evaporation* seems to be more appropriate [168–172] than the *electron-beam evaporation* which has received little attention [173]. As-deposited V_2O_5 films obtained by thermal evaporation of a pure oxide powder are amorphous. A 200-nm-thick film deposited on ITO substrate is reported to exhibit an important capacity fading when operating between 4 and 1.8 V in conventional liquid electrolyte [168]. This strongly contrasts with the cycling stability claimed in [145] for a sputtered amorphous film evaluated in liquid electrolyte. A protective $LiAlF_4$ layer of similar thickness also deposited by thermal evaporation on the cathode material allows a considerable improvement of the cycle life. Indeed, a high capacity of 60 μA h cm^{-2} μm^{-1} (i.e., 12 μA h cm^{-2}) is reached over 800 cycles. A deep insight into the kinetics of Li insertion has been performed for crystallized films, 1.6 and 3.6 μm thick, obtained after annealing at 400 °C in air and the high reversibility of the reaction has been established [169, 170].

Two original examples of producing a TFB device with the idea of avoiding the presence of the lithium metal anode for many reasons explained in the introduction section are known. A rocking chair TFB device has been proposed with a V_2O_5 cathode (100 nm) on Au and an Li-rich V_2O_5 anode (8 Li/V_2O_5), 300 nm thick, prepared by thermal evaporation of lithium onto V_2O_5 layers at room temperature [171]. Only the LiPON solid electrolyte was deposited by RF magnetron sputtering and a stable but low capacity of 3.33 μA h cm^{-2} is obtained between 3.5 and 1 V

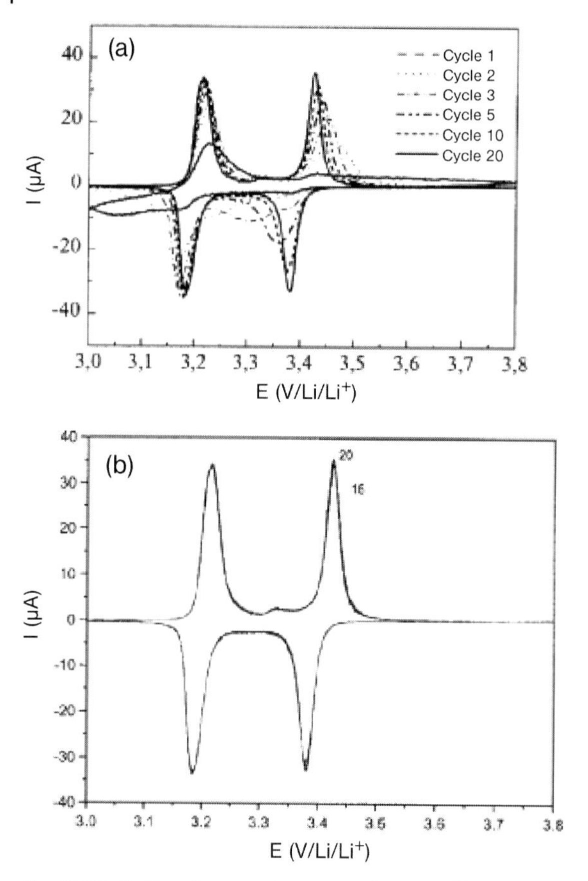

Fig. 10.36 Cyclic voltammetric curves of c-V_2O_5 films 700 nm thick deposited by atomic layer epitaxy and heat-treated at 400 °C. From [166].

over a few thousands of cycles at 10 μA cm^{-2}. A TFB with an initial configuration stainless steel/LiPON, 1.5 μm/$Li_{1.3}V_2O_5$, 500 nm/Cu avoids the presence of Li anode since metallic Li is electroplated at the Cu current collector during the first charge [172]; a stable capacity of 40 μA h cm^{-2} μm^{-1} is obtained between 3.8 and 1.8 V at 0.1 mA cm^{-2}.

10.5.5 V_2O_5 Films Prepared by Electrostatic Spray Deposition

The ESD technique has been applied to synthesize V_2O_5 films evaluated as cathode materials [174]. Such films are amorphous and require a heat treatment in the range 200–275 °C to crystallize. Good cyclability and high capacity (270 mA h g^{-1}) are claimed at C/5 rate in liquid electrolyte but their discharge/charge profile does

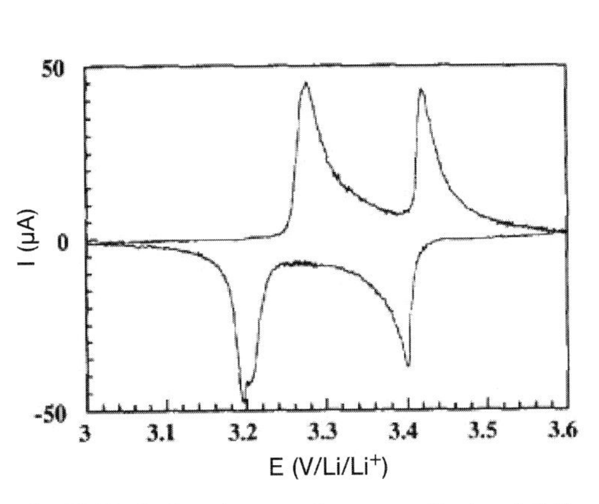

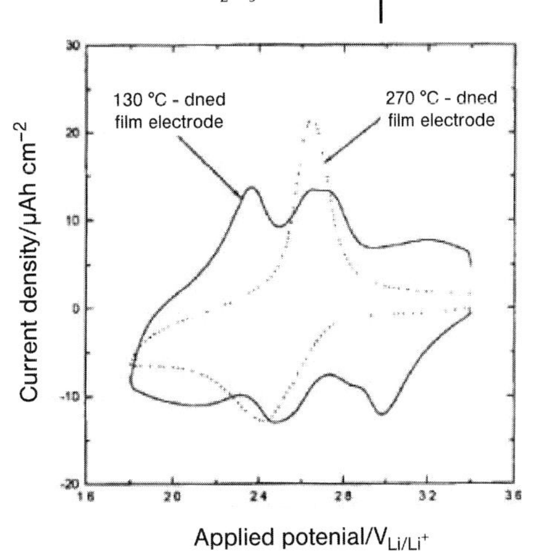

Fig. 10.37 Cyclic voltammetric curves of sol–gel V_2O_5 films obtained from the heat treatment of VXG at 400 °C (10 μV s^{-1}) from [139], 270 °C and 130 °C (at 1 mV s^{-1}). From [176].

not correspond to that expected for a c-V_2O_5. The thickness not being indicated, any comparison with other films made by other techniques is difficult.

10.5.6
V_2O_5 Films Prepared via Solution Techniques

The sol–gel process allows to get films of the xerogel V_2O_5, 1.6 H_2O (VXG) from the dehydration of the corresponding gel at room temperature. The thermodynamic, structural and kinetic study of the Li-insertion reaction in the VXG [175] and heat-treated forms [176] has been reported, but the low electronic conductivity of the former hinders the use of high current densities. This method, using aqueous or organic medium, has been demonstrated to be suitable for producing thin films of crystalline V_2O_5 with attractive properties. The crystallinity of the films is dependent on the temperature of heat treatment applied to the VXG film as illustrated by the cyclic voltammetric curves in Figure 10.37. Structural changes in the $Li_xV_2O_5$ films have been characterized in detail by XRD and XPS studies [139, 140, 177].

Films of $V_{1.8}Mo_{0.2}O_{5.1}$ xerogel were fabricated by spin coating or casting technique using a solution of molybdenum-doped polyvanadate [178]. This material (0.4 μm thick) shows a capacity of 14 μA h cm^{-2} at the current density of 25 μA cm^{-2} between 3.5 and 2.2 V and cast films allow to prepare thick films up to 18 μm for high specific capacities of about 480 μA h cm^{-2} stable over 30 cycles.

The oxidative way has been used to produce thin films of vanadium pentoxide from a vanadium foil [179–182]. The electrochemical oxidation of vanadium metal performed in aqueous medium first at high constant current density and at controlled potential in a second step leads to an amorphous compound and a heat treatment at 400 °C is required to get c-V_2O_5 films, 100–200 nm thick [179]. Li-insertion kinetics in such films has been studied in detail. The chemical oxidation of V at 500 °C at a high oxygen pressure results in the growth of vanadium oxide (200 nm thick) with an outer layer made of c-V_2O_5, 100 nm thick, and the inner oxide layer being constituted of VO_2 and V_6O_{13} [179–182]. An extensive XPS study of the Li-insertion mechanism has been performed and the reasons for capacity fading in long-term cycling experiments between 3.8 and 2.8 V are examined. Li insertion is shown to be partially irreversible with an considerable amount of vanadium ions remaining in the V^{4+} state after charging at 3.8 V [182] and which increases with cycles. The formation of a SEI layer including lithium carbonate and Li-alkoxides and/or Li-alkyl carbonates is pointed out. Hence cycling properties of such films are affected by Li trapping in the film at the grain boundaries combined with a loss of material by grain exfoliation.

10.5.7
Conclusion

From the wide range of techniques applied for V_2O_5 thin-film deposition, such as ESD, PLD, solution techniques, thermal evaporation, etc, a number of relevant results have been obtained by using DC or RF sputtering. All the works performed with other techniques suffer from a thickness ≤ 0.5 μm, which greatly limits the maximum capacity that is reached and this is associated with poor cycling properties. Conversely, by controlling the deposition rate, RF and DC sputtering allows deposition of (*h*00) or (110) oriented crystalline films in the range 0.6–3.6 μm with a high electrochemical efficiency and rate capability that does not depend on the thickness. Thus without any postannealing treatment, a stable capacity of 100 μA h cm^{-2} is achieved in the 3.8–2.8 V range at 100 μA cm^{-2} and 150 μA h cm^{-2} are recovered at 1 mA cm^{-2} in the enlarged 3.8–2.15 V region. Kinetic and structural data combining an XRD and Raman investigation of structural changes induced by Li insertion show an electrochemical and structural response that differs from that of bulk samples used in composite electrodes. This finding can be explained by their specific microstructure. However, the capacity decrease in the range 3.8–2.15 V has not been elucidated as yet even when some works suggest the growth of an SEI layer as the main factor. The use of this large potential window is a real challenge which opens, if met successfully, the possibility to get higher capacity values. Some attempts must be also made to establish the interest of the amorphous material versus the crystalline film owing to contradictory published data. In addition to relatively old data reported in TFBs using LiPON electrolyte, more recent work shows the possibility to get stable capacities, for instance 26 μA h cm^{-2} with a c-V_2O_5 film, 1 μm thick, between 3.8–2.15 V at 100 μA cm^{-2}.

10.6 MoO_3 Thin Films

MoO_3 is an attractive candidate from several standpoints. Indeed, the MoO_3 orthorhombic α phase exhibits a unique layered structure characterized by weak van der Waals attraction between the layers. This oxide can reversibly insert via a topotactic reaction up to 1.5 Li ions per mole of oxide corresponding to a specific capacity of 280 mA h g^{-1} for a discharge cutoff voltage of 1.5 V. Assuming fully densified films, a theoretical specific capacity of $\approx$130 μA h cm^{-2} μm can be expected compared with 69 μA h cm^{-2} $μm^{-1}$ for $LiCoO_2$ available at higher potential. However, in spite of these interesting properties, MoO_3 thin films have been mainly investigated for their electrochromic properties. The synthesis of MoO_3 thin-film cathodes combined with an electrochemical evaluation as cathode material in a liquid or solid electrolyte is reported using PLD [183], sputtering [10, 184–187], and flash evaporation [12, 188, 189].

10.6.1 Liquid Electrolyte

PLD MoO_3 films in the range 300–500 nm thick have been recently prepared on silicon wafers heated between room temperature to 500 °C [183]. The crystalline α phase is obtained from 200 °C. There is a great influence of the substrate temperature (T_s) on the shape of the discharge/charge curves and the capacity; for $T_s = 400$ °C, the first discharge/charge profile in liquid electrolyte is close that of the bulk oxide and the specific capacity changes from 53 μA h cm^{-2} μm for $T_s = 200$ °C to 90 μA h cm^{-2} $μm^{-1}$ for $T_s = 400$ °C.

Interesting cycling data are available for sputtered thin films, 300 nm thick, containing Pt nanoparticles [184]. In fact MoO_3–Pt composite films have been deposited on Pt-coated Si substrate by RF cosputtering MoO_3 and Pt targets simultaneously. From the shape of their discharge curves, it can be thought that amorphous phases are obtained. Such composite films show a high capacity of 100 μA h cm^{-2} μm after 30 cycles between 3.5 and 1.2 V at a current density of 20 μA cm^{-2} and a more stable cycling behavior than the single MoO_3 film without Pt particles, for which a sharp capacity decline is observed, to reach 25 μA h cm^{-2} μm after 30 cycles. This improvement of MoO_3 electrochemistry can be explained by the presence of Pt nanophases which might serve as spacers to reduce volume change by alleviating the mechanical stress and could also decrease the sheet resistance by a factor 10.

The feasibility of using flash-evaporated MoO_3 films, 500–600 nm thick, on heated substrate as Li-insertion material has been shown with, of course, an influence of the growth temperature [188, 189]. At 250 °C, the synthesis of the stable crystalline α phase is allowed, whereas a mixture of both molybdenum oxides MoO_3 and MoO_2 is obtained at 120 °C and an amorphous compound at lower temperature. A surprising flat discharge curve is reported at 3 V for the crystalline form obtained at 250 °C which contrasts with most of the data available on bulk and film electrodes; a maximum Li uptake of 1.5 Li per mole of oxide is obtained. However, no useful related cycling data are available for flash-evaporated films.

10.6.2 Solid State Electrolyte

In the case of MoO_3, three different solid electrolytes have been used to fabricate a microbattery: $Li_2O/V_2O_5/SiO_2$, LiPON and $Li_{1.4}B_{2.5}S_{0.1}O_{4.9}$.

In the case of all solid-state batteries, the most promising results have been obtained with $Li_2O/V_2O_5/SiO_2$ as solid electrolyte [10, 185, 186]. The electrical conductivity of the film electrolyte was high and equal to 10^{-4} S m^{-1}. The sputtered cathode, 1 μm thick, is reported to be a reduced form of the oxide, $MoO_{3-x} = MoO_{2.75}$, deposited in an Ar atmosphere on a heated substrate of stainless steel. Cycling between 3 and 1 V led to a stable capacity of 60 μA h cm^{-2} μm^{-1} over 250 cycles but the as-prepared cells suffer from a high rate of self-discharge [10, 185]. A significant gain of capacity is possible by using thicker RF-sputtered MoO_3 films [186]. A 4.66-μm-thick film of molybdenum oxide, probably $MoO_{2.875}$, exhibits S-shaped discharge curves upon cycling between 3.5 and 1.5 V with an average working voltage of 2.3 V. The capacity rapidly decreases during the first 10 cycles to about 80% of the initial capacity, then reaching 300 μA h cm^{-2}. Thereafter, the capacity fading is slow and 260 μA h cm^{-2} (i.e., 56 μA h cm^{-2} μm^{-1}) are achieved after 40 cycles at 10 μA cm^{-2} (Figure 10.38). Such capacity values are the highest among those reported for TFBs.

For some specific applications like batteries on low earth orbiting (LEO) spacecraft, a possible critical limitation of TFBs could be their sensitivity to long-term elevated temperature cycling (around 120 °C). Therefore, sputtered MoO_3, 0.3 μm

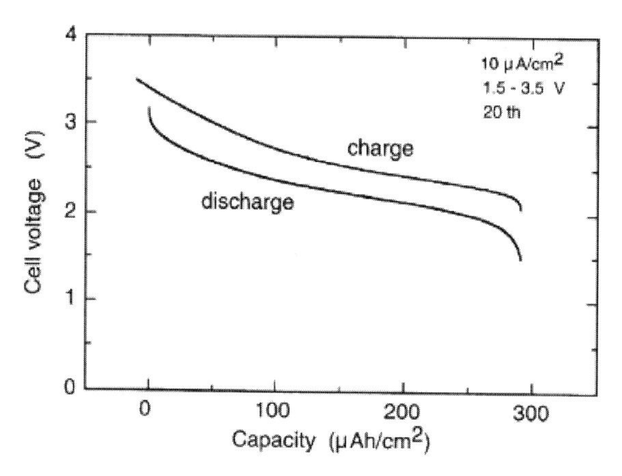

Fig. 10.38 Discharge and charge curves of a $Li/Li_2O/V_2O_5/SiO_2/MoO_{3-x}$ TFB at the 20th cycle. From [186].

thick, have been examined in all solid-state batteries using an LiPON electrolyte ($Li_{3.3}PO_{3.8}N_{0.22}$) in the potential range 5–1 V at 150 °C at a current density of 0.7 mA cm^{-2} [187]. The MoO_3 films, heat-treated in air at 280 °C reveal a mixture of the α and β phases. The practical discharge capacity of 140 μA h cm^{-2} μm^{-1} for thousands of cycles is reported to be twice that of $LiCoO_2$ cells. However, it is worth noting that the films examined were relatively thin and that the results described above have to be extended to thickness in the range 1–2 μm in order to significantly exceed 40 μA h cm^{-2}.

A thin-film microbattery has been constructed with a glassy $Li_{1.4}B_{2.5}S_{0.1}O_{4.9}$ electrolyte film and a flash-evaporated MoO_3 cathode film, which, in fact, is a mixture of MoO_3 and MoO_2 [12]. A discharge capacity of 75 μA h cm^{-2} μm^{-1} is given at a 1-V cutoff voltage for the positive electrode, but no data are given on the reliability and the cycling performance of the cell.

10.6.3 Conclusion

In spite of a limited number of reports, MoO_3 raises some interest as thin-film cathode owing to a good compromise between the working voltage around 2.5 V and a high capacity, practically twice that of $LiCoO_2$. Best results, obtained with sputtered films in liquid electrolyte, indicate a stable capacity as high as 33 μA h cm^{-2} (100 μA h cm^{-2} μm for a 0.3-μm-thick film) over 30 cycles with composite Pt–MoO_3 films. With a $Li_2O/V_2O_5/SiO_2$ solid electrolyte, a capacity of 60 μA h cm^{-2} is demonstrated to be stable over 250 cycles, while the capacity can be improved by using thicker films up to 4.66 μm with 260 μA h cm^{-2} after 40 cycles. These high capacities prompt us to intensively investigate the deposition and the electrochemical properties of high-performance MoO_3 thin films in liquid and solid electrolytes.

10.7 General Conclusions

In this chapter, we have tried to illustrate the recent achievements with materials with corresponding electrochemical properties for thin-film metal oxides that are usable as electrode materials in TFBs. Several trends of research can be drawn from this analysis. Before implementation of an electrode thin-film material in a TFB, a systematic approach of its electrochemical behavior in liquid electrolyte is required. Whatever the deposition method used, most of the studies deal with the effect of experimental conditions on the structure, morphology, chemistry, and electrochemistry of the deposited film. The sputtering technique and PLD are the most widely applied methods to get thin-film metal oxides with optimized properties, the sputtering method being by far the most appropriate and consistent for fabricating TFBs with the sputtered LiPON solid electrolyte.

The well-defined composition, structure, microstructure, and dimensions of the deposited layers without any binding or electronic additive make the thin films suitable for fundamental electrochemical and structural studies of the lithium-intercalation process. This approach to investigate the intrinsic properties of thin-film materials constitutes a very exciting field of research with possible applications for TFBs and also for the understanding of conventional composite electrodes of secondary Li batteries and LIBs. For instance, the demonstration of the influence of the preferred orientation on the electrochemical behavior of $LiCoO_2$ and V_2O_5 thin films has been allowed by the thin-film configuration. A key factor for highly reversible Li-insertion reaction into the host lattice of the metal oxide is contained in the nature and the magnitude of its structural response. The thin-film geometry using pure active material promotes high-quality structural investigations using *ex situ* and *in situ* approaches. Of course, the electrode–electrolyte interface plays a major role in the behavior of the film in terms of discharge capacity, chemical stability, and cycle life. It appears clear that the reactivity of thin-film cathode in usual liquid electrolytes is higher than with a solid electrolyte like the LiPON electrolyte, thus implying a lower cycle life for the film oxide tested in liquid electrolyte. This specific reactivity of 4- and 3-V materials originates a significant capacity loss with the possible occurrence of dissolution phenomenon and/or the growth of an SEI layer. Another challenge consists in the deposition of films thick enough to be characterized by a homogeneous electrochemical efficiency to benefit from a significant enhancement of the capacity combined with a satisfactory rate capability.

The best results are obtained with $LiCoO_2$ films for which the theoretical capacity (0.5 Li) of 69 μA h cm^{-2} μm^{-1} can be reached for relatively thick films of a few μm, its cycling behavior being stable in solid-state electrolyte. However, the use of well-crystallized thin film with a high-temperature treatment after deposition can induce severe interfacial reaction between a cathode, a current collector film, and an Si substrate. Therefore, new processes (RTA, bias, optimization of the power, lower annealing temperature etc.) have still to be developed and improved to solve this problem. The stability of the cycling behavior in liquid electrolyte must be addressed. The other 4-V cathode materials like $LiNiO_2$ and its derivative compounds $LiNi_{1-x}MO_2$, $LiMn_2O_4$, and its substituted forms, offer good opportunities for improving the discharge capacity, but basic data on their electrochemical and structural properties are still lacking. In addition, a considerable increase of film thickness, at present limited to about 0.3–0.5 μm at the best, is needed. The PLD technique has proved to be particularly attractive for obtaining $LiNi_{1-x}MO_2$ with capacity in the range 60–90 μA h cm^{-2} μm^{-1} in liquid electrolyte. In the case of $LiMn_2O_4$, the spin-coating and other solution techniques compete with sputtered and PLD-prepared films, leading to interesting capacities in the range 50 μA h cm^{-2} μm^{-1}, but the demonstration of a stability guaranteed over a large number of cycles in liquid electrolyte has not been performed.

Low crystallization temperature, as low as 300 °C, makes the V_2O_5 film oxides very attractive since they are prepared without any postannealing treatment via sputtering and PLD techniques, the sputtering method giving the best results. The sputtered V_2O_5 film oxides have been clearly shown to offer large capacities (100 μA h cm^{-2} for a 3.6-μm-thick film, i.e., 30 μA h cm^{-2} $μm^{-1}$), good adherence to the substrate, high rate capability, and good cycle life at high potential (3.8–2.8 V) in liquid electrolyte and larger capacities up to 150 μA h cm^{-2} in the extended range 3.8–2.15 V. Their cycling stability in this potential range needs to be improved and greater effort in the understanding of the Li-insertion process is needed for obtaining enlarged voltage cycling limits. Even when amorphous compounds are reported to be more suited to ensure a stable cycling behavior in both types of electrolytes, crystallized V_2O_5 thin films are thought to be considered with interest even in all solid-state TFBs. Amorphous and crystalline V_2O_5 films with thickness >1 μm are still under investigation for application in all solid-state TFBs. Advances in discharge capacity can be also expected if other cathode materials like MoO_3 are considered.

Owing to the solder reflow assembly in which the IC is heated to 250–260 °C for a short time causing all of the components to be soldered at once, some attempts must be made to replace the evaporated Li negative electrode by stable intercalation compounds. Hence low-voltage materials such as $Li_4Ti_5O_{12}$ [60, 67], TiO_2 [190, 191], $LiNiVO_4$ [192], or SnO_2 [62] can be considered with interest that opens new attractive targets for solid-state chemists searching for better thin-film positive and negative electrodes for TFBs.

References

1 Granqvist, C.G. (1995, 2002) *Handbook of Inorganic Electrochromic Materials*, Elsevier, Amsterdam, The Netherlands.

2 Monk, P.M.S., Mortimer, R.J., and Rosseinsky, D.R. (1995) *Electrochromism: Fundamentals and Applications*, Wiley-VCH Verlag GmbH, Weinheim, Germany.

3 Avendano, E., Berggren, L., Niklasson, G.A., Granqvist, C.G., and Azens, A. (2006) *Thin Solid Films*, **496**, 30.

4 Kanehori, K., Matsumoto, K., Miyauchi, K., and Kudo, T. (1983) *Solid State Ionics*, **9–10**, 1445.

5 Jones, S.D. and Akridge, J.R. (1992) *Solid State Ionics*, **53–56**, 628.

6 Meunier, G., Dormoy, R., and Levasseur, A. (1989) *Mater. Sci. Eng.*, **B3**, 19.

7 Lindic, M.H., Pecquenard, B., Vinatier, P., Levasseur, A., Martinez, H., Gonbeau, D., Petit, P.E., and Ouvrard, G. (2005) *J. Electrochem. Soc.*, **152**, A141.

8 Martins-Litas, I., Vinatier, P., Levasseur, A., Dupin, J.C., and Gonbeau, D. (2001) *J. Power Sources*, **97–98**, 545.

9 Levasseur, A., Schmidt, E., Meunier, G., Gonbeau, D., Benoist, L., and Pfister-Guillouzo, G. (1995) *J. Power Sources*, **54**, 352.

10 Ohtsuka, H. and Yamaki, J. (1989) *Solid State Ionics*, **35**, 201.

11 Ohtsuka, H., Okada, S., and Yamaki, J. (1990) *Solid State Ionics*, **40–41**, 964.

12 Julien, C.M. (2000) *Materials for Lithium-ion Batteries*, NATO Science Series, High technology, Vol. **85**, Kluwer Academic Publishers, p. 381.

13 Julien, C.M. (1994) in *Lithium Batteries: New materials, Developments and Perspectives* (ed. G. Pistoia), Elsevier, p. 167.

14 Bouwman, P.J., Boukamp, B.A., Bouwmeester, H.J.M., Wondergem, H.J., and Notten, P.H.L. (2001) *J. Electrochem. Soc.*, **148**, A311.

15 Bouwman, P.J., Boukamp, B.A., Bouwmeester, H.J.M., and Notten, P.H.L. (2002) *J. Electrochem. Soc.*, **149**, A699.
16 Whitacre, J.F., West, W.C., Brandon, E., and Ratnakumar, B.V. (2001) *J. Electrochem. Soc.*, **148**, A1078.
17 Chiu, K.F., Hsu, F.C., Chen, G.S., and Wu, M.K. (2003) *J. Electrochem. Soc.*, **150**, A503.
18 Pracharova, J., Pridal, J., Bludska, J., Jakubec, I., Vorlicek, V., Malkova, Z., Dikonimos Makris, Th., Giorgi, R., and Jastrabik, L. (2002) *J. Power Sources*, **108**, 204.
19 Jeon, S.W., Lim, J.K., Lim, S.H., and Lee, S.M. (2005) *Electrochim. Acta*, **51**, 268.
20 Park, H.Y., Lee, S.R., Lee, Y.J., Cho, B.W., and Cho, W.I. (2005) *Mater. Chem. and Phys.*, **93**, 70.
21 Whitacre, J.F., West, W.C., and Ratnakumar, B.V. (2001) *J. Power Sources*, **103**, 134.
22 Lee, J.K., Lee, S.J., Baik, H.K., Lee, H.Y., Jang, S.W., and Lee, S.M. (1999) *Electrochem. Solid State Lett.*, **2**, 512.
23 Cho, J., Kim, Y.Y., Kim, T.J., and Park, B. (2001) *Angew. Chem. Int. Ed.*, **40**, 3367.
24 Kim, Y.J., Kim, H., Kim, B., Ahn, D., Lee, J.G., Kim, T.J., Son, D., Cho, J., Kim, Y.W., and Park, B. (2003) *Chem. Mater.*, **15**, 1505.
25 Kim, B., Kim, C., Ahn, D., Moon, T., Ahn, J., Park, Y., and Park, B. (2007) *Electrochem. Solid State Lett.*, **10**, A32.
26 Kim, Y.J., Lee, E.K., Kim, H., Cho, J., Cho, Y.W., Park, B., Oh, S.M., and Yoon, J.K. (2004) *J. Electrochem Soc.*, **151**, A1063.
27 Alamgir, F.M., Strauss, E., denBoer, M., Greenbaum, S., Whitacre, J.F., Kao, C.C., and Neih, S. (2005) *J. Electrochem. Soc.*, **152**, A845.
28 Dupin, J.C., Gonbeau, D., Benqlilou-Moudden, H., Vinatier, Ph., and Levasseur, A. (2001) *Thin Solid Films*, **384**, 23.
29 Benqlilou-Moudden, H., Blondiaux, G., Vinatier, P., and Levasseur, A. (1998) *Thin Solid Films*, **333**, 16.
30 Rao, K.J., Benqlilou-Moudden, H., Desbat, B., Vinatier, P., and Levasseur, A. (2002) *J. Solid State Chem.*, **165**, 42.
31 Itoh, T., Anzue, N., Mohamedi, M., Hisamitsu, Y., Umeda, M., and Uchida, I. (2000) *Electrochem. Commun.*, **2**, 743.
32 Go, J.Y. and Pyun, S.I. (2003) *J. Electrochem. Soc.*, **150**, A1037.
33 Wang, B., Bates, J.B., Hart, F.X., Sales, B.C., Zuhr, R.A., and Robertson, J.D. (1996) *J. Electrochem. Soc.*, **143**, 3203.
34 Dudney, N.J. and Jang, Y.I. (2003) *J. Power Sources*, **119–121**, 300.
35 Bates, J.B., Dudney, N.J., Neudecker, B.J., Hart, F.X., Jun, H.P., and Hackney, S.A. (2000) *J. Electrochem. Soc.*, **147**, 59.
36 Bates, J.B., Dudney, N.J., Neudecker, B., Ueda, A., and Evans, C.D. (2000) *Solid State Ionics*, **135**, 33.
37 Hayashi, M., Takahashi, M., and Sakurai, Y. (2007) *J. Power Sources*. doi:10.1016/j.jpowsour.2007.06.081.
38 Neudecker, B.J., Dudney, N.J., and Bates, J.B. (2000) *J. Electrochem. Soc.*, **147**, 517.
39 Park, H.Y., Nam, S.C., Lim, Y.C., Choi, K.G., Lee, K.C., Park, G.B., Kim, J.B., Kim, H.P., and Cho, S.B. (2007) *Electrochim. Acta*, **52**, 2062.
40 Jang, Y.I., Dudney, N.J., Blom, D.A., and Allard, L.F. (2002) *J. Electrochem Soc.*, **149**, A1442.
41 Li, C.N., Yang, J.M., Krasnov, V., Arias, J., and Nieh, K.W. (2007) *Appl. Phys. Lett.*, **90**, 263102.
42 Kuwata, N., Kawamura, J., Toribami, K., Hattori, T., and Sata, N. (2004) *Electrochem. Commun.*, **6**, 417.
43 Kuwata, N., Kawamura, J., Toribami, K., Hattori, T., and Sata, N. in (eds B.V.R. Chowdari) *et al.* (2004) "Solid State Ionics: The Science and Technology of Ions in Motion", Proceedings of the Asian Conference 9th (2004) Jeju Island, Republic of Korea, p. 637.
44 Antaya, M., Dahn, J.R., Preston, J.S., Rossen, E., and Reimers, J.N. (1993) *J. Electrochem. Soc.*, **140**, 575.

45 Striebel, K.A., Deng, C.Z., Wen, S.J., and Cairns, E.J. (1996) *J. Electrochem. Soc.*, **143**, 1821.

46 Perkins, J.D., Bahn, C.S., Mc Graw, J.M., Parilla, P.A., and Ginley, D.S. (2001) *J. Electrochem. Soc.*, **148**, A1302.

47 Julien, C., Camacho-lopez, M.A., Escobar-Alarcon, L., and Haro-Poniatowski, E. (2001) *Mater. Chem. Phys.*, **68**, 210.

48 Tang, S.B., Lai, M.O., and Lu, L. (2006) *J. Alloys Compd.*, **424**, 342.

49 Xia, H., Lu, L., and Ceder, G. (2006) *J. Power Sources*, **159**, 1422.

50 Xia, H., Lu, L. and Ceder, G. (2006) *J. Alloys Compd.*, **417**, 304.

51 Xia, H. and Lu, L. (2007) *Electrochim. Acta.*, **52**, 7014.

52 Xia, H., Lu, L., Meng, Y.S., and Ceder, G. (2007) *J. Electrochem. Soc.*, **154**, A337.

53 Iriyama, Y., Kako, T., Yada, C., Abe, T., and Ogumi, Z. (2005) *Solid State Ionics*, **176**, 2371.

54 Cho, S.I., and Yoon, S.G. (2002) *J. Electrochem. Soc.*, **149**, A1584.

55 Cho, S.I. and Yoon, S.G. (2003) *Appl. Phys. Letters*, **82**, 3345.

56 Fragnaud, P., Brousse, T., and Schleich, D.M. (1996) *J. Power Sources*, **63**, 187.

57 Fragnaud, P., Nagarajan, R., Schleich, D.M., and Vujic, D. (1995) *J. Power Sources*, **54**, 362.

58 Yu, Y., Shui, J.L., Jin, Y., and Chen, C.H. (2006) *Electrochim. Acta*, **51**, 3292.

59 Rho, Y.H., Kanamura, K., and Umegaki, T. (2003) *J. Electrochem. Soc.*, **150**, A107.

60 Rho, Y.H., Kanamura, K., Fujisaki, M., Hamagami, J.I., Suda, S.I., and Umegaki, T. (2002) *Solid State Ionics*, **151**, 151.

61 Kim, M.K., Chung, H.T., Park, Y.J., Kim, J.G., Son, J.T., Park, K.S., and Kim, H.G. (2001) *J. Power Sources*, **99**, 34.

62 Maranchi, J.P., Hepp, A.F., and Kumta, P.N. (2005) *Mater. Sci. Eng. B*, **116**, 327.

63 Tao, Y., Chen, Z., Zhu, B., and Huang, W. (2003) *Solid State Ionics*, **161**, 187.

64 Uchida, I. and Sato, H. (1995) *J. Electrochem. Soc.*, **142**, L139.

65 Lee, S.T., Jeon, S.W., Yoo, B.J., Choi, S.D., Kim, H.J., and Lee, S.M. (2006) *J. Power Sources*, **155**, 375.

66 Park, M.S., Hyun, S.H., and Nam, S.C. (2006) *J. Power Sources*, **159**, 1416.

67 Brousse, T., Fragnaud, P., Marchand, R., Schleich, D.M., Bohnke, O., and West, K. (1997) *J. Power Sources*, **68**, 412.

68 Zhao, S.L., Fu, F.W., and Qin, Q.Z. (2002) *Thin Solid Films*, **415**, 108.

69 Zhao, S.L. and Qin, Q.Z. (2002) *J. Power Sources*, **122**, 174.

70 Lee, S.J., Lee, J.K., Kim, D.W., and Baik, H.K. (1996) *J. Electrochem. Soc.*, **143**, L268.

71 Kim, H.K., Seong, T.Y., and Yoon, Y.S. (2002) *Electrochem. Solid State Lett.*, **5**, A252.

72 Kim, H.K., Seong, T.Y., and Yoon, Y.S. (2004) *Thin Solid Films*, **447–448**, 619.

73 Ramana, C.V., Zaghib, K., and Julien, C.M. (2006) *Chem. Mater.*, **18**, 1397.

74 Ramana, C.V., Zaghib, K., and Julien, C.M. (2006) *J. Power sources*, **159**, 1310.

75 Ramana, C.V., Zaghib, K., and Julien, C.M. (2007) *Appl. Phys. Letters*, **90**, 21916.

76 Wang, G.X., Lindsay, M.J., Ionescu, M., Bradhurst, D.H., Dou, S.X., and Liu, H.K. (2001) *J. Power Sources*, **97–98**, 298.

77 Yamada, K., Sato, N., Fujino, T., Lee, C.G., Uchida, I., and Selman, J.R. (1999) *J. Solid State Electrochem.*, **3**, 148.

78 Svegl, F., Orel, B., and Kaucic, V. (2000) *Solar Energy*, **68**, 523.

79 Han, K.S., Tsurimoto, S., and Yoshimura, M. (1999) *Solid State Ionics*, **121**, 229.

80 Neudecker, B.J., Zuhr, R.A., Kwak, B.S., Bates, J.B., and Robertson, J.D. (1998) *J. Electrochem. Soc.*, **145**, 4148.

81 Neudecker, B.J., Zuhr, R.A., Robertson, J.D., and Bates, J.B. (1998) *J. Electrochem. Soc.*, **145**, 4160.

82 Shokoohi, F.K., Tarascon, J.M., Wilkens, B.J., Guyomard, D., and Chang, C.C. (1992) *J. Electrochem. Soc.*, **139**, 1845.

83 Bates, J.B., Lubben, D., Dudney, N.J., and Hart, F.X. (1995) *J. Electrochem. Soc.*, **142**, L149.

84 Hwang, K.H., Lee, S.H., and Joo, S.K. (1994) *J. Electrochem. Soc.*, **141**, 3296.

85 Lee, K.L., Jung, J.Y., Lee, S.W., Moon, H.S., and Park, J.W. (2004) *J. Power Sources*, **130**, 241.

86 Chiu, K.F., Lin, H.C., Lin, K.M., and Tsai, C.H. (2005) *J. Electrochem. Soc.*, **152**, A2058.

87 Komaba, S., Kumagai, N., Baba, M., Miura, F., Fujita, N., Groult, H., Devilliers, D., and Kaplan, B. (2000) *J. Appl. Electrochem.*, **30**, 1179.

88 Bates, J.B., Dudney, N.J., Lubben, D.C., Gruzalski, G.R., Kwak, B.S., Yu, X., and Zuhr, R.A. (1995) *J. Power Sources*, **54**, 58.

89 Dudney, N.J., Bates, J.B., Zuhr, R.A., Young, S., Robertson, J.D., Jun, H.P., and Hackney, S.A. (1999) *J. Electrochem. Soc.*, **146**, 2455.

90 Park, Y.S., Lee, S.H., Lee, B.I., and Joo, S.K. (1998) *Electrochem. Solid State Lett.*, **2**, 58.

91 Park, C.H., Park, M., Yoo, S.I., and Joo, S.K. (2006) *J. Power Sources*, **158**, 1442.

92 Striebel, K.A., Deng, C.Z., Wen, S.J., and Cairns, J. (1996) *J. Electrochem. Soc.*, **143**, 1821.

93 Morcrette, M., Barboux, P., Perriere, J., and Brousse, T. (1998) *Solid State Ionics*, **112**, 249.

94 Morcrette, M., Barboux, P., Perriere, J., Brousse, T., Traverse, A., and Boilot, J.P. (2001) *Solid State Ionics*, **138**, 213.

95 Julien, C., Haro-Poniatowski, E., Camacho-Lopez, M.A., Escobar-Alarcon, L., and Jimenez-Jarquin, J. (2000) *Mater. Sci. Eng.*, **B72**, 36.

96 Tang, S.B., Lai, M.O., Lu, L., and Tripathy, S. (2006) *J. Solid State Chem.*, **179**, 3831.

97 Tang, S.B., Lai, M.O., and Lu, L. (2006) *Electrochim. Acta*, **52**, 1161.

98 Tang, S.B., Lai, M.O., and Lu, L. (2007) *J. Power Sources*, **164**, 372.

99 Yamada, I., Abe, T., Iriyama, Y., and Ogumi, Z. (2003) *Electrochem. Commun.*, **5**, 502.

100 Striebel, K.A., Sakai, E., and Cairns, E.J. (2002) *J. Electrochem. Soc.*, **149**, A61.

101 Iriyama, Y., Nishimoto, K., Yada, C., Abe, T., Ogumi, Z., and Kikuchi, K. (2006) *J. Electrochem. Soc.*, **153**, A821.

102 Kuwata, N., Kumar, R., Toribami, K., Suzuki, T., Hattori, T., and Kawamura, J. (2006) *Solid State Ionics*, **177**, 2827.

103 Nishizawa, M., Uchiyama, T., Itoh, T., Abe, T., and Uchida, I. (1999) *Langmuir*, **15**, 4949.

104 Uchiyama, T., Nishizawa, M., Itoh, T., and Uchida, I. (2000) *J. Electrochem. Soc.*, **147**, 2057.

105 Chung, K.Y., Shu, D., and Kim, K.B. (2004) *Electrochim. Acta*, **49**, 887.

106 Hjelm, A.K. and Lindbergh, G. (2002) *Electrochim. Acta*, **47**, 1747.

107 Hjelm, A.K., Lindbergh, G., and Lundqvist, A. (2001) *J. Electroanal. Chem.*, **509**, 139.

108 Mohamedi, M., Takahashi, D., Itoh, T., and Uchida, I. (2002) *Electrochim. Acta*, **47**, 3483.

109 Mohamedi, M., Takahashi, D., Uchiyama, U., Itoh, T., Nishizawa, M., and Uchida, I. (2001) *J. Power Sources*, **93**, 93.

110 Mohamedi, M., Takahashi, D., Itoh, T., Umeda, M., and Uchida, I. (2002) *J. Electrochem. Soc.*, **149**, A19.

111 Cao, F. and Prakash, J. (2002) *Electrochim. Acta*, **47**, 1607.

112 Chung, K.Y., Yoon, W.S., Kim, K.B., Yang, X.Q., and Oh, S.M. (2004) *J. Electrochem. Soc.*, **151**, A484.

113 Shu, D., Kumar, G., Kim, K.B., Ryu, K.S., and Chang, S.H. (2003) *Solid State Ionics*, **160**, 227.

114 Park, Y.J., Kim, J.G., Kim, M.K., Kim, H.G., Chung, H.T., Um,

W.S., Kim, M.H., and Kim, H.G. (1998) *J. Power Sources*, **76**, 41.
115 Park, Y.J., Kim, J.G., Kim, M.K., Kim, H.G., Chung, H.T., and Park, Y. (2000) *J. Power Sources*, **87**, 69.
116 Wu, X.M., Li, X.H., Xu, M.F., Zhang, Y.H., He, Z.Q., and Wang, Z. (2002) *Mater. Res. Bull.*, **37**, 2345.
117 Chiu, K.F., Lin, H.C., Lin, K.M., and Chen, C.C. (2006) *J. Electrochem. Soc.*, **153**, A1992.
118 Shih, F.Y. and Fung, K.Z. (2006) *J. Power Sources*, **159**, 179.
119 Rho, Y.H., Dokko, K., and Kanamura, K. (2006) *J. Power Sources*, **157**, 471.
120 Kim, K.W., Lee, S.W., Han, K.S., Chung, H.J., and Woo, S.I. (2003) *Electrochim. Acta*, **48**, 4223.
121 Lei, J., Li, L., Kostecki, R., Muller, R., and McLarnon, F. (2005) *J. Electrochem. Soc.*, **152**, A774.
122 Striebel, K.A., Rougier, A., Horne, C.R., Reade, R.P., and Cairns, E.J. (1999) *J. Electrochem. Soc.*, **146**, 4339.
123 Xia, H., Tang, S.B., Lu, L., Meng, Y.S., and Ceder, G. (2007) *Electrochim. Acta*, **52**, 2822.
124 Xia, H., Meng, Y.S., Lu, L., and Ceder, G. (2007) *J. Electrochem. Soc.*, **154**, A737.
125 Mohamedi, M., Makino, M., Dokko, K., Itoh, T., and Uchida, I. (2002) *Electrochim. Acta*, **48**, 79.
126 Soudan, P., Brousse, T., Taillades, G., and Sarradin, J. (2003) Proceedings Electrochemical Society 2003-20, Paris, p. 633.
127 Park, Y.J., Ryu, K.S., Kim, K.M., Park, N.G., Kang, M.G., and Chang, S.H. (2002) *Solid State Ionics*, **154–155**, 229.
128 Park, Y.J., Ryu, K.S., Park, N.G., Hong, Y.S., and Chang, S.H. (2002) *J. Electrochem. Soc.*, **149**, A597.
129 Miyazaki, H., Sakamura, H., Kamei, M., and Yasui, I. (1999) *Solid State Ionics*, **122**, 223.
130 Kumagai, N., Kitamoto, H., Baba, M., Durand-Vidal, S., Devilliers, D., and Groult, H. (1998) *J. Appl. Electrochem.*, **28**, 41.
131 Koike, S., Fujieda, T., Sakai, T., and Higuchi, S. (1999) *J. Power Sources*, **81–82**, 581.
132 Navone, C., Pereira-Ramos, J.P., Baddour-Hadjean, R., and Salot, R. (2005) *J. Power Sources*, **146**, 327.
133 Navone, C., Baddour-Hadjean, R., Pereira-Ramos, J.P., and Salot, R. (2005) *J. Electrochem., Soc.*, **152**, A1790.
134 Navone, C., Pereira-Ramos, J.P., Baddour-Hadjean, R., and Salot, R. (2006) *J. Electrochem. Soc.*, **153**, A2287.
135 Mui, S.C., Jasinski, J., Leppert, V.J., Mitome, M., Sadoway, D.R., and Mayes, A.M. (2006) *J. Electrochem. Soc.*, **153**, A1372.
136 Talledo, A. and Granqvist, C.G. (1995) *J. Appl. Phys.*, **77**, 4655.
137 Cocciantelli, J.M., Doumerc, J.P., Pouchard, M., Broussely, M., and Labat, J. (1991) *J. Power Sources*, **34**, 103.
138 Scarminio, J., Talledo, A., Andersson, A.A., Passerini, S., and Decker, F. (1993) *Electrochim. Acta*, **38**, 1637.
139 Vivier, V., Farcy, J., and Pereira-Ramos, J.P. (1998) *Electrochim. Acta*, **44**, 831.
140 Meulenkamp, E.A., van Klinlen, W., and Schlatmann, A.R. (1999) *Solid State Ionics*, **126**, 235.
141 Zhang, J.G., McGraw, J.M., Turner, J., and Ginley, D. (1997) *J. Electrochem. Soc.*, **144**, 1630.
142 Kumagai, N., Komaba, S., Nakano, O., Baba, M., Groult, H., and Devilliers, D. (2004) *Electrochemistry*, **72**, 261.
143 Garcia, M.E. and Garofalini, S.H. (1999) *J. Electrochem. Soc.*, **146**, 840.
144 Gies, A., Pecquenard, B., Benayad, A., Martinez, H., Gonbeau, D., Fuess, H., and Levasseur, A. (2005) *Solid State Ionics*, **176**, 1627.
145 Benayad, A., Martinez, H., Gies, A., Pecquenard, B., Levasseur, A., and Gonbeau, D. (2006) *J. Phys. Chem. Solids*, **67**, 1320.
146 Benayad, A., Martinez, H., Gies, A., Pecquenard, B., Levasseur, A., and Gonbeau, D. (2006) *J. Electron Spectrosc. Relat. Phenom.*, **150**, 1.

147 Kim, Y.S., Ahn, H.J., Shim, H.S., and Seong, T.Y. (2006) *Solid State Ionics*, **177**, 1323.

148 Artuso, F., Decker, F., Krasilnikova, A., Liberatore, M., Lourenco, A., Masetti, E., Pennisi, A., and Simone, F. (2002) *Chem. Mater.*, **14**, 636.

149 Dudney, N.J. and Neudecker, B.J. (1999) *Curr. Opin. Solid State Mater. Sci.*, **4**, 479.

150 Bates, J.B., Dudney, N.J., Gruzalski, G.R., Zuhr, R.A., Choudhury, A., Luck, C.F., and Robertson, J.D. (1993) *J. Power Sources*, **43–44**, 103.

151 West, K., Zachau-Christiansen, B., Skaarup, S.V., and Poulsen, F.W. (1992) *Solid State Ionics*, **57**, 41.

152 Kim, H.K., Seon, T.Y., and Yoon, Y.S. (2003) *J. Vac. Sci. Technol., B*, **21**, 754.

153 Hwang, H.S., Oh, S.H., Kim, H.S., Cho, W.I., and Lee, D.Y. (2004) *Electrochim. Acta*, **50**, 485.

154 Lee, J.M., Hwang, H.S., Cho, W.I., Cho, B.W., and Kim, K.Y. (2004) *J. Power Sources*, **136**, 122.

155 Jeon, E.J., Shin, Y.W., Nam, S.C., Cho, W.I., and Yoon, Y.S. (2001) *J. Electrochem. Soc.*, **148**, A318.

156 Baba, M., Kumagai, N., Kobayashi, H., Nakano, O., and Nishidate, K. (1999) *Electrochem. Solid State Lett.*, **2**, 320.

157 Nakazawa, H., Sano, K., and Baba, M. (2005) *J. Power Sources*, **146**, 758.

158 McGraw, J.M., Perkins, J.D., Zhang, J.G., Liu, P., Parilla, P.A., Turner, J., Schulz, D.L., Curtis, C.J., and Ginley, D.S. (1998) *Solid State Ionics*, **113–115**, 407.

159 Mc Graw, J.M., Bahn, C.S., Parilla, P.A., Perkins, J.D., Readey, D.W., and Ginley, D.S. (1999) *Electrochim. Acta*, **45**, 187.

160 Julien, C., Haro-Poniatowski, E., Camacho-lopez, M.A., Escobar-Alarcon, L., and Jimenez-Jarquin, J. (1999) *Mater. Sci. Eng. B*, **65**, 170.

161 Ramana, C.V., Smith, R.J., Hussain, O.M., Massot, M., and Julien, C.M. (2005) *Surf. Interface Anal.*, **37**, 406.

162 Huang, F., Fu, Z.W., Chu, W.Y. Q., Liu, W.Y., and Qin, Q.Z. (2004) *Electrochem. Solid State Lett.*, **7**, A180.

163 Barreca, D., Armelao, L., Caccavale, F., Noto, V.D., Gregori, A., Rizzi, G.A., and Tondello, E. (2000) *Chem. Mater.*, **12**, 98.

164 Liu, P., Zhang, J.G., Turner, J.A., Tracy, C.E., Benson, D.K., and Bhattacharya, R.N. (1998) *Solid State Ionics*, **111**, 145.

165 Mantoux, A., Groult, H., Balnois, E., Doppelt, P., and Gueroudji, L. (2004) *J. Electrochem. Soc.*, **151**, A368.

166 Badot, J.C., Ribes, S., Yousfi, E.B., Vivier, V., Pereira-Ramos, J.P., Baffier, N., and Lincot, D. (2000) *Electrochem. Solid State Lett.*, **3**, 485.

167 Baddour-Hadjean, R., Golabkan, V., Pereira-Ramos, J.P., Mantoux, A., and Lincot, D. (2001) *J. Phys. IV*, **11**, 85.

168 Lee, S.H., Cheong, H.M., Liu, P., Tracy, C.E., Pitts, J.R., and Deb, S.K. (2003) *Solid State Ionics*, **165**, 81.

169 Levi, M.D., Lu, Z., and Aurbach, D. (2001) *Solid State Ionics*, **143**, 309.

170 Lu, Z., Levi, M.D., Salitra, G., Gofer, Y., Levi, E., and Aurbach, D. (2000) *J. Electroanal. Chem.*, **491**, 211.

171 Lee, S.H., Liu, P., Tracy, C.E., and Benson, D.K. (1999) *Electrochem. Solid State Lett.*, **2**, 425.

172 Lee, S.H., Liu, P., and Tracy, C.E. (2003) *Electrochem. Solid State Lett.*, **6**, A275.

173 Ramana, C.V., Hussain, O.M., and Srinivasulu Naidu, B. (1997) *Mater. Chem. Phys.*, **50**, 195.

174 Kim, Y.T., Gopukumar, S., Kim, K.B., and Cho, B.W. (2003) *J. Power Sources*, **117**, 110.

175 Baddour, R., Pereira-Ramos, J.P., Messina, R., and Perichon, J. (1991) *J. Electroanal. Chem.*, **314**, 81.

176 Pyun, S.I. and Bae, J.S. (1997) *J. Power Sources*, **68**, 669.

177 Ibris, N., Salvi, A.M., Liberatore, M., Decker, F., and Surca, A. (2005) *Surf. Interface Anal.*, **37**, 1092.

178 Li, Y.M., Hibino, M., Tanaka, Y., Wada, Y., Noguchi, Y., Takano, S., and Kudo, T. (2001) *Solid State Ionics*, **143**, 67.

179 Bae, J.S. and Pyun, S.I. (1996) *Solid State Ionics*, **90**, 251.
180 Lindström, R., Maurice, V., Zanna, S., Klein, L., Groult, H., Perrigaud, L., Cohen, C., and Marcus, P. (2006) *Surf. Interf. Anal.*, **38**, 6.
181 Swiatowska-Mrowiecka, J., Maurice, V., Zanna, S., Klein, L., and Marcus, P. (2007) *Electrochim. Acta*, **52**, 5644.
182 Swiatowska-Mrowiecka, J., Maurice, V., Zanna, S., Klein, L., Briand, E., Vickridge, I., and Marcus, P. (2007) *J. Power Sources*, **170**, 160.
183 Ramana, C.V. and Julien, C.M. (2006) *Chem. Phys. Lett.*, **428**, 114.
184 Kim, Y.S., Ahn, H.J., Shim, H.S., Nam, S.H., Seong, T.Y., and Kim, W.B. (2007) *Electrochem. Solid State Lett.*, **10**, A180.
185 Yamaki, J.I., Ohtsuka, H., and Shodai, T. (1996) *Solid State Ionics*, **86–88**, 1279.
186 Ohtsuka, H. and Sakurai, Y. (2001) *Solid State Ionics*, **144**, 59.
187 West, W.C. and Whitacre, J.F. (2005) *J. Electrochem. Soc.*, **152**, A966.
188 Julien, C., Hussain, O.M., El-Farh, L., and Balkanski, M. (1992) *Solid State Ionics*, **53–56**, 400.
189 Julien, C., Nazri, G.A., Guesdon, J.P., Gorenstein, A., Khelfa, A., and Hussain, O.M. (1994) *Solid State Ionics*, **73**, 319.
190 Natarajan, C., Fukunaga, N., and Nogami, G. (1998) *Thin Solid Films*, **322**, 6.
191 Liu, W.Y., Fu, Z.W., and Qin, Q.Z. (2007) *Thin Solid Films*, **515**, 4045.
192 Reddy, M.V., Pecquenard, B., Vinatier, P., and Levasseur, A. (2007) *Electrochem. Commun.*, **9**, 409.

11
Research and Development Work on Advanced Lithium-Ion Batteries for High-Performance Environmental Vehicles

Hideaki Horie

11.1
Introduction

The possibility of applying lithium-ion batteries to environmental vehicles, especially electric vehicles (EVs) and hybrid electric vehicles (HEVs), has been actively and continuously studied since the early 1990s from the standpoint of fully utilizing the high-performance potential of these batteries. This chapter discusses the characteristics required of a power source for such environmental vehicles and considers the configuration of an environmental vehicle system based on the distinctive features of lithium-ion batteries.

11.2
Energy Needed to Power an EV

EVs have been expected to be a very effective means of preserving the environment, but several serious issues have prevented their widespread use. The first major issue is the high cost of the vehicle itself, and a second major issue is the short driving range of EVs on a single battery charge. Various other problems have also been pointed out. Vehicle performance and interior space are insufficient. Batteries lack sufficient reliability, and the battery service life is sometimes shorter than expected. The charging operation is troublesome and time consuming. Moreover, the number of EV charging stations is severely limited.

The amount of energy obtainable by burning gasoline is approximately $450 \times 10^5\ \mathrm{J\,kg^{-1}}$. By comparison, the energy density of a typical lead-acid battery is approximately $1.26 \times 10^5\ \mathrm{J\,kg^{-1}}$. There is roughly a 50-fold difference, even though the average energy conversion efficiency of an internal combustion engine is only 14%. Accordingly, an EV must be fitted with a large number of batteries to travel a long distance. This results in a vicious circle, however, in that more batteries not only increase the vehicle weight, which degrades power performance, but they also reduce the available interior space and increase the battery cost.

Lithium Ion Rechargeable Batteries. Edited by Kazunori Ozawa

ISBN: 978-3-527-31983-1

With regard to vehicle propulsion, the resistance component can be expressed as

$$F_{\text{resist}} = \mu_{\text{r}} \cdot g \cdot M + \frac{1}{2}\rho \cdot C_{\text{D}} \cdot A \cdot V^2 + g \cdot M \cdot \sin\theta$$

where M is the vehicle weight in kilograms and V is the vehicle velocity. For example, if we consider the equation in terms of the basic units of meters, kilograms, and seconds, a figure of 17.7 m s^{-1} would be substituted for V at a driving speed of 60 km h^{-1}. The first term on the right-hand side of the equation is the rolling friction resistance of the vehicle, including the tires. The friction coefficient μ typically has a value of around 0.025. The gravitational constant g has a value of 9.8 m s^{-2}. Assuming a vehicle weight of 1500 kg, the calculation becomes $0.025 \times 9.8 \times 1500 = 367.5$ N (or $0.025 \times 1500 = 37.5$ kg f). The second term is the air resistance, which is proportional to the projection area (A expressed in meters squared) of a plane perpendicular to the vehicle's longitudinal direction and it is also proportional to the square of the vehicle velocity. Air density ρ has a value of 1.2 kg m^{-3} and the drag coefficient (C_{D}) of an ordinary vehicle is around 0.35. For example, assuming a vehicle projection area of 2 m^2 and the same vehicle velocity of 16.7 m s^{-1}, the calculation becomes $0.5 \times 1.2 \times 0.35 \times 2\times$ $(16.7 \times 16.7) = 117$ N (or if converted to the unit of kg f, 117 N/9.8 = 11.9 kg f).

The third term represents the hill-climbing resistance relative to the vehicle weight when the vehicle is traveling up an incline having a gradient of θ. While the first and second terms are energy dissipation terms, the third term is conserved as positional energy. Adding up these terms yields the vehicle's total resistance F_{resist} expressed in newtons. Letting $F_{\text{powertrain}}$ denote the energy input by the vehicle's power source, the balance obtained by subtracting the total resistance F_{resist} from $F_{\text{powertrain}}$ can be regarded as the actual power expended to accelerate the vehicle. The vehicle is accelerated if the value is positive and decelerated if it is negative.

$$\begin{aligned}\frac{\mathrm{d}V}{\mathrm{d}t} &= \frac{F_{\text{accel}}}{M} = \frac{F_{\text{powertrain}} - F_{\text{resist}}}{M} \\ &= \frac{F_{\text{powertrain}} - (\mu_{\text{r}} \cdot g \cdot M + \frac{1}{2}\rho \cdot C_{\text{D}} \cdot A \cdot V^2 + g \cdot M \cdot \sin\theta)}{M}\end{aligned}$$

We let L denote the distance the vehicle has moved in the direction of force F by the energy (in joules) input from the power source. From their product, we can express the power output as the energy per unit time as follows:

$$\begin{aligned}P &= \int F_{\text{powertrain}} \cdot \mathrm{d}L = \int F_{\text{powertrain}} \cdot \left(\frac{\mathrm{d}L}{\mathrm{d}t}\right) \mathrm{d}t \\ &= \int F_{\text{powertrain}} \cdot V\mathrm{d}t = \overline{F_{\text{powertrain}} \cdot V}\end{aligned}$$

$$\therefore P = V \cdot \left(\mu_{\text{r}} \cdot g \cdot M + \frac{1}{2}\rho \cdot C_{\text{D}} \cdot A \cdot V^2 + g \cdot M \cdot \sin\theta + M \cdot \frac{\mathrm{d}V}{\mathrm{d}t}\right)$$

If the acceleration pattern $V(t)$ is set, vehicle acceleration can also be found. Substituting that value into the equation above yields the necessary power output. In terms of the example above we obtain the following:

1. *Power output consumed in overcoming rolling friction:.*

$$367.5\ \mathrm{N} \times 16.7\ \mathrm{m\ s^{-1}} = 6137\ \mathrm{W} = 6.1\ \mathrm{kW}$$

2. *Power output consumed in overcoming air resistance:.*

$$117\ \mathrm{N} \times 16.7\ \mathrm{m\ s^{-1}} = 1954\ \mathrm{W} = 1.9\ \mathrm{kW}$$

At a steady driving speed of 60 km h^{-1}, power output of approximately 8 kW is needed to propel the vehicle. In other words, the amount of energy consumed in one hour of driving is 8 kW h (= 8000 × 3600J = 28.8 MJ). Since the distance traveled in that time is 60 km, the energy loss by the vehicle in traveling a unit distance of 1 km can be calculated as

$$8000\ \mathrm{W\ h}/60 = 133\ \mathrm{W\ h\ km^{-1}}$$

From the energy that can be extracted from the battery, it is necessary to consider the energy losses that occur in the power transmission system, the motor controller system, and in battery charging/discharging operations. Considered from the standpoint of power output, rolling friction resistance is proportional to the vehicle velocity V and air resistance is proportional to V^3. Accordingly, rolling friction resistance is the main power-consuming factor at low speeds, and the share of power consumption accounted for by air resistance increases at high speeds. By performing the calculations noted above, the number of batteries that must be fitted on a vehicle can be determined from the single-charge driving range required of the vehicle.

11.3 Quest for a High-Power Characteristic in Lithium-Ion Batteries

HEVs specifically require batteries with a high power output characteristic. If the output characteristic of the battery could be further improved, the number of batteries required could be reduced by a corresponding extent and a substantial cost reduction could be expected as a result. In the past, the power density of a battery was around 0.2 k W kg^{-1} at the most. Taking an HEV fitted with a 30-kW motor as an example, a battery weight of approximately 150 kg would be needed to provide the power for propelling the vehicle. In terms of the actual vehicle design, the battery pack alone would account for as much as or more weight and volume than a conventional internal combustion engine. For this reason, there was a long period of time when batteries were not considered to be practical for vehicle propulsion systems. However, if power density could be markedly improved to provide higher performance as a result of improvements made to new types of batteries, the number of batteries needed could be reduced, providing a way to reduce the weight and size of the battery pack.

Before discussing the improvement of power density, let us briefly consider the relationship between the internal resistance of a cell and (i) maximum power output, (ii) energy efficiency in charging/discharging operations, and (iii) reduction of cell heat generation. The reason behind this is that the absolute value of battery output in an environmental vehicle is exceptionally large, so if the energy efficiency of the battery is low, the vehicle's energy efficiency will decline sharply. Moreover, as is described later, the large size of the battery means that heat generation becomes a much more critical issue than in the case of small batteries, making it necessary to suppress the amount of heat generated by the battery itself.

Figure 11.1 illustrates cell voltage behavior during charging and discharging. Cell voltage falls during discharging and rises during charging in proportion to the cell's open-circuit voltage. The difference relative to the open-circuit voltage during charging/discharging is thought to be due to the internal resistance intrinsic to the cell. In other words, the voltage difference ΔV is determined by the current $(I) \times$ internal resistance (R) in accordance with Ohm's law.

1. *Maximum power output:* The maximum dischargeable current value is determined such that the cell voltage during discharging does not fall below the lower limit voltage set for each type of cell. Assuming that overvoltage accompanying cell reactions can be ignored, the maximum current IMAX is

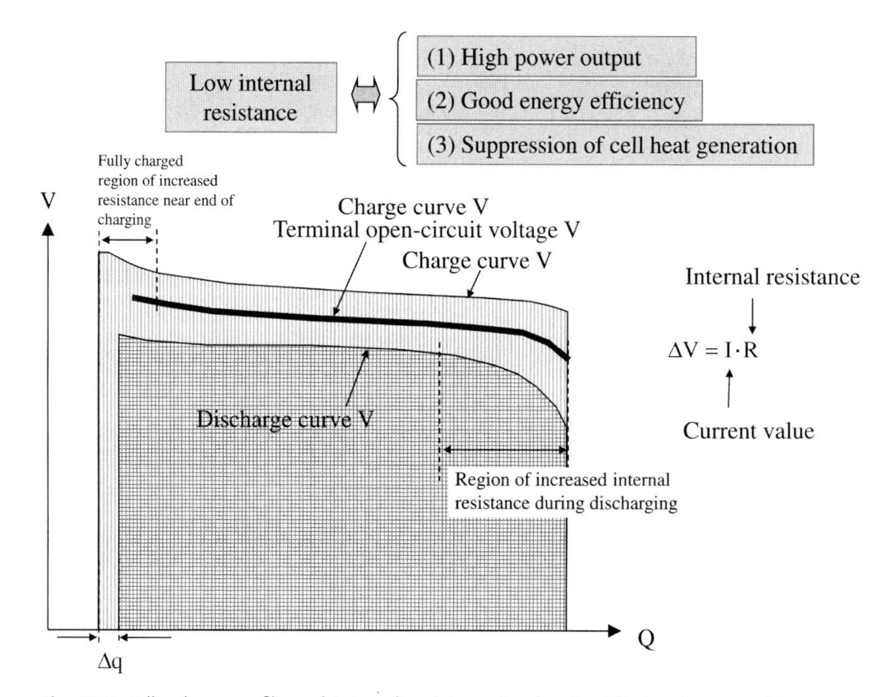

Fig. 11.1 Cell voltage profile and internal resistance in charging/discharging operations.

determined by Ohm's law as follows:

$$\text{IMAX} = \left[\left\{\text{cell open-circuit voltage}\right\} - \left\{\text{lower limit voltage}\right\}\right] / R$$

Hence, the maximum power output of a cell can be calculated as IMAX × lower limit voltage. Increasing the maximum power output requires a reduction of internal resistance R.

2. *Cell charging/discharging efficiency:* The electrical energy transferred during charging/discharging can be calculated as $\Delta Q \times V$, where ΔQ is the transferred charge and V is the terminal voltage at that time. Accordingly, the electrical energy expended externally during charging is the area between the charge curve and the horizontal axis (Q axis) in Figure 11.1. Similarly, during discharging, it is the area between the discharge curve and the horizontal axis. Charging/discharging efficiency is simply the ratio of this area during discharging to the area during charging. Therefore, minimizing the difference in these areas along the horizontal and vertical axes is effective in improving charging/discharging efficiency. In other words, it is necessary (i) to reduce internal resistance and (ii) to bring Coulomb efficiency (the ratio of the discharge to the charge) close to 100%.
3. Charging/discharging operations cause the cell temperature to rise. It is essential to suppress this heat generation because high temperatures above 50 °C generally induce cell degradation. Since cell reactions are reversible processes, they are not considered here. (The heat generated and heat absorbed accompanying cell reactions in one charging/discharging cycle balance out to be zero, and the values themselves are not particularly large.) It is assumed that the entire portion lost from 100% charging/discharging efficiency becomes heat in the end and is represented in Figure 11.1 as the difference in area between the charge-curve area and the discharge-curve area. Stated the other way around, internal resistance must be reduced to suppress battery heat generation.

To sum up the foregoing discussion, reducing internal resistance is necessary to improve battery power output, to increase charging/discharging efficiency, and to suppress heat generation; and there is a one-to-one correspondence between them. In short, reducing internal resistance is equivalent to improving cell performance characteristics.

Let us briefly consider the internal resistance of a cell. The resistance component is related to ion conductivity, electron conductivity, and energy exchanges in cell reactions, among other things. For example, electron conductivity is related to the materials and geometrical structures of the metal parts such as the collectors and terminals, the network between active-material particles, and electron conductivity within the crystals of the active materials. With regard to ion conductivity, the flow of ions can be given by the following diffusion equation:

$$\frac{\partial c}{\partial t} = \mathrm{D}\left(\frac{\partial^2}{\partial x^2} + \frac{\partial^2}{\partial y^2} + \frac{\partial^2}{\partial z^2}\right) c$$

Because lithium-ion batteries use an organic solvent having low ion conductivity, it was previously believed for a long time even among learned societies and the battery industry worldwide that these batteries were intrinsically incapable of high power output. However, the author was very dubious of that view, because if it were true, then the lithium-ion diffusion process would be considered the principal rate-determining factor. Accordingly, it should be possible to improve power density substantially by taking the opposite approach of promoting lithium-ion diffusion.

This line of thinking provided the motivation for initiating a theoretical study around 1993 with the aim of reducing internal resistance. To pursue all sorts of conceivable possibilities, Mathematica was used to make several-dozen simulation programs that were run on a personal computer in a process of trail and error. In this way, success was achieved in identifying the factors that seemed to contribute the most to performance. As a result, it became clear that lithium-ion transport was the key factor, as had originally been suspected.

The ion current, representing the total quantity of lithium ions transported, can be expressed by the following equation:

$$I = \int j \mathrm{d}S = \int \mathrm{D} \frac{\partial c}{\partial x} \mathrm{d}S$$

Calculations are performed at three locations for the separate media that the lithium ions pass through:

1. ion conduction in the separator electrolyte;
2. ion conduction in the electrolyte between the active materials; and
3. ion conduction within the solid particles of the active materials.

The simultaneous equations for these locations can be solved by determining a specific diffusion coefficient and diffusion equation for each one and by assuming an equation for continuity at the boundary layers.

The results of this study revealed that much greater power output could be obtained by adjusting lithium-ion transport in the electrolyte and in the active materials. A lithium-ion cell operates on the basic principle of intercalation in the crystalline structure of the positive and negative electrodes. Therefore, the changes in volume and strain that occur in the cell materials accompanying charging and discharging can be substantially suppressed to allow the use of an exceptionally thin electrode structure. Using a broader electrode surface area and a narrower gap between the electrodes can be expected to promote much greater lithium-ion diffusion.

Figure 11.2 shows an example of the simulation results. The vertical axis indicates the power density ratio, letting the power density of conventional lithium-ion batteries equal one. In this simulation, the material properties and geometrical structures were arranged such that the ion current in each cross section would basically have the same value and the rate-determining parts restricting lithium-ion transport would be eliminated. Specifically, the quantity of lithium ions transported though the electrolyte and the quantity transported through the active materials

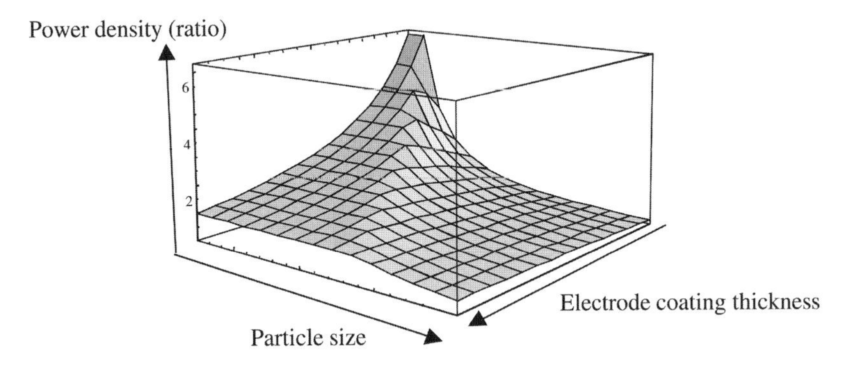

Fig. 11.2 Potential of lithium-ion cells for higher power output. (Source: Horie, H. (1998) *Development of a PHEV Source System using High-power Lithium-ion Batteries.* Preprints of the Autumn Scientific Lecture Series of JSAE, No. 80, pp. 5–8 (Oct. 1998) (in Japanese)).

were adjusted by reducing the size of the active-material particles and expanding the total solid-surface area. The simulation results made it clear that power density could be significantly improved by optimizing the parameters in this way.

Naturally, there are many factors that hamper the attainment of higher power output. If the things that were learnt through much repeated trial and error in the mid-1990s were to be summarized, it would be that an especially crucial point is to adjust lithium-ion diffusion between the electrolyte and the active materials (Figure 11.3). Essentially, there is an enormous difference in diffusion constants between the active materials and the electrolyte. In spite of that huge difference, the lithium-ion flux can be brought to the same or nearly the same order in the end by optimizing the electrode thickness and the particle size of the active materials. Since ion movement is governed by the diffusion equations, the large difference in diffusion constants can be compensated for by the effect of squaring the inherent dimensions.

In line with the foregoing concept, we pushed ahead with our research and constructed a cell simulation program for examining cell performance more quantitatively. The basic concept of this cell performance simulation program is outlined in Figure 11.4. This simulation program consists of two main elements:

1. lithium-ion transport equations: diffusion equations; and
2. electron transport equations: electric field and electron conduction equations.

Let us take a brief look at some of the calculation results obtained with this simulation program. In Figure 11.5, the vertical axis shows the obtainable capacity (percent) in a continuous discharge for three different positive electrode thicknesses. The experimental data (indicated by the triangles, squares, and diamonds) and the simulation results (shown by the curves) coincide closely.

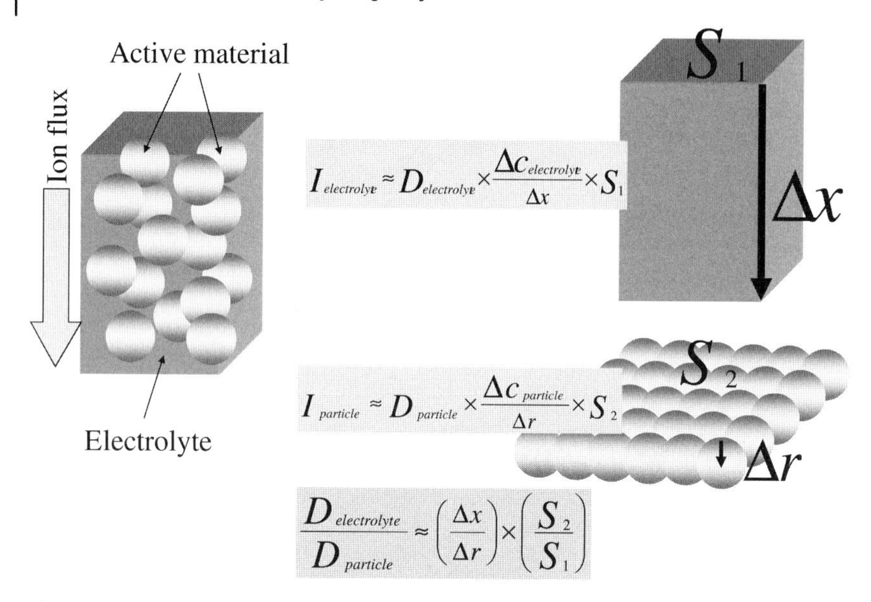

Fig. 11.3 Attainment of higher power output by adjusting lithium-ion diffusion constants. (Source: Horie, H. (2006) *Development of High Power Lithium-ion Batteries*. The 23rd Battery Seminar, Fort Lauderdale, FL, p. 3).

Figure 11.6 shows the obtainable capacity (percent) in large-current discharges as a function of the thickness of the composite positive electrode. These results also indicate that the simulation program can calculate cell discharge behavior with remarkably high accuracy.

An attempt was also made to simulate voltage drop behavior at various current values for different electrode thicknesses, and the results are shown in Figure 11.7. As the data presented here indicate, this simulation program can accurately predict cell performance.

To validate the concept explained here, a prototype cell of an ordinary size was fabricated and evaluated. Figure 11.8 shows the data that were made public in 2004 for this ultrahigh-power prototype cell. When fully charged, the cell displayed specific power of more than 10 k W kg^{-1}, thus breaking the 10 k W kg^{-1} barrier for the first time anywhere in the world.

Figure 11.9 shows the specific power displayed by the prototype cell in laboratory evaluations. The results unmistakably indicate that a significant improvement in power output was achieved with this cell. Although it was previously believed that lithium-ion batteries were intrinsically incapable of producing high power, we were the first to discover that lithium-ion batteries have the potential to provide exceptionally high power output. In validation tests conducted on actual cells during the last 10 years, we have continued to break our own world record for the highest power output of lithium-ion cells and have created and led this research field almost independently.

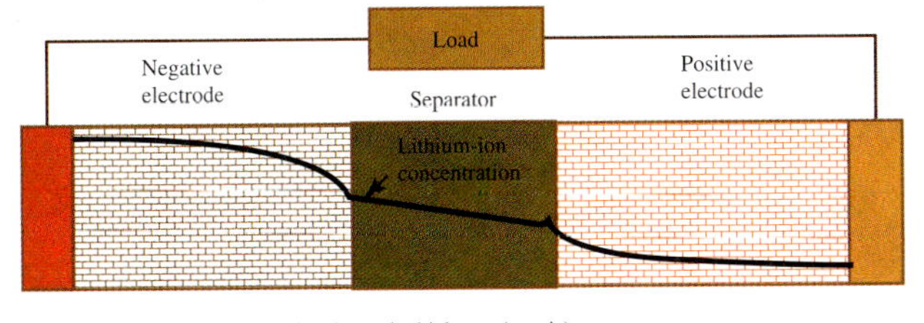

• Equation for lithium-ion transport in active materials

$$\frac{\partial C_s}{\partial t} = D_s \nabla_r^{2\bullet} C_s \qquad \text{(one dimension in the radial direction)}$$

• Equation for lithium-ion transport in electrolyte

$$\varepsilon \frac{\partial C_e}{\partial t} = \varepsilon D_e \nabla_x^{2\bullet} C_e + \frac{(1-t_s)}{F} a\, i_n$$

(one dimension along the electrode thickness)

(i) Lithium-ion Transport Model

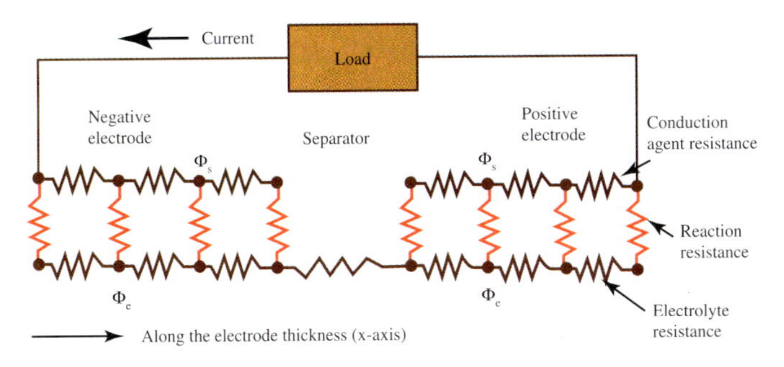

• Equation for current transfer in active materials

$$\sigma_s \nabla_x^{2\bullet} \Phi_s - a i_n = 0 \qquad \text{(one dimension along the electrode thickness)}$$

• Equation for current transfer in electrolyte

$$\sigma_e \nabla_x^{2\bullet} \Phi_e + a i_n - \sigma_e \nabla_x^{2\bullet} (\mathrm{InCe}) \bullet 2RT(1-t) = 0$$

$$i_n = k\, Ce^{\alpha s} (Ct - Cs)^{\alpha s} Cs^{\alpha c} [\exp(\alpha F/RT^* \eta) - \exp(-\alpha F/RT^* \eta)]$$

$$\eta = \Phi_s - \Phi_e - U(Cs) \qquad \text{(one dimension along the electrode thickness)}$$

(ii) Current Transfer Equation

Fig. 11.4 Configuration of cell performance simulation program. (Source: Abe, T. *et al.* (2001) *Simulation of a High-Power Lithium-ion Battery*. ECS 2001 Joint International Meeting, San Francisco, California).

Using the cell performance simulation program, predictions were made of the lithium-ion cell performance needed to satisfy the performance requirements of EVs, plug-in HEVs, and HEVs. As the results in Figure 11.10 indicate, lithium-ion cells can be optimized over a wide performance range to provide the energy density and power density required by each type of vehicle.

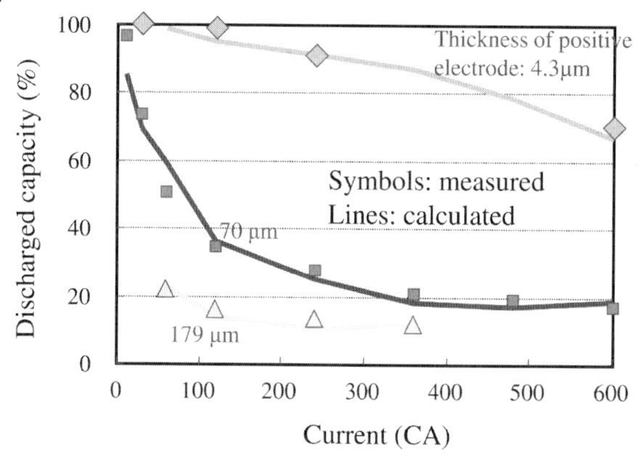

Fig. 11.5 Obtainable capacity in large-current discharges.

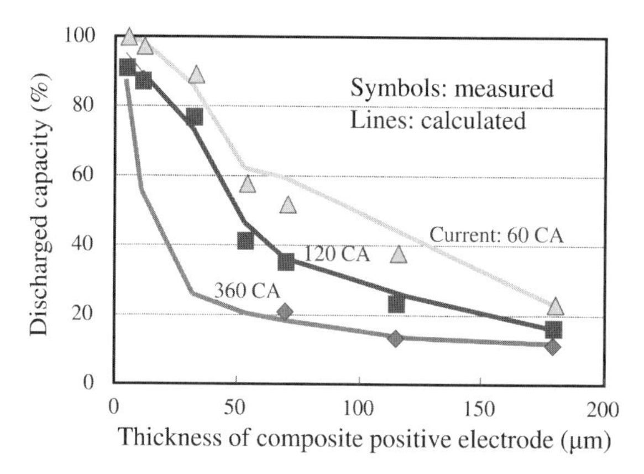

Fig. 11.6 Obtainable discharge capacity as a function of positive electrode thickness.

11.4
Cell Thermal Behavior and Cell System Stability

In this section, we will calculate the heat generation behavior of a lithium-ion cell for environmental vehicle use. Heat generation by the cell can be divided between joule heat and the heat of reaction produced by the cell reactions.

$$\begin{aligned}
\omega &= \lfloor \mathrm{w_{joule}} + \mathrm{w_{react}} \rfloor_{\mathrm{T=const}} \\
&= \int \mathrm{R_{direct}} \cdot \mathrm{I}^2 dt + \int \mathrm{T} dS \Big|_{\mathrm{T=const}} \\
&= \int \mathrm{R_{direct}} \cdot \mathrm{I}^2 dt - \int \mathrm{T} \frac{\mathrm{dV}}{\mathrm{dT}} dq
\end{aligned}$$

In the quasi-static limit, thermal values are identical during charging and discharging but have opposite signs. In principle, heat generation and absorption

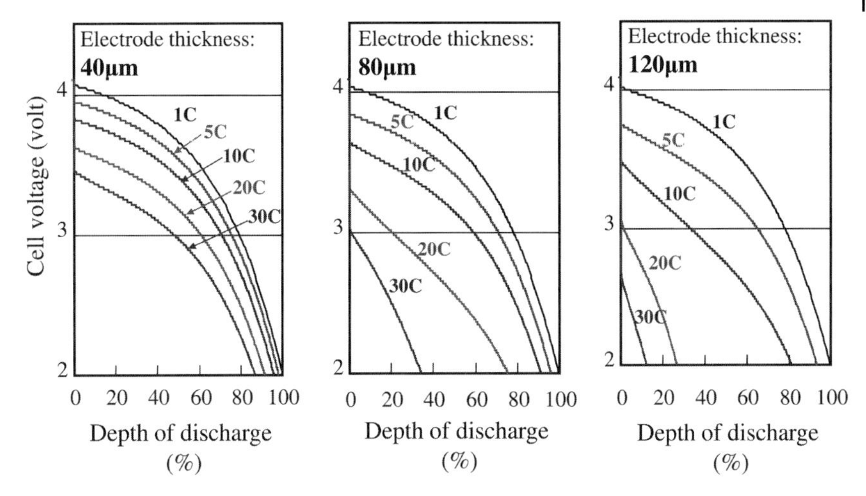

Fig. 11.7 Voltage drop characteristics for different discharge current values. (Source: Shimamura, O., Abe, T., Watanabe, K., Ohsawa, Y., and Horie, H. (2006) *Research and Development Work on Lithium-ion Batteries for Environmental Vehicles.* Electric Vehicle Symposium 22, Yokohama, p. 10).

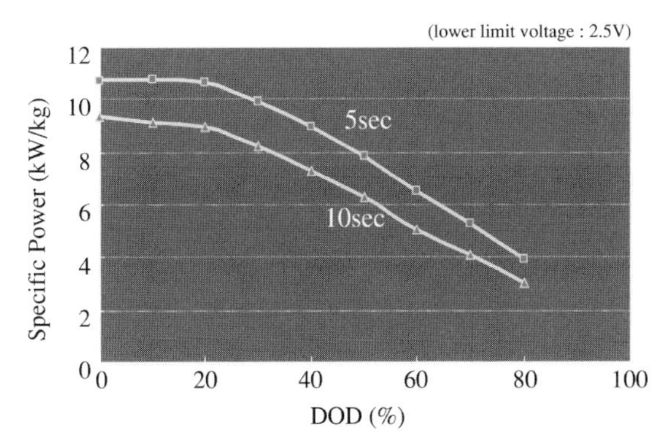

Fig. 11.8 Large-current discharge characteristics. (Source: Horie, H., Shimamura, O., Saito, T., Abe, T., Ohsawa, Y., Kawai, M., and Sugawara, H. (2004) *Development of Ultra-high Power Lithium-ion Batteries.* 12th International Meeting on Lithium Batteries, Nara, Japan, June 27-July 2, 2004).

cancel each other out in one charging/discharging cycle, as was mentioned earlier. Joule heat is determined by internal resistance and the current in relation to the demanded output. The reaction volume per unit of time can be derived from the current value, making it possible to determine the quantity of heat generated. In the case of an EV or HEV, the level of power demanded is high, the cell size is large, and the materials making up the cell have a small heat transfer coefficient. For these reasons, the cell temperature is apt to rise owing to heat that stays inside the

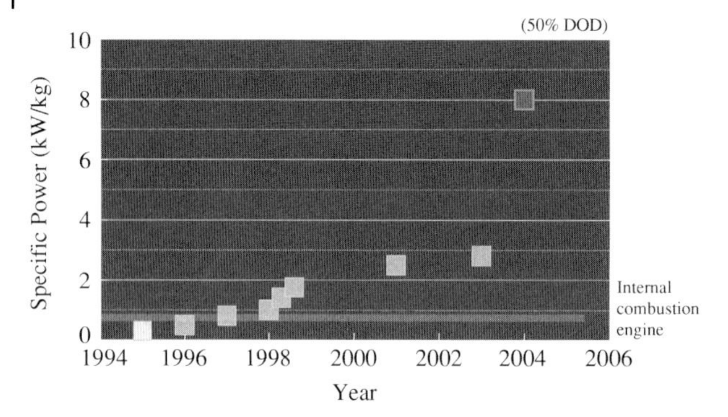

Fig. 11.9 Improvement of power output of lithium-ion cells. (Source: Horie, H., Shimamura, O., Saito, T., Abe, T., Ohsawa, Y., Kawai, M., and Sugawara, H. (2004) *Development of Ultra-high Power Lithium-ion Batteries*. 12th International Meeting on Lithium Batteries, Nara, Japan, June 27-July 2, 2004).

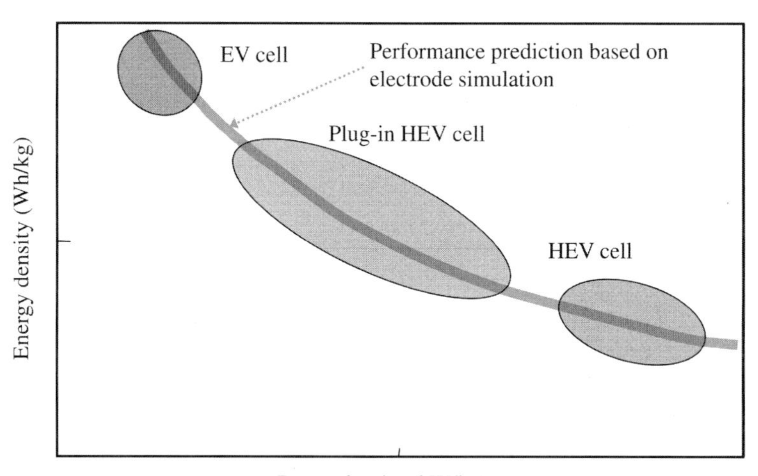

Fig. 11.10 Cell performance simulation results. (Source: Shimamura, O., Abe, T., Watanabe, K., Ohsawa, Y., and Horie, H. (2006) *Research and Development Work on Lithium-ion Batteries for Environmental Vehicles*. Electric Vehicle Symposium 22, Yokohama, p. 10).

cell. Since cell degradation generally occurs under exposure to high temperatures, steps must be taken to keep the cell temperature at a low level.

Figure 11.11 shows the maximum temperature inside a cell in relation to the cell thickness. Doubling the cell thickness results in a fourfold, not a twofold, increase in the maximum temperature inside the cell. This indicates that even a slight increase in cell thickness can cause the maximum temperature inside the

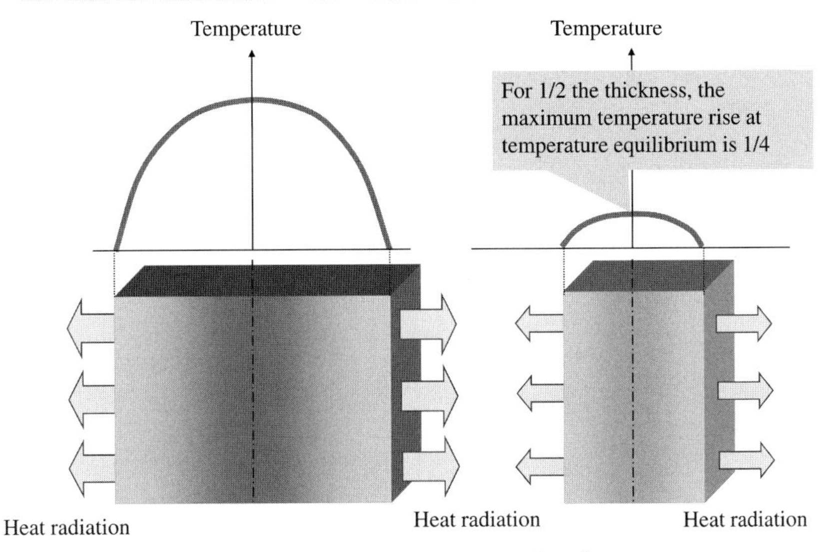

Fig. 11.11 Cell thickness versus maximum temperature inside cell.

cell to rise considerably. This temperature rise is a potential problem that must be addressed in developing batteries for environmental vehicles.

Finally, let us consider the stability of a cell system in terms of thermal performance. This issue is explained here using a simple toy model that has been devised by the author. In the case of cells with an aqueous solvent electrolyte, electrolysis of water also accompanies charging/discharging reactions (Figure 11.12). A small cell has only one equilibrium solution for heat generation and heat radiation from the cell surface, so it is possible to tell if the system has thermal stability. However, as seen in the graph in Figure 11.12(a), medium-size to large cells have three points where there is an equilibrium solution for heat generation and heat radiation from the cell surface. Among them, the middle intersection point is an unstable one. This means that if the cell temperature rises, it may jump suddenly to the equilibrium solution in the high-temperature region. Cell systems with an aqueous solvent electrolyte thus have thermally stable and unstable internal states. The author was the first researcher to discover the relationship between the branching of the equation solution and this jumping behavior of the equilibrium temperature.

In contrast, lithium-ion cells do not have such thermal instability because the organic solvent does not undergo electrolysis, and only the cell reactions consume electrical charge. This is true even for medium-size to large cells for plug-in HEVs and also for EV cells. The thermal stability of lithium-ion cells gives them a superior advantage for constructing battery systems. Therefore, in terms of thermal stability as well, lithium-ion cells can be used to build optimal battery systems for HEVs, plug-in HEVs, and EVs.

Issue in cell systems where charging efficiency declines at high temperature
Temp. rise → Decline in charging efficiency →Increased heat generation →Temp. rise

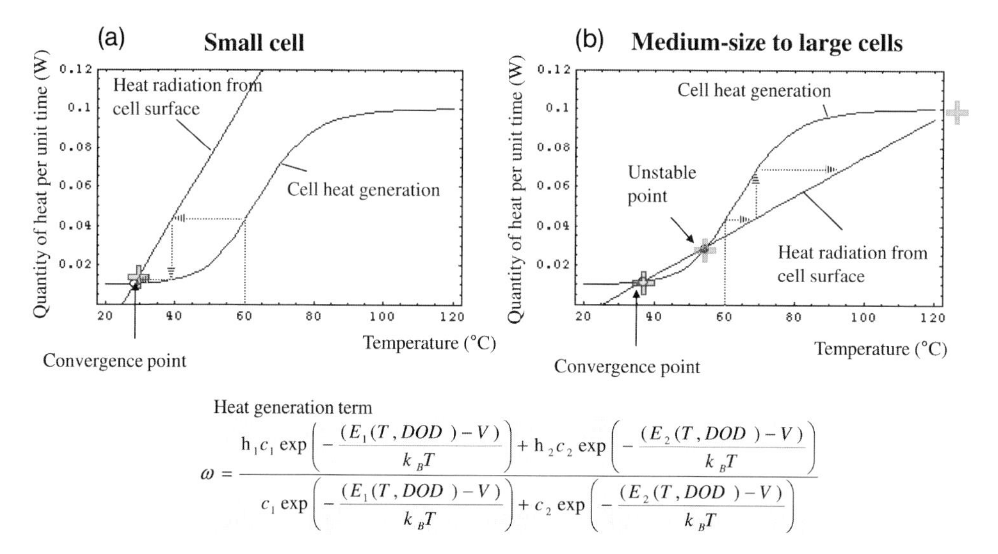

Fig. 11.12 Thermal instability of medium-size to large cell systems with an aqueous solvent electrolyte.

Further Reading

Abe, T. *et al.* (2001) "Simulation of a High-Power Lithium-ion Battery". Electrochemical Society 2001 Joint International Meeting – San Francisco, California.

Abe, T., Shimamura, O., Watanabe, K., Ohsawa, O., and Horie, H. (2006) "Development of a Lithium-ion Battery System for Environmental Vehicles". Autumn Scientific Lecture Series of JSAE, No. 132-06, pp. 1–5, 9/27/ Japan (in Japanese).

Horie, H., Tanjo, Y., Abe, T., Katayama, K., and Shigetomi, J. (1997) *Development of A High Power Lithium-Ion Battery System for HEV*. EVS14, Orlando, Dec.

Horie, H., Touda, M., Tanjo, Y., and Shigetomi, J. (1997) "Development of a High-power Lithium-ion Battery System for HEV Application". Preprints of the Spring Scientific Lecture Series of JSAE 971, p. 53 (in Japanese).

Horie, H., Shimamura, O., Saito, T., Abe, T., Ohsawa, Y., Kawai, M., and Sugawara, H. (2004) "Development of Ultra-high Power Lithium-ion Batteries". 12th International Meeting on Lithium Batteries, Nara, Japan.

Horie, H. *et al.* (2005) "Development of High Power Lithium-ion Batteries". The 46th Battery Symposium, Japan.

Horie, H. (2006) "Development of High Power Lithium-ion Batteries". 23th Battery Seminar, Fort Lauderdale, Fl.

Kitada, S. and Horie, H. (1996) *Development of Nissan HEV*. EVS13.

Miyamoto, T., Tohda, M., and Katayama, K. (1996) "Advanced Battery System for Electric Vehicle (FEV-II)".

Electric Vehicle Symposium13 (EVS13), Osaka, 13–16 Sept.
Nishi, Y. (1997) *The Story of Lithium-ion Secondary Batteries*. Shokabo, Tokyo, (in Japanese).
Nishi, Y., Katayama, K., Shigetomi, J., and Horie, H. (1998) "The Development of Lithium-Ion Secondary Battery Systems for EV and HEV". 13th Annual Battery Conference, Long Beach, CA, Jan.
Origuchi, M., Miyamoto, T., Horie, H., and Katayama, K. (1997) "Development of a Lithium-ion Battery System for Evs". SAE Technical Paper 970238, Detroit.
Shigetomi, J., Katayama, K., Tanjo, Y., Shimamura, O., Abe, T., Horie, H., Ohsawa, Y., and Kawai, M. (1988) "Increasing the Power Output of Lithium-ion Batteries". Abstracts of the 39th Battery Forum, p. 221 (in Japanese).
Shimamura, O., Abe, T., Saito, T., Ito, T., Kinoshita, T., Watanabe, K., Horie, H., Sugawara, H., Makita, H. and Miyamoto, N. (2004) "Development of a High-power, Compact Lithium-ion Battery System". Autumn Scientific Lecture Series of JSAE, Japan (in Japanese).
Shimamura, O. *et al.* (2005) *Development of a High Power Compact Lithium-ion Battery System*. EVS21.
Shimamura, O., Abe, T., Watanabe, K., Ohsawa, Y., and Horie, H. (2006) *Research and Development Work on Lithium-ion Batteries for Environmental Vehicles*. EVS23.

Index

a
additive surface film 164, 168f
additives
– for suppression of manganese (II) dissolution 188f
– functions of 180
admetal doping 125
aging see conditioning
air resistance of electro vehicle 314
alignment of lithium diffusion planes 259
alkali salt coating 207f
alkyl carbonate-type solvent 7, 29f, 33f, 73, 164, 217
alloying of lithium 165
all-solid-state lithium polymer battery 231f
– comparison with liquid electrolyte battery 232f
– electrochemical behavior of 240f
– electrode surface chemistry 247
– fabrication of nonflammable 243f, 246
– fabrication of plasticizer-containing 232f
α-$LiFeO_2$-type layered structure 56
α-lithium manganese dioxide 125f
α-$NaFeO_2$-type layered structure 14, 39, 56
alternating current impedance analysis 223f, 227, 238f, 246, 279
aluminum phosphate 245f, 262
ammonium iodide 189
analysis of cell performance 319f, 324
analysis of local structure see Raman microspectrometry
anatase
– crystal structure of 143f
– Raman-active modes of 146
– structure variations induced by lithium insertion 148f
anion conduction 227
anode can 7
aqueous electrolyte 2f
as-deposited film 260

b
battery
– functions 1
– general structure 1f
battery cycling system 71
bias sputtering 260
binder 232
biphenyl (BP) 172
bulk resistance 227, 231, 239

c
calcite-type borate cathode 62f
calculation of maximum power output 316f
capacity fading 122f, 125
carbon anode 4, 6
carbon coating 64
– effect on electrochemistry of cathode-active material 154f
– nanocoating 58
carbon-lithium cobalt oxide cell 184
carbon-LMO cell 180f
– charge capacity of 182
– potential relation of reactions in 181
carbonaceous anode material 70, 74f, 163f
– artificial graphite "MAG" 165
– effect of heat treatment on 74f
– graphitization degree of 75f, 84
– heat-treated material 79f
– highly crystalline graphite anode 170f
– interlayer spacing 84
– natural graphite 82f, 85, 181
– pre-coke material 77f
– surface charge distribution on 176
carbonization 74f
catalytic electrolyte decomposition 163
– effect of additives on 164

Lithium Ion Rechargeable Batteries. Edited by Kazunori Ozawa

ISBN: 978-3-527-31983-1

cathode capacity 165
cathode protection agent 180
cathodic additive 168, 172
cation dissolution 27f
cation mixing 40, 54, 70
cation ordering 15f, 70, 125f
cation vacancy 15
cell heat generation 316f, 322f
– dependence on cell thickness 324f
cell resistance 263
– relationship to energy efficiency in charge/discharge process 316
– relationship to maximum power output 316
cell voltage profile
ceramic-coated cathode 236
charge/discharge curve
– of a graphite anode 171, 182, 192f, 199, 204, 208
– of anatase 144
– of cobalt-containing LiNiMO cathodes 46
– of composite lithiated vanadium pentoxide electrode 132
– of LiNiMO cathodes 30, 44, 50
– of LMO cathode 26, 28
– of LMO thin film 280
– of LTO cathode 32f, 144
– of olivine-type iron phosphate cathode 62
– of perovskite-type fluoride cathode 64
– of vandium pentoxide thin film 138
charge/discharge process 1f
charge/discharge-end voltages of 12-V batteries 31
charging voltage limitation 165, 172
chemical lithiation 62
chemical vapour deposition (CVD) 269
Clarke numbers 54
Cole-Cole plot (Nyquist plot) 200f, 223, 238f, 291
conditioning 174
conductive membrane 165
control of SEI film formation 174f
corrosion inhibitor 180, 193
C-rate (charge/discharge-current rate) 67
crystallinity of polymers 220
crystal-lattice vibration 104
crystal structure 13
– of α-$NaFeO_2$-type layered structure 13
– of anatase 144
– of ferric sulfate 57f
– of lithium cobalt oxide 107
– of olivine 13, 244
– of spinel-derived superstructures 16f
– of spinel $MgAl_2O_4$ 12f
– of vandium pentoxide 127f, 285
current transfer equation 321
cutoff voltage 165, 265
cyclability
– of additive-containing lithium cobalt oxide-graphite cells 172f
– of lithiated manganese oxides 124
– of lithium iron phosphates 62
– of overlithiated materials 49f
cycle life 67
cyclic voltammetry (CV) of SPE films containing B-PEG plasticizer 229f
cycling, effect on thermodynamics 93
cyclohexylbenzene (CHB) 173

d

Daniel cell 2
depolarizing ability 3
deposition temperature 263
diffusion length of lithium ions in electrodes 232
dilute Stage I intercalation compound 82
direct-current (DC) sputtering 259
discharge capacity
– dependence on cathode thickness 322f
discharge curve
– of lithium cobalt oxide thin film 268
– of lithium/LiPOB/lithium cobalt oxide thin cell 263
dissolution of cobalt ions 262
dissolution of manganese ions 122, 182, 191, 281
doping/undoping 4
DSC analysis 220f

e

electric vehicle (EV)
– calculation of the resistance of 314
– energy demand for powering 313f
electrochemical cell for themodynamic studies 73
electrochemical cobalt intercalation 184
electrochemical extraction of lithium ions 122, 137f, 145
electrochemical insertion of lithium ions 122, 137f, 145
electrochemical quartz crystal microbalance (EQCM) technique 281
electrochemical thermodynamics measurement system (ETMS) 71f, 95
electrochromic material 257
electrode additive 164
electrode cavity 232

electrode composition current dependence 72
electrode morphology
– effect on electrochemistry of cathode-active material 137f
electrode-electrolyte interaction 232f
electrodeposition of metals on electrode 182f, 193
electrode reaction 67
electrolyte additive 166f
– effect of chloride anions as 208
– effect of cobalt (II) ion as 184f
– effect of manganese (II) ion as 182f
– effect of nickel (II) ion as 186f
– effect of potassium ion as 199, 204f
– effect of sodium chloride coating 207
– effect of sodium ion as 197f
– effect of 2-vinylpyridine as 189f
– forming protective surface film on graphite anodes 167, 171
– strategy for development of 170
electrolyte decomposition 184, 188f
electrolyte/electrode interface 179
– impedance induced by 223
– in all-solid-state lithium polymer batteries 233f
– in liquid-electrolyte batteries 233f
– structural modification of 197, 200f
electrolyte salt 6
electrolyte solvent 6, 164, 166
– decomposition of 167f
electron-beam evaporation 273, 275
electron cyclotron resonance (ECR) plasma sputtering 263
electron source-limited capacity 26, 28f
electroreductive polymerization 190f
electroreductive SEI formation 181, 190, 199
electrostatic spray deposition (ESD) 258
energy density of lithium-ion cells, increase of 164f, 172f
energy efficiency of EV 316f
enhancement of power output 320f, 324
ESD see electrostatic spray deposition
exfoliation 232, 234, 300

f
ferric fluoride cathode, carbon coated 64
ferric oxide cathode 55
ferric sulfate 57f
flame retardant additive 173, 217
flooded electrolyte 31
fluorinated cyclotriphosphazene 173
formation of passivation layer 233
free volume theory of polymers 224
functional electrolyte 170

g
γ-Al_2O_3spinel type compounds 14
Galvanic cell 1f
galvanostatic lithiation 139
gas evolution on electrodes 170, 174
gas lattice model 82
gel polymer electrolyte 213, 217
– thermal stability of 219f
glass transition temperature of polymers 220f, 226
graphite anode
– capacity of 63, 163f
– dependence of surface morphology on electrolyte additive 196f
– intercalation of cobalt into 184
– intercalation of potassium ions into 204f
– intercalation of sodium ions into 197
graphite conductor 6
graphitization 74f
– degree of 84
– entropy of 85f
group-subgroup relation of spinel space groups 19f

h
high-capacity thin film 290
highest occupied molecular orbital (HOMO) energy of additives 166f
high-performance environmental vehicle 313
hybride electric vehicle (HEV) 313
hysteresis 74

i
ignition test with plasticizer-containg SPE 243f
inductive effect 62
in-plane lithium concentration 82
in situ bending beam method (BBM) 281
in situ phase transition 137
in situ plated lithium anode 264
interface resistance 227
interlayer expansion 206
internal cell resistance see cell resistance
intramolecular vibration 104
ion current 318
ionic conductivity of SPE films 231
– Arrhenius plots for polymer electrolytes 224
– containing B-PEG plasticizers 223f
– effect of lithium salt concentration on 226f

ionic conductivity of SPE films (*contd.*)
– temperature dependence of 224f
ionic radii 23
iron-based cathode-active materials, capacity of 63

j

Jahn-Teller distortion, in lithiated manganese oxides 120, 123f

l

lattice energy 215f
lattice parameters, variation of 43, 48
layered lithium cobalt oxide 4
layered rocksalt structure 54
lead acid storage battery 3, 30f
lead-free accumulator 29f
– application of 31
Leclanche cell 3
Li-$AlPO_4$composite PEG-BE-$LiFePO_4$cell 246f
$LiCoO_2$-graphite battery
– crystal structure 12f
limiting frequency 293
LiNiMO see lithium nickel manganese oxide
LiNiMO thin film 274f
liquid-liquid junction see also separator 2
lithiated anatase titanium dioxide 143f
– lattice dynamics simulation of 147f
– phase transitions of 145
– Raman spectra of 145f, 149
– structural investigations of 147f
lithiated vandium pentoxide
– comparison of bulk material and sputtered thin film 139f
– deformation of vanadium pentoxide layer 135
– phase structure of 130, 133
– phase transitions of 132, 135
– Raman spectra of 134f, 136f, 139, 142
– structural investigations of 133f
– thin film of 138f
lithiation/delithiation process 70f
lithium aluminum manganese oxide (LAMO) 32f
lithium bis-pentafluoroethanesulfonimide (LiBETI) 216
lithium bis-trifluoromethanesulfonimide (LiTFSI) 216, 219
lithium bromide 188f
lithium cell 4f
– anode-active materials for 4f
– cathode-active materials for 4
– laminated film-packaged 165
lithium cobalt oxide 39, 70, 86f, 107
– as cathode in TFB 262
– crystal structure of 107f
– microstructural analysis of 111
– phase diagramm of 87
– Raman spectrum of 108f, 112
lithium cobalt oxide thin film 259
– CVD film 269
– lithium loss in 266
– morphology of 266f
– nano-silica modified 270
– PLD film 265
– sputtered on stainless-steel substrate 265f
lithium diffusion 293, 318
lithium diffusion layer 54
lithium diffusion model 251
lithium-free battery 264
lithium hexafluoro phosphate 6, 29f, 33f, 73
lithium hexafluorophosphate stabilizer 180
lithium insertion electrode 4
lithium insertion material 7
– lithium titanate 24
– with layer structure 7f
– with olivine structure 7f
– with spinel-frame work structure 7f
– zero-strain 27f
lithium insertion/deinsertion
– into vanadium pentoxide 131
lithium intercalation/deintercalation 4, 70
– detection of local structure variations induced by 107, 288
– entropy of 78, 80f, 91
– into carbonaceous material 77f, 170f
– into lithium cobalt oxide cathode 87f, 109
– into LMO cathode 90f
lithium iodide 188f
18650 lithium-ion cell 5f
– capacity of 164
lithium-ion cell discharge reaction 68
– enthalpy of 68
– entropy of 68
– free energy of 68
lithium-ion exchange 67
lithium-ion rechargeable battery 4f
– cathode-active materials for 7f
– potential windows of 229
– reactions in 4f
lithium iron phosphates 58
lithium-iron-phosphorous ternary system 61
lithium/LiPON/a-vanadium pentoxide-TFB 294f
lithium-LMO cell 4, 30, 70
lithium-LTO cell 29

lithium manganese oxide (LMO)
24, 28, 32f, 70, 90f, 114f
– Birnessite-type, Raman spectrum of 115
– crystallographic data of 115
– layered rocksalt phases of 125f
– Raman spectra of 114f, 121, 126
– spinel phases of 117f
– ternary lithiated derivatives 117
lithium-manganese-oxygen ternary system 117f
– phase diagramm of 118
– Raman-active modes of 120f
lithium metal anode 4
lithium microbattery see thin-film battery
lithium nickel cobalt oxided cathode film 273
lithium nickel manganese oxide (LiNiMO) 29, 39f, 113f
– Raman spectrum of 113f
– structural characterization of cobalt-containing 44f
– structural characterization of cobalt-free 40f
lithium nickel oxide 54, 272
– Raman spectrum of 113f
lithium oxide – vanadium pentoxide – silica (LVSO) solid electrolyte 265
lithium-oxygen interaction
– effect on electrochemistry of cathode-active material 147f
lithium phosphorus oxynitride (LiPON) 262, 267, 273, 275, 284, 294
lithium polymer secondary battery (LPB) 213, 231
lithium-silicum alloy 165
lithium titanium oxide (LTO) 27, 32f, 143f
lithium tris(trifluoromethanesulfonyl) methanide 216
lithium-vanadium pentoxide cell 181
LMO see lithium manganese oxide
LMO thin film 275f
– characteristics of 283
– electrochemical behaviour of TFB with 276
– ESD film 281f
– PLD film 277f
– spin-coated film 282
– substituted spinel film 283
localized vibration analysis 119
lowest unoccupied molecular orbital (LUMO) energy of additives 166f
LTO see lithium titanium oxide
LTO-LAMO cell 33
LTO-LiNiMO cell 34f

m
manganese metal electrode 182f
metallic lithium anode improver 180
molecular orbital (MO) calculation 166f, 170
molybdenum trioxide thin film
– electrochemical characteristics of 301, 303
– flash-evaporated film 301
– PLD film 301
multilayer system 2

n
$Na/NaFeO_2$cell 54f
NASICON-type sulfate cathode 55f
– discharge profile 57
nickel-cadmium battery 3
nickel-metal hydride battery 3
non-aqueous electrolyte 4f

o
OCV see open-circuit voltage
olivine ($LiFePO_4$)
– crystal structure 13, 152
– Raman spectrum of 153
– Raman-active modes of 151f
olivine-type phosphate cathode 58f
– composition of 61
– conductivity of 58f
– electrochemical properties of 149f
– for ignition test of all-solid-state lithium polymer battery 245
– lithiation/delithiation of 152
– phase transformation upon lithium extraction 154
– structural investigations of 150f
– synthesis of 59f
open-circuit voltage (OCV) 68, 95, 121, 316
– entropy dependence of 86, 88
– measurement of 71f, 78, 81f, 85, 92
– temperature dependence of 70
– time dependence of 72
overcharge protection additive 173
overlithiation 40
– effect on cyclability 43f, 49
– effect on oxidation state of transition metals 43f
– effect on structure of LiNiMO-based material 42f
oxygen swing 27

p

PEG see poly(ethylen glycol) as plasticizer
perovskite-type fluoride cathode 64f
phase transition in electrodes 70f, 87f, 92, 122f
phosphate solvents 217
phospho-olivine compound see olivine-type phosphate cathode
plasticization of polymer chains 222, 226
plasticizer 217
– effect on lithium ion conductivity 217
– side reaction with electrodes 240f
PLD see pulsed laser deposition
polarization 227, 231, 294
poly(ethylene glycol) (PEG) as plasticizer 217
– aluminate derivative (Al-PEG) 217f
– borate derivative (B-PEG) 217f, 228
– synthesis of derivatives of 218
poly(ethylene oxide) (PEO) 213f, 294
– crosslinking of 215
– helix coil structure 215
– lithium ion conductivity in 215
polyamine matrix 214
polyanionic cathode 58, 63
polyether matrix 214f
polyethylene (PE) separator 6
polymer-ceramic interface see solid-solid interface
polymer coating 190f
polymer electrolyte 213f
– lithium salts for 215f
polymer matrix 213f
– copolymers 218
– influence of lithium salts on thermal properties of 221f
– permittivity of 215f
polypropylene (PP) separator 29f, 33f
polysulfide matrix 214
postannealing treatment 260f, 266, 270
potassium ion 199, 204f
potentiostatic polarization measurement 227f
power output characteristic 315
predoping 62, 64
promotion of lithium-ion diffusion 318
1,3-propanesulton (PS) 164, 170
propylene carbonate (PC) 8
pulsed laser deposition (PLD) 258
pulverization of cathode particles 240

q

qualitative microstructural analysis of lithium-ion cathode 110f

r

radio frequency (RF) magnetron sputtering 259, 262, 273f, 277
Raman activity of lithiated transition metal oxides 118, 120
Raman effect 104
Raman microspectrometer 105f
Raman microspectrometry 104f, 138, 143, 289
– of metal-oxide-based compounds 106f
Raman shift 104
Raman spectroscopy 76f, 103f
– *in situ* 110, 112, 123, 133, 142
rapid thermal annealing (RTA) 264
rare-metal-free cathode 54f
rate capability
– of lithium iron phosphates 58f
– of LMO thin film electrode 284
– of overlithiated materials 43, 47f
redox potential 3
rolling friction resistance of electric vehicles 314

s

safety protection agent 180
screen-printing 271
secondary solid phase
segmental motion of polymer chains 224f
SEI see solid electrolyte interface 2f
self discharge of active material 1f, 7f, 69
separator 6
site-limited capacity 27
sodium chloride coating 207
sodium ion 197f
sol-gel coating 261
sol-gel spin-coating 270, 282f
solid electrolyte interface (SEI) 1f, 164, 174f, 248
solid electrolyte interface (SEI) forming improver 180
solid polymer electrolyte (SPE) film 213, 216
– containing plasticizer see gel polymer electrolyte
– electrochemical stability of 229f
– evaluation of thermal properties of 219f
– in thin film battery 277
– temperature dependence of ionic conductivity of 224f
– transport number of lithium ions in 227f
solid-solid interface 231

solid solution 83, 117
solvation enhancer 180
solvation of lithium ions in polymers 214, 226
space group symmetry 16
– of lithiated transition metal oxides 118
SPE see solid polymer electrolyte
specific capacity of electrode active material 72, 125
spinel 14f
– defect 14
– inverse 14
– normal 14
– phases in lithium-manganese-oxygen system 117f
– spinel oxide lattice structure 119
– transformation of 14
spinel-type cathode 181f
spray pyrolysis 270
sputtering
– for preparation of lithium cobalt oxide thin film 259
state of charge (SOC) 67, 93
state of discharge (SOD) 67
Stokes Raman scattering 104
storage/stop process 7, 24, 28
substitutional doping 58, 63
sulfuric acid 30f
superlattice 15f, 23
superlattice structures derived from spinel 15f
– characterization by XRD 21f
– chemical composition of 19
– oxidation state of cations in 19f
– preparation of 20
supression of manganese (II) dissolution 188f, 192
symmetric atomic displacement combinations for Pmmn 129
synthesis
– of crystalline thin film 264, 266, 269, 276
– of lithium cobalt oxide thin film 259f
synthesis of lithium iron phosphate cathodes 59f
synthesis of polymer SPE film 218

t
TFB see thin-film battery
thermal stability of lithium-ion cells 326
thermodynamic data acquisition 73f
thermodynamic studies of lithium-ion cells 69f
thermogravimetric analysis 219
thin oxide films 137f
– compressive stress in 140
– insertion of lithium ions into 138f
– Raman microspectrometry of 138f
– single phase behaviour of 143
thin-film battery (TFB)
– components of 258
– influence of film thickness on the capacity 290f
– with solid state electrolyte 262f, 267, 273, 294, 302
thin-film transition metal oxide (TMO) 257
– effect of film thickness on discharge capacity 261
– methods for preparation of 258, 269f, 274
– temperature effect on formation of thin film 278
three-dimensional NASICON-type structure 56
titanium disulfide 257
topotactic reaction 1
transport number of lithium ions in SPE films 227f
trioctyl phosphate 173
two-dimensional layered rocksalt-type oxide cathode 54f, 56
two-phase reaction 25, 150, 152

u
unit-cell orientation of sputtered thin films 259
UV-irradiation 246

v
vanadium pentoxide
– as cathode-active material 127
– crystal structure of 127f, 285
– Raman spectrum of 130
– Raman-active modes of 128f
– sputtered crystalline film 137f, 286
vanadium pentoxide thin film
– composite films 294
– crystal structure of 286
– CVD film 297
– doped with silver 294f
– effect of crystal orientation on TFB characteristics 288f
– ESD film 298
– influence of film thickness on the electrochemical behaviour of 290f
– morphology of 288
– PLD film 296f
– prepared by evaporation techniques 297
– spin-coated film 299

vanadium pentoxide thin film (*contd.*)
– sputtered films 286
– synthesis of 285
12-V battery 31f
vinylene carbonate (VC) 164, 192f
2-vinylpyridine 189f
Volta cell 11
voltamperometric curve 111

w
wettability 173
wetting agent 180
Williams-Landel-Ferry (WLF) equation 224f
working electrode for thermodynamic studies 73
Wulff's theorem 24f

x
X-ray absorption near edge spectroscopy (XANES) 41f
X-ray diffraction (XRD) 84

z
zero-strain insertion mechanism 28f